T0238341

Near-Rings and Near-Fields

Near-Rings and Near-Fields

Proceedings of the Conference on Near-Rings and Near-Fields,
Stellenbosch, South Africa, July 9–16, 1997

edited by

Yuen Fong

Department of Mathematics,
National Cheng Kung University,
Tainan, Taiwan, Republic of China

Carl Maxson

Department of Mathematics,
Texas A & M University,
College Station, Texas, U.S.A.

John Meldrum

Department of Mathematics,
University of Edinburgh,
Edinburgh, Scotland

Günter Pilz

Institute for Mathematics,
Johannes Kepler Universität Linz,
Linz, Austria

Andries van der Walt

and

Leon van Wyk

Department of Mathematics,
University of Stellenbosch,
Stellenbosch, South Africa

Springer-Science+Business Media, B.V.

A C.I.P. Catalogue record for this book is available from the Library of Congress.

ISBN 978-94-010-3802-7   ISBN 978-94-010-0954-6 (eBook)
DOI 10.1007/978-94-010-0954-6

Printed on acid-free paper

All Rights Reserved
© 2001 Springer Science+Business Media Dordrecht
Originally published by Kluwer Academic Publishers in 2001
Softcover reprint of the hardcover 1st edition 2001

No part of the material protected by this copyright notice may be reproduced or
utilized in any form or by any means, electronic or mechanical,
including photocopying, recording or by any information storage and
retrieval system, without written permission from the copyright owner.

CONTENTS

FOREWORD

This volume contains three invited lectures and sixteen other papers which were presented at the 14th International Conference on Nearrings and Nearfields held in Stellenbosch, South Africa, July 9-16 1997.

It was also the first nearring conference to be held after the untimely death of James R Clay, who over the years had been an inspiration to many algebraists interested in nearring theory. The occasion was marked by the invited talk of Gerhard Betsch, which was devoted to an overview of Clay's contributions to nearring and nearfield theory.

This book is affectionately dedicated to the memory of James R Clay.

All the papers presented here have been refereed under the supervision of the Editorial Board: Fong Yuen, Carl Maxson, John Meldrum, Günter Pilz, Leon van Wyk and Andries van der Walt. Thanks are due to the referees and to the Editorial Board.

A special word of thanks is due to Wen-fong Ke for preparing the final version of the TEX files, and to Fong Yuen for his pains in arranging for the publication of the volume with Kluwer Academic Publishers.

Andries van der Walt
Stellenbosch, August 1999

COMBINATORIAL ASPECTS OF NEARRING THEORY
TO THE MEMORY OF JAMES RAY CLAY

GERHARD BETSCH

A brief curriculum vitae of James Ray (Jim) Clay

Born November 5, 1938 at Burley (Idaho).
Died January 16, 1996 at Tucson (Arizona).
Married since 1959 to Carol Cline BURGE,
    "a truly beautiful daughter of Zion" (Dedication of Jim's 1992 book).
Three daughters, ten grand-children.

Training and professional career:

1956	US Naval Academy (class 1960); Jim studied engineering.
1959 (February)	University of Utah
1960	B. Sc. University of Utah
1962	M. Sc. in Mathematics University of Washington in Seattle
1966	Ph. D. in Mathematics University of Washington in Seattle
	(Supervisor: Ross A. Beaumont)
1965/66	Mathematician for the CIA,
	Part-time Instructor US Dept. of Agriculture
	Part-time Assistant Professor George Washington University
1966	Assistant Professor University of Arizona in Tucson
1969	Associate Professor University of Arizona
	1969–1972 Associate Head of Department
1974	Full Professor University of Arizona

Visiting Professor at: Tubingen, London (King's College), Munchen/Munich (TU), Edinburgh, Stellenbosch, Hamburg (Univ. der Bundeswehr), Tainan (Nat. Cheng Kung University), Linz (Joh. Kepler Universitat).

In addition to visiting professorships: 40 International lectures at Universities in Bulgaria, Hungary, Italy, Ireland, India, Hong Kong, Thailand, Singapore, and China.

1

Y. Fong et al. (eds.), Near-Rings and Near-Fields, 1–9.
© 2001 *Kluwer Academic Publishers. Printed in the Netherlands.*

Award:

1972/73   Humboldt Foundation's Distinguished Senior U.S. Scientist Award

(Jim was only 34 at the time!)

• **Publications:** Three Books, over fifty articles in journals.

Cf. also the Obituary by C. R. MAXSON in Results in Math. **30** (1996) and a fine picture of Jim Clay in the same volume of this journal.

1. INTRODUCTION

On 3rd August 1995, at the Hamburg Nearring Conference, Jim Clay gave a survey lecture on "Recent Developments, Discoveries, and Directions for Planar Nearrings" [sic!]. Nobody could foresee, that this excellent lecture was the last public presentation which our friend Jim would deliver to the nearring community. On 16 January 1996, Jim died of a sudden death at the age of 57, while riding his bicycle home from work.-

Jim attended all international nearring meetings so far. Andries van der Walt and I decided that a proper way to honour our deceased friend and colleague would be

to give **a survey on combinatorial aspects of nearring theory.**

Jim himself substantially and decisively contributed to those branches of nearring theory, which have combinatorial aspects. He did like this line of research very much.

According to Peter DEMBOWSKI (1928–1971), Combinatorics is the Theory (or enumeration) of subsets of finite sets. I claim:

1) Speaking of combinatorial aspects of nearring theory almost always involves finite *planar nearrings*, possibly generalizations of these structures;
2) the central combinatorial concept we have to deal with in this context is the concept of *balanced incomplete block designs* (BIBDs), possibly partially balanced incomplete block designs (PBIBDs), or other generalizations.

Let me give you the definition.

Definition. A *balanced incomplete block design* (BIBD) is a pair $(P, \mathcal{B})$ with the following properties: P is a set and $\mathcal{B} \subseteq 2^P$, $v := |P| > 0$, $b := |\mathcal{B}| > 0$. The elements of P are called *points*, and the elements of $\mathcal{B}$ are called *blocks*. We assume that the following axioms are satisfied:

- For any $B \in \mathcal{B}$, $|B| = k$.
- Every $p \in P$ belongs to exactly r blocks.
- Every two distinct blocks have $\lambda > 0$ points in common.

(The integers r and λ are supposed to be constants: r is independent of p, λ is independent of the chosen blocks.)

Examples are abundant: Affine planes; projective planes; Moebius planes; "good" experimental designs, with good symmetries.

In order to explain how Jim Clay came to study planarity, let me give you some historical information.

2. PREHISTORY OF THE SUBJECT

1905 Leonard Eugene DICKSON constructed and investigated finite near-fields (see his paper in TAMS 6). In particular, he exhibited a proper near-field with 9 elements, which is in fact the smallest proper near-field.

1905/1907 O. VEBLEN and J. H. MACLAGAN-WEDDERBURN, in their paper on "Non-desarguesian and non-pascalian geometries" (TAMS **8** (1907)), applied DICKSON's finite near-fields to construct finite projective planes, which are non-desarguesian and non-pascalian (today we would say "Pappian" instead of "pascalian"). It follows from a theorem by G. HESSENBERG of 1905(!), that a non-desarguesian plane has to be automatically non-pascalian (Beweis des Desarguesschen Satzes aus dem Pascalschen. Math. Ann. **61** (1905)). Using DICKSON's proper near-field of order 9, VEBLEN and WEDDERBURN constructed in particular two non-desarguesian and non-pascalian projective planes of order 9.

1931 CARMICHAEL discovered that the finite sharply 2-transitive permutation groups are precisely the groups of affine transformations

$$x \mapsto ax + b \quad (a \neq 0)$$

of a suitable finite near-field into itself (Amer. J. Math. **53** (1931), 631–644). Hence, from a group-theoretical point of view, it was important to determine all finite near-fields.

1934 In his dissertation, Hans ZASSENHAUS determined all (finite) sharply 3-transitive permutation groups. Of course, the characterization involved near-fields. (The dissertation was published in Abh. Math. Sem. Univ. Hamburg **11** (1936), 17–40).

1936 In a famous paper, ZASSENHAUS determined all finite near-fields (Abh. Math. Sem. Univ. Hamburg **11** (1936), 132–145). He proved: Up to seven exceptional near-fields (which he described precisely), any finite near-field may be derived from a finite field by DICKSON's method of 1905.

1943 Marshall HALL Jr. established the coordinatization of projective planes by a ternary ring. This involved the introduction of **Planarity:** It was required that in the ternary ring of coordinates the equation

$$ax = bx + c \text{ has a unique solution, if } a \neq b.$$

The geometrical meaning is clear: We aim at the unique point of intersection of two distinct lines.

All finite near-fields are automatically planar. This is due to the fact that any injective map of a finite set into itself is bijective. But what about the infinite case?

1959   In his book "The Theory of Groups," Marshall Hall Jr. discusses in great
            detail the correspondence between "Doubly Transitive Groups and Near-
            Fields" (Section 20.7). In constructing a near-field from a strictly 2-
            transitive group G the author needs an additional hypothesis ("which
            may not be necessary but is required for our proof", page 382). This is a
            transitivity condition (3) or the condition (3′), that G is finite.

1964   J. L. ZEMMER constructed infinite non-planar near-fields (*Near-fields,
            planar and non-planar.* The Math. Student **31** (1964), 145–150).

At this state of the art, Michael ANSHEL and Jim CLAY started their investigation of
planarity.

> "In 1967, I was wanting to find some geometric applications of nearrings.
> Nearfields for which each equation $ax = bx + c$, $a \neq b$, had unique solu-
> tions were exactly the ones which were used successfully in geometry, so
> I focused on this equation for nearrings. It was not surprising to various
> experts at the time [HUGHES, ZEMMER] that nearrings in which each
> equation $ax = bx + c$, $a \neq b$, had a unique solution would also be near-
> fields. So a different point of view was needed" (Jim CLAY, The equation
> $ax = bx + c$, Preface; see also his contribution to the Oberwolfach 1968
> meeting).

Now let me start a systematic survey.

3. SYSTEMATIC SURVEY

Let $(N, +, .)$ be a left nearring.

Definitions and Remarks. If $a, b \in N$, then $a \equiv_m b$ is defined by $ax = bx$ for all $x \in N$.
Obviously, $\equiv_m$ is an equivalence relation on N.

N is called *planar* if

(i) $|N/\equiv_m| \geq 3$, and
(ii) if a is not equivalent to b under $\equiv_m$, then $ax = bx + c$ has a unique solution.

N is called *integral planar* iff $\{a \mid a \equiv_m 0\} = \{0\}$.

There are some **Main Examples**, due to Michael ANSHEL, which Jim CLAY referred
to quite frequently:

> "I keep getting inspiration from these examples" (first sentence of Jim's
> lecture at Fredericton).

Take $(\mathbb{C}, +, .)$, the field of complex numbers. Now define

(i) $a \times b := |a| \cdot b$;
(ii) $a \bullet b := \Theta(a) \cdot b$, where $\Theta(a) := |a|^{-1} \cdot a$, if $a \neq 0$, and $\Theta(0) := 0$.

Then $(\mathbb{C}, +, \times)$ and $(\mathbb{C}, +, \bullet)$ are integral planar nearrings.

What is the equivalence class of a with respect to $\equiv_m$ in these nearrings?

(i) $\{z \in \mathbb{C} \mid |z| = |a|\}$, the *circle* around 0 with radius $|a|$;
(ii) $\{0\}$ and $\{z \in \mathbb{C} \mid z = \lambda a, \lambda \in \mathbb{R}, \lambda > 0\}$, the *ray* from 0 in the direction of a.

Let us look at certain "distinguished subsets" (blocks):

- $\mathbb{C} \times e + d$ is the ray starting from d in the direction of e;
- $\mathbb{C}^* \bullet e + d$ is the circle with midpoint d and radius $|e|$, (here $\mathbb{C}^*$ is the set of nonzero complex numbers);

Furthermore:

$$(\mathbb{C} \times u + m) \cup (\mathbb{C} \times (-u) + m) = \mathbb{C} \times \{u, -u\} + m$$

is the line through point m in the direction of u. Moreover,

$$(\mathbb{C} \times (s-t) + t) \cap (\mathbb{C} \times (t-s) + s)$$

is the segment from t to s of the line through t and s.

Let $(N, +, \cdot)$ be a planar nearring. We form

$$\mathcal{B} := \{Na + b \mid a, b \in N, a \neq 0\},$$
$$\mathcal{B}^* := \{N^*a + b \mid a, b \in N, a \neq 0\}, N^* := \{n \mid n \neq 0\},$$
$$\bar{\mathcal{B}} := \{N\{a, -a\} + b \mid a, b \in N, a \neq 0\},$$
$$S := \{\overline{a, b} \mid a, b \in N, a \neq b\}, \text{ where } \overline{a, b} := (N(a-b) + b) \cup (N(b-a) + a).$$

In his Hamburg 1995 lecture, Jim proposed the following **Program**:

(1) Study the structures $(N, \mathcal{B})$, $(N, \mathcal{B}^*)$, $(N, \bar{\mathcal{B}})$, (N, S). What is their geometric meaning?

(2) Study (N, α, β), where $\alpha, \beta \in \{\mathcal{B}, \mathcal{B}^*, \bar{\mathcal{B}}, S\}$.

To "study" these structures means: Try to answer the following questions.

(i) Is $(N, \mathcal{B})$ a BIBD? Or a PBIBD (partially balanced incomplete block design)? Or some incidence structure with a weaker type of symmetry?

(ii) Is $(N, \mathcal{B})$ a BIBD with interesting additional properties?

And then we have the analogous questions for the other structures mentioned above.

To attack these questions, we have to study in great detail the planar nearring N we start with.

Let me just mention some basic facts.

4. BASIC FACTS ON PLANAR NEARRINGS N

- N is always 0-symmetric. This is due to the fact, that for all $z \in N$, the elements 0 and $0z$ are both solutions of $ax = 0x + 0$.
- The *MAIN STRUCTURE THEOREM* on planar nearrings, due to ANSHEL-CLAY 1968 and CLAY 1971, describes N in great detail:

  Let $A := \{a \mid a \not\equiv_m 0\}$, and if $a \notin A$, let 1_a be the unique solution of $ax = 0x + a = a$. Now, we assume $a \notin A$.

  Then the following assertions are true:

  (a) $B_a := \{x \in N \setminus A \mid x1_a = x\}$ is a group with respect to the multiplication in N.

  (b) $N = A \cup \bigcup \{B_a \mid a \in N \setminus A\}$ (disjoint union of A and the different groups B_a).

  (c) $(N \setminus A)B_a = B_a$.

  (d) $\Phi : B_a \to B_c : x \mapsto x1_c$ is a group isomorphism.

(e) $\forall a \in N \setminus A : 1_a$ is a left identity of the nearring $(N,+,\cdot)$.

• $\forall a \in N \setminus A$: The mapping $\varphi_a : x \mapsto ax$ is a *fixed-point-free automorphism* of $(N,+)$.

A careful analysis of these facts finally leads to a universal method, due to G. FERRERO, of constructing planar nearrings. Jim jokingly called this method the *Ferrero planar nearring factory*, and he used to add, that this factory is not to be confused with the well-known Ferrero factory of sweets. To explain this method, we start with the following situation

(Frob)   Let (N,Φ) be a pair consisting of a finite additive group $(N,+)$ and $1 \neq \Phi$ be a fixed point free group of automorphisms of $(N,+)$.

In addition we assume: $\forall \varphi \in \Phi \setminus \{1\}$, $1 - \varphi$ is bijective. Then we form a set A which consists of 0 an a (possibly empty) union of orbits of Φ on N. From all the orbits of Φ not contained in A we select a representative. Let $a \in N \setminus A$, and let e be the representative of the orbit of Φ, to which a belongs. Since Φ is fixed point free, there exists exactly one $\varphi_a \in \Phi$ with the property $\varphi_a(e) = a$. And now we define a multiplication on N by

$$ax := \begin{cases} \varphi_a(x), & \text{if } a \in N \setminus A, \\ 0, & \text{if } a \in A. \end{cases}$$

This definition turns $(N,+,\cdot)$ into a planar nearring. The method is universal: Every (finite) planar nearring may be constructed by this method. Moreover, N is an integral planar nearring if and only if $A = \{0\}$.

Let me take a brief detour.

5. DETOUR IN SEVERAL STEPS

a) The situation designed by (Frob) is equivalent to the statement: The semidirect product of N with Φ is a Frobenius group with kernel N. (See e.g. D. GOREN-STEIN, Finite Groups. Chelsea ed. 1980. Page 37 ff.) Applying a famous theorem of John G. THOMPSON, we infer that $(N,+)$ must be a finite nilpotent group, which means: not too far from abelian. (J. G. THOMPSON, *Finite groups with fixed-point-free automorphisms of prime order.* Proc. Nat. Acad. Sci. **45** (1959), 578–581.)-

b) In our paper "Block Designs from Frobenius Groups and Planar nearrings" (Proc. Conference on Finite Groups 1976, 473–502), Jim and I explained the Ferrero construction method, and methods to obtain BIBDs from planar nearrings, in great detail. We discussed the question, "whether planar nearrings are really needed" in order to construct the combinatorial structures $(N,\mathcal{B})$. Furthermore, we looked at the "efficiency" $E := \lambda v / kr$ of the BIBDs obtained. And we could give a series of examples, which contained BIBDs with efficiency arbitrarily close to 1.-

c) Fixed point free automorphisms did arise in a very early stage of nearring theory:
Let M be a nearring which is 2-primitive on the M-group Γ. Then the group G of nonzero M-endomorphisms of Γ is a (possibly trivial) group of fixed point free automorphisms of Γ. This leads to the question: Is there any reasonable connection between the pair (Γ, M) and a planar nearring, or combinatorial structure derived from (Γ, G)?

d) In his 1972 paper "Generating BIBDs from planar nearrings" (J. Algebra **22**), Jim discussed methods to obtain a finite group with a non-trivial group of fixed point free automorphisms. In the case of abelian groups, fixed point free automorphisms are abundant. What about the non-abelian case?

This finishes our detour and leads back to our survey.

6. CONTINUATION OF THE SURVEY

Let $(F, +, \cdot)$ be a finite field, and Φ a group of automorphisms of the multiplicative group F^* of F; let $\Phi \neq \{id\}$. For $a \in F^*$ we define the automorphism a^* of $(F, +)$ by $a^*(x) := ax$. Then we form

$$\Phi' := \{a^* \mid a \in F^*\}.$$

The pair (F, Φ') satisfies condition (Frob) from section 4. Hence we may build up planar nearrings from (F, Φ) and form the combinatorial structures defined in the Hamburg program of Jim CLAY (see section 3).

Some results to be mentioned are

- $(F, \mathcal{B}^*)$ is always a BIBD;
- $(F, \mathcal{B})$ is "normally" (i.e., in most cases) a BIBD, and always a tactical configuration;
- $(F, \mathcal{B}^-)$ is "normally" a BIBD;
- (F, S) is always a BIBD.

7. CODES, CIRCULAR PLANAR NEARRINGS, DOUBLE PLANAR NEARRINGS

a) Peter FUCHS, Gerhard HOFER, and Guenter PILZ investigated "Codes from Planar Near-rings" (IEEE Transactions on Information Theory **36** (1990), 647–651). Each tactical configuration determines four different codes, the codewords being the rows or the columns of the incidence matrix A of the configuration, or the linear hulls generated by the rows or the columns of A. The paper stresses the particular case, that the tactical configuration is a BIBD. Several of these codes turn out to be "almost perfect." The authors also examine the case, where the BIBDs are constructed from planar nearrings, and from "circular" planar nearrings in particular. Jim CLAY had pointed out, that the class of circular planar nearrings might be especially useful in this context.

b) *Circular planar nearrings* are planar nearrings with the additional properties:
- 3 points belong to at most one block (circle);
- 2 points belong to at least 2 blocks.

The nearring $(\mathbb{C}, +, \bullet)$ (cf. section 3, main example ii)) is an example of a circular planar nearring.

In the dissertation by Wen-Fong KE on "Structures of Circular Planar Nearrings" (University of Arizona 1992, written under the supervision of Jim CLAY), the circular planar nearrings are thoroughly investigated.

Jim CLAY introduced the circular planar nearrings in his lecture at the Middles-brough conference 1987. I remember with pleasure his opening sentence, spoken with a charming smile: "Perhaps the most important thing I have to say is: My wife sends kind regards to everybody".

c) Finally, we have to mention *double planar nearrings*, which Jim CLAY intro-duced, inspired again by the main examples in section 3. An algebraic structure $(N, +, *, \bullet)$ is called a *double planar nearring* if each of $(N, +, *)$ and $(N, +, \bullet)$ is a planar nearring, and each of $*$ and $\bullet$ is left distributive over the other.

Double planar nearrings play a key role in one of Jim CLAY's last papers on "Geometry in Fields" (Algebra Colloquium **1** (1994), 289–306). This delightful paper builds a bridge from algebra to geometry. I feel, that this might be a promis-ing line of research.

In conclusion, I would like to add a few remarks.

8. REMARKS (MAINLY PERSONAL)

(1) Jim was a *family man*. He was a loving husband, a devoted father and grandfather, and a faithful and helpful friend. I mention the dedication of his book "To CAROL CLINE BURGE, a truly beautiful daughter of Zion".

(2) Jim had *personal experience in applications*, in particular in the fields of
  • Engineering,
  • Cryptography (from his work with the CIA),
  • Agriculture.
He was a scholar in pure mathematics, with a strong feeling for applications.

(3) Jim was an *outstanding teacher*. He had seven Ph. D. students.
"He took a great deal of pride in his teaching. He showed endless pa-tience with students, not giving quick answers but suggesting ways to explore the problem. He would always find a kind and positive com-ment to encourage students and colleagues as well. He was always in-terested in finding new ways to help students learn. His book 'Nearrings: Geneses and Applications' is a model of exposition, directed toward stu-dents and student learning" (C. J. MAXSON, in his obituary in Results in Math. **30**).
I would like to add: An exposition on a very high level! Also I refer to his book "The equation $ax = bx + c$".

(4) The pair of concepts "Geneses" versus "Applications" seems to be typical for our friend Jim. He would always start a lecture giving the concrete background and motivating examples, perhaps a visualization, if possible. And he would look for consequences in various fields, "real(!)" applications.

What will remain? The answer certainly depends on personal convictions. But we can say: There will remain

- A legacy of fine books and articles on the subject of nearrings, some of them outstanding, and decisive in shaping and promoting the theory;
- And fond memories of a kind and noble person, and of a very good personal friend.

ACKNOWLEDGEMENT

I am indebted to C. J. MAXSON. In preparing this article I frequently used his obituary on James Ray CLAY, published in Results in Math. **30** (1996).

FURTBRUNNEN 17, D71093 WEIL IM SCHOENBUCH, GERMANY
(FORMERLY: MATHEMATISCHES INSTITUT DER UNIVERSITÄT TÜBINGEN, AUF DER MORGENSTELLE 10, 72076 TUBINGEN, GERMANY)

LEFT SELF-DISTRIBUTIVE RINGS AND NEARRINGS

GARY F. BIRKENMEIER

ABSTRACT. A (near-) ring R is called left self-distributive, *LSD*, if $vxy = vxvy$ for all v, x, y in R. Right self-distributive (near-) rings, *RSD*, are defined similarly. A (near-) ring is called self-distributive, *SD*, if it is both *LSD* and *RSD*. Observe that the class of *LSD* (left near-) rings is exactly the class of (left near-) rings for which each left multiplication mapping (i.e., $x \to ax$) is a (left near-) ring endomorphism. Hence the class of *LSD* (left near-)rings includes the AE-(left near-) rings (i.e., those (left near-) rings for which every additive endomorphism is a (left near-) ring endomorphism). In this paper we will discuss the history and recent developments for the class of *LSD* and related (left near-) rings. Examples will be included to illustrate and delimit the theory.

INTRODUCTION

In this survey we focus on algebraic systems with a binary operation which is left (and/or right) self distributive. Thus if G is a nonempty set with a binary operation $\cdot$, then the groupoid $(G, \cdot)$ is called *left (right) self distributive, LSD (RSD)*, if $a(bc) = (ab)(ac)$ $((ab)c = (ac)(bc))$ for all $a, b, c \in G$. If $(G, \cdot)$ is both *LSD* and *RSD* then we say $(G, \cdot)$ is *self distributive, SD*. At first glance this notion may seem somewhat artificial, however it becomes quite intriguing once one realizes that a left (right) self distributive groupoid is exactly the type of groupoid for which left (right) multiplication maps (i.e., $x \to \alpha x$ ($x \to x\alpha$) for α a fixed element of G and x an arbitrary element of G) are endomorphisms. For an example of a *SD* groupoid $(G, *)$, let G be the set of real numbers; and let the binary operation $*$ be the weighted average of two real numbers (i.e., $a * b = pa + qb$, where p, q are fixed real numbers such that $p + q = 1$).

The *LSD*, *RSD* and *SD* groupoids have been studied extensively and applications have been made to universal algebra (see [5], [29], [35], [45]), semigroup theory (see [33], [41], [42]), combinatorics (see [26]), knot theory (see [31], [39]), and set theory (see [17], [19]). The above applications and cited references in no way form a complete list but only provide an appetizer to the feast. Also there is at least one monograph [30] on *SD*-groupoids.

Our main interest will be on rings and nearrings in which the multiplication operation is left (right) self distributive. Such a ring or nearring will be called a *LSD (RSD) ring* or *nearring*, respectively. If it is both *LSD* and *RSD* then we say it is an *SD ring* or *nearring*, respectively.

Throughout this paper R denotes an associative ring or a left nearring. $\mathbb{Z}$ and $\mathbb{Z}_n$ symbolize the ring of integers and the ring of integers modulo n, respectively. $C(n)$ denotes the cyclic group of order n. We use $I(R)$ and $\mathcal{N}(R)$ for the set of idempotent elements of

---

1991 *Mathematics Subject Classification.* 16S99, 16U99, 16Y30.

Y. Fong et al. (eds.), *Near-Rings and Near-Fields*, 10–22.
© 2001 *Kluwer Academic Publishers. Printed in the Netherlands.*

R and the set of nilpotent elements of R, respectively. Let X be a nonempty subset of R. We say X is *reduced* if X has no nonzero nilpotent elements of R. Also $\langle X \rangle$, $l(X)$, $r(X)$, **gp** (X), and **snr** (X) denote the ideal of R generated by X, the left annihilator of X in R, the right annihilator of X in R, the subgroup of $(R, +)$ generated by X, and the subnearring of R generated by X. Moreover we say X is *left (right) essential* in R if every nonzero left (right) ideal of R has nonzero intersection with X. Let $(S, \cdot)$ be a semigroup. Then $(S, \cdot)$ is called *left (right) permutable* if $abc = bac$ $(abc = acb)$ for all $a, b, c \in S$. $(S, \cdot)$ is called *permutable* if it is both left and right permutable. Let $(G, \cdot)$ be a groupoid, then

$$L(G) = \{\alpha \in G \mid \alpha(xy) = (\alpha x)(\alpha y), \text{ for all} x, y \in G\},$$
$$\mathcal{R}(G) = \{\alpha \in G \mid (xy)\alpha = (x\alpha)(y\alpha), \text{ for all} x, y \in G\},$$
$$\mathcal{D}(G) = L(G) \cap \mathcal{R}(G).$$

1. **LSD** Rings

Our story begins in 1969 with the following result of Petrich [42] in which he characterizes *SD* rings. From [30], one can see that this result is a specialization of a groupoid result to a ring result.

Theorem 1.1. *R is a SD ring if and only if $R = B \oplus N$ (ring direct sum), where B is a Boolean ring and N is a nilpotent ring of index at most three.*

In 1987 H. E. Heatherly and the author [5] introduced the concept of an *OI*-system. This system allows one to construct algebras with certain desired properties. In particular it generalizes the nearring construction of J. R. Clay [15]. The *OI*-system allows one to design *LSD* rings and nearrings as will be illustrated below.

In 1988 H. E. Heatherly, T. Kepka, and the author submitted the first paper [12] dealing exclusively with the investigation of *LSD* rings. Due to delays in publication this paper did not appear until 1992. They showed that every *LSD* ring is left permutable; however this is not true, in general, for *LSD* semigroups. Thus *LSD* semigroups which are not left permutable cannot be the multiplicative semigroup of a ring. The following results from [12] provide a characterization of the subdirectly irreducible *LSD* rings and a generalization of Petrich's decomposition theorem (Theorem 1.1).

Theorem 1.2 ([12]). *Every subdirectly irreducible LSD ring is either nilpotent of index at most three, isomorphic to $\mathbb{Z}_2$, or isomorphic to the four element ring in the following example.*

Example 1.3 ([12]). There exists a subdirectly irreducible *LSD* ring which is not *SD*. This ring can be formed by taking w and e as the generators for the group $C(2) \oplus C(2)$ (Klein four group) and defining multiplication via the following table:

	0	w	e	$w+e$
0	0	0	0	0
w	0	0	0	0
e	0	w	e	$w+e$
$w+e$	0	w	e	$w+e$

This example is isomorphic to the semigroup ring $\mathbb{Z}_2[S]$ where S is the semigroup such that $|S| = 2$ and $xy = y$, for all $x, y \in S$.

Theorem 1.4 ([12]). *Let R be a LSD ring.*

(i) $\mathcal{N}(R)$ *is an ideal of R,* $[\mathcal{N}(R)]^3 = 0$, *and $R/\mathcal{N}(R)$ is a Boolean ring.*

(ii) $\mathcal{N}(R) = \langle K \rangle$, *where $K = \{a - a^2 \mid a \in R\}$.*

(iii) *If P is a prime ideal of R, then $R/P \simeq \mathbb{Z}_2$, hence P is completely prime and maximal.*

Theorem 1.5 ([12]). *Let R be an LSD ring and A a maximal reduced left ideal of R (such an A exists by Zorn's Lemma).*

(i) $l(A) = l(I(R)) = \mathcal{N}(R)$ *and $r(I(R)) \subseteq r(A) \subseteq \mathcal{N}(R)$.*

(ii) *A contains every reduced right ideal of R.*

(iii) *$B = A \oplus \mathcal{N}(R)$ is an ideal of R which is both left and right essential in R.*

(iv) *B is a completely semiprime ideal of R.*

(v) *If $I(R) \subseteq B$, then $R = B$.*

It is shown in [12] that Petrich's result, Theorem 1.1, is a corollary of Theorem 1.5.

Theorem 1.6 ([12]). *Let R be an LSD ring and $\bar{R} = R/\mathcal{N}(R)$. If $\bar{R} = \bar{R}\bar{e}_1 + \ldots + \bar{R}\bar{e}_n$, for some $\bar{e}_i \in \bar{R}$, then $R = A \oplus \mathcal{N}(R)$, where A is a left ideal of R which is a Boolean ring with unity.*

Corollary 1.7 ([12]). *Let R be an LSD ring. If R satisfies any of the following conditions, then $R = A \oplus \mathcal{N}(R)$, where A is a left ideal of R which is a finite Boolean ring.*

(i) *$R/\mathcal{N}(R)$ is finite.*

(ii) *R has no infinite set of orthogonal idempotents.*

(iii) *R has ACC or DCC on ideals.*

A ring R is *medial* if $abcd = acbd$, for all $a, b, c, d \in R$. Observe that since every *LSD* ring is left permutable, then it is medial. Our next result is a partial converse to Theorem 1.6.

Theorem 1.8 ([12]). *Let R be a medial ring. If $R = A \oplus \mathcal{N}(R)$, where A is a left ideal of R and $[\mathcal{N}(R)]^3 = 0$, then R is an LSD ring.*

Our next example shows that the medial condition in Theorem 1.8 is not superfluous.

Example 1.9. Let

$$R = \begin{pmatrix} \mathbb{Z}_2 & \mathbb{Z}_2 & \mathbb{Z}_2 \\ 0 & 0 & \mathbb{Z}_2 \\ 0 & 0 & 0 \end{pmatrix}.$$

Then $R = A \oplus \mathcal{N}(R)$ where A is a left ideal of R and $[\mathcal{N}(R)]^3 = 0$, but R is not a *LSD* ring.

It is easy to construct *LSD* rings from free rings.

Example 1.10. Let F be a free ring and let X be the ideal of F generated by

$$\{abc - abac \mid a, b, c \in F\}.$$

Then F/X is an *LSD* ring.

The following two examples where obtained in [12] using *OI*-systems.

Example 1.11. Let T be a ring, M a left T-module and $f : M \to T$, $h : M \to M$ be T-homomorphisms satisfying $fh = f$ and $h^2 = h$. Define $x * y = f(x)h(y)$ for each $x, y \in M$. Then $(M, +, *)$ is a ring. If $f(x)f(y) = f(x)f(y)f(x)$ for each $x, y \in M$, then $(M, +, *)$ is an *LSD* ring.

One concrete realization of this given by taking M to be the full set of n by n matrices, $n > 1$, over a Boolean ring T and use $f = \text{trace}$, $h = 1_M$. The ring so formed will always contain nonzero nilpotent and idempotent elements. Another realization arises by using the same M, T, f but taking $h(a_{ij}) = (a'_{ij})$ where $a'_{ii} = a_{ii}$ and $a'_{ij} = 0$ for $i \neq j$.

Other similar constructions come readily to mind.

Example 1.12. Let S be a ring and f, h endomorphisms of the additive group $(S, +)$ such that:

(i) for each $y \in f(S)$, $x \in h(S)$, $f(yx) = yf(x)$ and $h(yx) = yh(x)$;
(ii) $h^2 = h$, $f^2 = f$ and $fh = f$.

Define $a * b = f(a)h(b)$ for each $a, b \in S$. Then $(S, +, *)$ is a ring. If

$$f(a)f(b)f(a)h(c) = f(a)f(b)h(c) \quad \text{for all } a, b, c \in S$$

and if $f(S)$ is a subsemigroup of the multiplicative semigroup of S, then $(S, +, *)$ is an *LSD* ring.

As a particular example of this take R to be an *LSD* ring, S the full ring of n by n matrices over R, $n > 1$, and define f and h via $f(a_{ij}) = (b_{ij})$, where $b_{ij} = 0$ for $i \neq j$ and $b_{ii} = a_{ii}$, $h(a_{ij}) = (c_{ij})$, where $c_{ij} = 0$ for $i > j$ and $c_{ij} = a_{ij}$ otherwise. Then $(S, +, *)$ is an *LSD* ring which is not right permutable.

The results in [12] motivated the authors to ask: if R is an *LSD* ring, does $R = A \oplus \mathcal{N}(R)$, where A is a left ideal of R which is a Boolean ring? In 1996, A. V. Kelarev answered this question in the negative using a certain contracted semigroup algebra [32].

The self distributive concept can be generalized as follows: a ring R is said to be a *left* (*right*) *n-distributive ring*, $n > 1$ a positive integer, if

$$aa_1 a_2 ... a_n = aa_1 aa_2 ... aa_n \ (a_1 a_2 ... a_n a = a_1 aa_2 a ... a_n a),$$

for all $a, a_1, a_2, ..., a_n \in R$. If R is both left and right n-distributive, then it is said to be an *n-distributive ring*. In 1990 M. Ćirić and S. Bogdanović [13] and again in 1992 S. Feigelstock and R. Raphael [25] obtained the following result.

Theorem 1.13. *A n-distributive ring is an extension of a nilpotent ring N satisfying $N^{n+1} = 0$ by a generalized Boolean ring B (i.e., $a^n = a$, for all $a \in B$.*

This result and many others for rings satisfying various semigroup identities were generalized in 1995 by M. Ćirić, S. Bogdanović, and T. Petković in [14].

2. THE SULLIVAN PROBLEM

In 1977 R. P. Sullivan [47] posed the following problem: characterize those rings in which every additive endomorphism is a ring endomorphism. Using S. Feigelstock's terminology [22], we call such a ring an *AE-ring*. In 1981 K. H. Kim and F. W. Roush [36] characterized all finite *AE*-rings. In 1987 S. Dhompongsa and J. Sanwong [18] obtained several elementary results of *AE*-rings and *AE*-nearrings. The result of K. H. Kim and F.

W. Roush was generalized in 1988 when S. Feigelstock characterized all AE-rings whose additive groups are torsion [22].

The author observed that since every multiplication map is an additive endomorphism then an AE-ring is a SD ring. With this observation in mind and their results on SD and LSD rings H. E. Heatherly and the author began work on the Sullivan problem in 1989. About the same time several other investigators including Y. Hirano, M. Dugas, J. Hausen, and J. Johnson began attacking the problem. In 1990 H. E. Heatherly and the author [9] characterized all AE-rings except those AE-rings R which satisfy all of the following conditions: $R^3 = 0$, $R^2 \neq 0$, $(R, +)$ is not a torsion group but R has a element of order 2. Their characterization included the previous results of K. H. Kim and F. W. Roush and S. Feigelstock. In 1991 Y. Hirano in [28] and M. Dugas, J Hausen, and J. Johnson [20] determined characterizations of AE-rings equivalent to those of H. E. Heatherly and the author. The following result gives the flavor of solutions to the Sullivan Problem while avoiding the technicalities of the more complete results. Observe that it is a special case of Petrich's result.

Theorem 2.1 ([9]). *Let R be a ring such that $R^3 \neq 0$. Then R is an AE-ring if and only if all of the following conditions are satisfied:*

(i) $R = \mathbb{Z}_2 \oplus \mathcal{N}(R)$ *(ring direct sum);*
(ii) $[\mathcal{N}(R)]^2 = 0 = \mathcal{N}_2(R)$, *where $\mathcal{N}_2(R)$ is the 2-component of $(\mathcal{N}(R), +)$; and*
(iii) $(\mathcal{N}(R), +)$ *is 2-divisible.*

In 1992 M. Dugas, J. Hausen, and J. Johnson [21] answered four questions posed in [9] by producing an AE-ring whose 2-component is not a direct summand. This example is quite amazing and shows there is little hope for a more complete solution to the Sullivian Problem.

In 1989 S. Feigelstock [23] generalized the concept of an AE-ring. He said a ring R is an AE_n-ring, $n > 1$, if every additive endomorphism ϕ satisfies

$$\phi(a_1 a_2 ... a_n) = \phi(a_1)\phi(a_2)...\phi(a_n)$$

for all $a_1, a_2, ..., a_n \in R$. In [23] and [24], he extends several results for AE-rings to AE_n-rings.

In 1993 T. Kepka [34] showed that for an algebra $(S, +, \cdot)$, where $(S, +)$ is a commutative semigroup and the multiplication is both left and right distributive over addition, then the condition that every additive endomorphism is also a multiplicative endomorphism (AE) guarantees that multiplication is associative if $(S, +, \cdot)$ satisfies either: (1) $(S, +)$ is an abelian group, or (2) $x + x = x$ for all $x \in S$. Thus surprisingly the AE condition guarantees the associativity of multiplication in the class of not necessarily associative rings.

Recently Y. U. Cho and the author have defined a ring R to be an AGE-ring if $End(R, +)$ is equal to the group generated by the ring endomorphisms on R. Similarly a nearring R is called an AGE-nearring if $\mathcal{E}(R)$ is equal to the group generated by the nearring endomorphisms on R. Their results appear in [4].

3. **LSD** Nearrings

C. Ferrero Cotti in 1972 [16] seems to be the first nearringer to investigate nearrings with self distributive multiplication (i.e., *SD* nearrings). She attempted to generalize Petrich's result (Theorem 1.1) as follows: a distributive nearring R is *SD* if and only if $R = B + N$ is a semidirect sum of a Boolean ring B and a nearring N such that $N^3 = 0$. However Example 1.3 provides a ring R which is a semidirect sum of Boolean ring and an ideal N such that $N^3 = 0$, however R is not a *RSD* ring. Thus her statement is incomplete.

In 1984 R. Scapellato [46] considered *SD* nearrings (he called them "autodistributive" near-rings). He showed the following result.

Theorem 3.1. *If R is a SD nearring then $\mathcal{N}(R)$ is an ideal of R such that $[\mathcal{N}(R)]^3 = 0$, and $R/\mathcal{N}(R)$ is a Boolean nearring.*

Also he showed that the class of β-nearrings (i.e., Boolean left permutable nearrings) coincides with the class of Boolean *SD* nearrings, and he gave a characterization of these nearrings in terms of subdirect products of subdirectly irreducible ones.

The author in 1989 [3] investigated the interaction between the circle operation "∘" (i.e., $x \circ y = x + y - xy$) and multiplication "·" in a distributive nearring $(R, +, \cdot)$. Using the following observation and other results a new class of nontrivial distributive nearrings was obtained.

Lemma 3.2. *Let R be a nearring and $v, x, y \in R$.*

   (i) *$v(x \circ y) = vx \circ vy$ if and only if $vxy = vxvy$.*

   (ii) *If v is a distributive element (in the nearring sense) of R, then $(x \circ y)v = xv \circ yv$ if and only if $xyv = xvyv$.*

In 1990 H. E. Heatherly and the author [8] investigated *LSD* nearrings and *RSD* near-rings. They were able to complete C. Ferrero Cotti's generalization of Petrich's results, generalize results of R. Scapellato, and provided examples to show the abundance of *LSD* nearrings. The following results and examples are some of the highlights of their paper [8]. Since the class of *LSD* nearrings is a variety (and hence is closed under the formation of subdirect products), further examples immediately arise from the ones explicitly given here.

Example 3.3 (Malone's trivial nearrings [37]). These are the nearrings R such that either (i) $0 \cdot R = R$ or (ii) $0 \cdot R \neq R$ and every element is either a left identity or a left annihilator of R. These near-rings are always *LSD* and left permutable. If $0 \cdot R \neq R$ and $R^2 \neq 0$, then R cannot be *RSD* or right permutable.

While Malone's trivial nearrings are in some sense just that-trivial-we discussed them here because they play a key role in classifying *LSD* nearrings.

A perusal of the standard lists of examples of nearrings given via Cayley tables, Clay representations, or Cayley tables for generating sets [1, 15, 43, 44, 49] reveals a considerable number of these near-rings are *LSD*. This is due only in part to the presence of Malone trivial ones. We next give a brief summary of our findings from said lists. For a more detailed listing of the nearrings of small order which are *LSD* and/or *RSD* see [7].

Summary 3.4.   (i) On the Klein four group there are twenty-three nonisomorphic near-rings [15]. Fifteen of these are *LSD*, of which five are Malone trivial.

(ii) On C_6 there are sixty non-isomorphic nearrings [15]. Forty-seven of these are
     LSD, of which twenty-one are Malone trivial.

(iii) On S_3 there are thirty-nine non-isomorphic nearrings [15]. Twenty-nine are LSD,
     of which thirteen are Malone trivial.

(iv) On the dihedral group of order eight there are twenty non-isomorphic d.g. near-
     rings [48]. Eighteen of these are LSD, of which three are Malone trivial.

(v) On the infinite dihedral group there are six non-isomorphic d.g. nearrings [38].
     All of these are LSD and three of them are Malone trivial.

Example 3.5. All Boolean nearrings are SD nearrings [8, p. 294].

The following example was obtained in [6] and [8] using an *OI*-system [5].

Example 3.6. Let $(R, +, \cdot)$ be a Boolean ring, M a left R-module, and $f : M \to R, h : M \to M$
R-homomorphisms satisfying $fh = f$ and $h^2 = h$. Define $a * b = f(a)h(b)$ for each $a, b \in M$.
Then $(M, +, *)$ is a (left) nearring which is LSD and left permutable.

A concrete realization of this occurs when $M = M_n(R)$, the full set of n by n matrices
over R, $n > 1$, and $f(\alpha) = \det \alpha$, $h = 1_M$. Then $\alpha \in M$ is nilpotent in $(M, *)$ if and only
if $\det \alpha = 0$ and each nonzero nilpotent element has index two. Next suppose R has unity;
then the left identity elements in $(M, *)$ are exactly the matrices with determinant one.
Also, $(M, *)$ is neither right permutable nor right self distributive. Using $R = \mathbb{Z}_2$ we have
an example where the set of nilpotent elements of $(M, +, *)$ is not closed under addition.

Example 3.7 (Clay construction [15]). Let $(G, +)$ be a group (not necessarily abelian)
and let $f : G \to End(G, +)$. Define $a * b = f(a)(b)$, for each $a, b \in G$. Then $(G, *)$ is
associative and $(G, +, *)$ is a nearring if and only if $f(a) \circ f(b) = f(a * b)$ [15, Theorem
1.2]. The nearring $(G, +, *)$ is LSD if $f(a) \circ f(b) = f(a) \circ f(b) \circ f(a)$, for each $a, b \in G$.

Example 3.8. Let F be a free nearring [44] and J the ideal of F generated (as an ideal) by
$\{abc - abac \mid a, b, c \in F\}$. Then F/J is an LSD near-ring.

Theorem 3.9 ([8]). *Let R be an LSD nearring.*

(i) *If $K = \{x - x^2 \mid x \in r(0)\}$, then $\langle \mathcal{N}(R) \rangle = \langle K \rangle$ and $\langle \mathcal{N}(R) \rangle$ is a completely semi-
     prime ideal.*

(ii) *If R is zero symmetric, then $R / \langle \mathcal{N}(R) \rangle$ is a Boolean nearring.*

Theorem 3.10 ([8]). *Let R be an RSD nearring.*

(i) *$\mathcal{N}(R)$ is a right ideal.*

(ii) *If R is zero symmetric, then $\mathcal{N}(R)$ is an ideal of R and $R/\mathcal{N}(R)$ is a Boolean
     nearring.*

As indicated in [8], Theorem 3.9 and Theorem 3.10 generalize Theorem 3.1.

Theorem 3.11 ([8]). *Let R be a zero symmetric nearring. Then the following conditions
are equivalent.*

(i) *R is a reduced LSD nearring.*

(ii) *R is a reduced SD nearring.*

(iii) *R is a Boolean nearring.*

(iv) *R is a β-nearring.*

An ideal P is called 3-*prime* (3-*semiprime*) if $xRy \subseteq P$ ($xRx \subseteq P$) implies $x \in P$ or
$y \in P$ ($x \in P$).

Theorem 3.12 ([8]). *Let R be a zero symmetric nearring and either LSD or RSD. Then P is a completely prime (completely semiprime) if and only if P is a 3-prime (3-semiprime) ideal.*

Theorem 3.13 ([8]). *Let $R \neq 0$ be a LSD subdirectly irreducible nearring with heart H. Then exactly one of the following conditions is satisfied:*

(i) $RH = 0$ and $l(R) \neq 0$.

(ii) $0 \cdot R = R$ and H contains no nonzero normal subgroups of $(R, +)$ other than itself.

(iii) *Each element of R is either a left unity for R or a left annihilator for H (at least one element of each type exists); and either H contains a left unity of R and $H = r(0)$, or $H^2 = 0$ and $HR = 0 \cdot R$.*

Theorem 3.14 ([8]). *Let R be an LSD zero symmetric nearring and A a left ideal of R which is maximal among left ideals having zero intersection with $\langle \mathcal{N}(R) \rangle$. Then A contains every reduced right ideal of R. Hence $A + \langle \mathcal{N}(R) \rangle$ is a semidirect sum which is left and right essential in R and A is a Boolean nearring.*

Theorem 3.15 ([8]). *Let R be a nearring. The following conditions are equivalent:*

(i) *R is LSD, right permutable, and $\mathbf{gp}(I(R))$ is reduced.*

(ii) *R is right permutable and $R = B + \mathcal{N}(R)$ is a semidirect sum, where $\mathcal{N}(R)$ is an ideal, B is a Boolean ring and a two-sided R-subgroup which is contained in the center of R, and $R \cdot [\mathcal{N}(R)]^2 = 0$.*

(iii) *R is SD and permutable.*

The following corollary completes C. Ferrero Cotti's result which was stated in the opening paragraph of this section.

Corollary 3.16 ([8]). *Let R be a distributive nearring. The following conditions are equivalent:*

(i) *R is SD.*

(ii) *$R = B + \mathcal{N}(R)$ is a semidirect sum, where B is a Boolean ring, $\mathcal{N}(R)$ is an ideal, and $[\mathcal{N}(R)]^3 = 0 = B \cdot [\mathcal{N}(R)] = [\mathcal{N}(R)] \cdot B$.*

4. LSD-GENERATED ALGEBRAS AND NEARRINGS

Recently, H. E. Heatherly and the author have begun the investigation of *SD*-generated and *LSD*-generated algebras [10], [11]. These classes of algebras effectively generalize the self distributive theory to an extensive class of algebras as evidenced by the following examples. Moreover these algebras which are generated by elements which exhibit some type of self distributively seem to have a much deeper structure theory than one obtains in algebras generated by other types of elements (e.g., units).

In this section K will denote a commutative ring with unity and A a K-algebra (i.e., A is a unital K-module which is also a ring and which satisfies $\alpha(xy) = (\alpha x)y = x(\alpha y)$, for each $\alpha \in K$ and $x, y \in A$). Let

$$L(A) = \{a \in A \mid axy = axay, \text{ for each} x, y \in A\},$$

the set of *left self distributive* elements of A. Let

$$\mathcal{R}(A) = \{a \in A \mid xya = xaya, \text{ for each} x, y \in A\},$$

the set of *right self distributive* elements of A. $\mathcal{D}(A) = L(A) \cap \mathcal{R}(A)$ is the set of *self distributive* elements of A. Observe that $l(A^2) \subseteq L(A)$, $r(A^2) \subseteq \mathcal{R}(A)$, all central idempotents are in $\mathcal{D}(A)$, and all one-sided unity elements are in $\mathcal{D}(A)$. Also $L(A)$, $\mathcal{R}(A)$ and $\mathcal{D}(A)$ are subsemigroups of $(A, \cdot)$.

If X is a nonempty subset of A, then $\text{mod}_K(X)$ denotes the unital K-submodule of A generated by X. Observe that $\text{mod}_K(L(A))$, $\text{mod}_K(\mathcal{R}(A))$, and $\text{mod}_K(\mathcal{D}(A))$ are subalgebras of A. We say A is *LSD-generated*, *RSD-generated*, or *SD-generated*, if $A = \text{mod}_K(L(A))$, $A = \text{mod}_K(\mathcal{R}(A))$, or $A = \text{mod}_K(\mathcal{D}(A))$, respectively. Note if K is considered as a K-algebra, then K is SD-generated with the unity of K as its generator. Also every LSD, RSD, or SD ring is a LSD, RSD, or SD generated $\mathbb{Z}$-algebra, respectively.

Example 4.1. A K-algebra A is called *SCI-generated* if $A = \text{mod}_K(S_l(A))$, where $S_l(A) = \{e = e^2 \in A \mid Ae = eAe\}$. Observe $S_l(A) = \{e = e^2 \in A \mid xe = exe, \text{ for all } x \in A\}$. Since $S_l(A) \subseteq \mathcal{R}(A)$, then the class of *SCI-generated* algebras is a subclass of the class of *RSD-generated* algebras. For a particular example, let

$$A = \begin{pmatrix} K & V \\ 0 & K \end{pmatrix},$$

where V is a (K, K)-bimodule. A generating set for A is

$$\left\{ \begin{pmatrix} 1 & 0 \\ 0 & 1 \end{pmatrix} \right\} \cup \left\{ \begin{pmatrix} 1 & v \\ 0 & 0 \end{pmatrix} \middle| v \in V \right\} \subseteq S_l(A).$$

Observe that any commutative ring generated by idempotents is *SCI-generated* $\mathbb{Z}$-algebra. These rings were studied in [2], [27], and [40].

Example 4.2. Let G be a multiplicatively closed subset of A with $A = \text{mod}_K(G)$. Let I be the algebra ideal of A generated by

$$\{x_1 x_2 x_3 - x_1 x_2 x_1 x_3 \mid x_1, x_2, x_3 \in G\}.$$

Then A/I is an *LSD-generated* K-algebra. Similarly one obtains *RSD-generated* and *SD-generated* K-algebras.

Example 4.3. Let S be a semigroup. Form the semigroup algebra $K[S]$. If S is LSD, RSD, or SD then $K[S]$ is *LSD-generated*, *RSD-generated*, or *SD-generated*, respectively.

Theorem 4.4 ([10]). *Let A be SD-generated. If A is subdirectly irreducible with heart H, then exactly one of the following holds:*

(i) *A has a unity 1, $A = K1$ (hence A is a homomorphic image of K, so A is commutative).*

(ii) *$AH = 0 = HA$ and H is a simple K-module.*

(iii) *$HA = 0$, A has a left unity, and H is a simple K-module.*

(iv) *$AH = 0$, A has a right unity, and H is a simple K-module.*

Theorem 4.5 ([11]). *Let A be LSD-generated.*

(i) *Every prime ideal of A is completely prime.*

(ii) *If A is prime then A has a unity 1, $A = K1$, and A is a commutative domain.*

Theorem 4.6 ([11]). *Let A be LSD-generated with no infinite set of orthogonal idempotents. Then either $A^3 = 0$, or there exists $0 \neq e = e^2 \in L(A)$ such that:*

(i) *$A = eA \oplus r(e)$ and $r(e)$ is an ideal of A;*
(ii) *$[r(e)]^3 = 0$;*
(iii) *$eA = \bigoplus_{i=1}^{n} e_i A$, where $\{e_1, ..., e_n\}$ is a set of orthogonal primitive idempotents;*
(iv) *$e_i A = K e_i \oplus e_i l(e_i)$ is a direct sum of left ideals of $e_i A$ and $e_i l(e_i)$ is a nilpotent right ideal of A.*

By comparing Theorem 4.6 with Theorem 1.6, one may view Theorem 4.6 as an extension of Theorem 1.6 from *LSD* rings to *LSD*-generated algebras.

Example 4.7 ([11]). The following *LSD*-generated algebra A, illustrates the decomposition of Theorem 4.6.

$$A = \begin{pmatrix} K & B & 0 & 0 \\ 0 & K & 0 & 0 \\ 0 & C & 0 & 0 \\ 0 & D & K & 0 \end{pmatrix}$$

where B and C are (K,K)-bimodules, and D is a right K-module containing C, and K is an indecomposable commutative ring with 1. Then A is generated by

$$\left\{ \begin{pmatrix} 1 & 0 & 0 & 0 \\ 0 & 1 & 0 & 0 \\ 0 & 0 & 0 & 0 \\ 0 & 0 & 0 & 0 \end{pmatrix} \right\} \cup \left\{ \begin{pmatrix} 0 & b & 0 & 0 \\ 0 & 1 & 0 & 0 \\ 0 & c & 0 & 0 \\ 0 & d & k & 0 \end{pmatrix} \ \middle| \ b \in B, c \in C, d \in D, k \in K \right\} \subseteq L(A).$$

Then take

$$e = \begin{pmatrix} 1 & 0 & 0 & 0 \\ 0 & 1 & 0 & 0 \\ 0 & 0 & 0 & 0 \\ 0 & 0 & 0 & 0 \end{pmatrix}, e_1 = \begin{pmatrix} 1 & 0 & 0 & 0 \\ 0 & 0 & 0 & 0 \\ 0 & 0 & 0 & 0 \\ 0 & 0 & 0 & 0 \end{pmatrix}, e_2 = \begin{pmatrix} 0 & 0 & 0 & 0 \\ 0 & 1 & 0 & 0 \\ 0 & 0 & 0 & 0 \\ 0 & 0 & 0 & 0 \end{pmatrix}.$$

Then

$$r(e) = \begin{pmatrix} 0 & 0 & 0 & 0 \\ 0 & 0 & 0 & 0 \\ 0 & C & 0 & 0 \\ 0 & D & K & 0 \end{pmatrix},$$

$[r(e)]^2 \neq 0$ (if $C \neq 0$), but $[r(e)]^3 = 0$. Observe eA is not a left ideal of A,

$$l(e_1) = \begin{pmatrix} 0 & B & 0 & 0 \\ 0 & K & 0 & 0 \\ 0 & C & 0 & 0 \\ 0 & D & K & 0 \end{pmatrix}, \quad \text{and} \quad l(e_2) = \begin{pmatrix} K & 0 & 0 & 0 \\ 0 & 0 & 0 & 0 \\ 0 & 0 & 0 & 0 \\ 0 & 0 & K & 0 \end{pmatrix}.$$

K can be taken to be a commutative ring with unity which has no infinite set of orthogonal idempotents. Then there exists a complete set of primitive idempotents $\{c_1, ..., c_m\}$ of K.

The e_i of Theorem 4.6 are of the form

$$\begin{pmatrix} c_j & 0 & 0 & 0 \\ 0 & 0 & 0 & 0 \\ 0 & 0 & 0 & 0 \\ 0 & 0 & 0 & 0 \end{pmatrix} \quad \text{or} \quad \begin{pmatrix} 0 & 0 & 0 & 0 \\ 0 & c_j & 0 & 0 \\ 0 & 0 & 0 & 0 \\ 0 & 0 & 0 & 0 \end{pmatrix}.$$

We say a nearring R is *LSD-generated* if $R = \mathbf{snr}(L(R))$. Similarly we define a nearring R to be *RSD-generated* or *SD-generated* if $R = \mathbf{snr}(\mathcal{R}(R))$ or $R = \mathbf{snr}(\mathcal{D}(R))$, respectively. Since preliminary work on *LSD*-generated nearrings is currently in progress by the author, I will only present a few examples to illustrate the existence of these nearrings and one elementary result to give the flavor of this research.

Example 4.8. Let $(G, +)$ be a group and K a subgroup of G. Let

$$R = \{\alpha \in M(G) \mid K\alpha \subseteq K\}.$$

Then $L(R) = \{1\} \cup T$, where

$$T = \{\alpha \in M(G) \mid G\alpha \subseteq K \text{ and } \alpha \text{ restricted to } K \text{ is the identity map on } K\}.$$

Hence $\mathbf{snr}(L(R))$ and T are *LSD*-generated subnearrings of R. Observe if K is fully invariant, then $\mathcal{E}(G) \subseteq R$. Also if $M(G)$ is replaced by $M_o(G)$ in the definition of R, then the zero map is in $L(R)$.

Example 4.9. Let $(G, +)$ be a group and K a subgroup of G. Let

$$W = \{\alpha \in M(G) \mid G\alpha \subseteq K\}.$$

Then

$$\mathcal{D}(W) = \{\alpha \in M(G) \mid G\alpha \subseteq K \text{ and } \alpha \text{ restricted to } K \text{ is the identity map on } K\}.$$

Hence $\mathbf{gp}(\mathcal{D}(W))$ is an *SD*-generated d.g. subnearring of W. Moreover, if K is abelian, then $\mathbf{gp}(\mathcal{D}(W))$ is an *SD*-generated ring.

Example 4.10. Let C be a ring such that $x^n = x$ for all $x \in C$ and some fixed $n > 1$. Then C is a commutative ring and

$$L(C) = \mathcal{D}(C) = \{\text{idempotents}\} = \{0\} \cup \{x^{n-1} \mid x \in C\}.$$

Let $(R, +, *)$ be the nearring defined on the set of n-by-n matrices over C with usual addition and multiplication defined by

$$\alpha * \beta = (\det(\alpha))\beta$$

for $\alpha, \beta \in R$ where $\det(\alpha)$ denotes the determinant of α [6]. Then

$$L(R) = \{\alpha \in R \mid \det(\alpha) \in \mathcal{D}(C)\}.$$

So $\mathbf{snr}(L(R))$ is a *LSD*-generated subnearring of R. If $C = \mathbb{Z}_p$ (p a prime), then $R = \mathbf{gp}(L(R))$ so R is an abelian, zero symmetric, *LSD*-generated nearring which is not *SD*-generated (and not a ring). Moreover, R is a completely prime nearring (i.e., $xy = 0$ implies $x = 0$ or $y = 0$).

Proposition 4.11. *If R is a 3-prime nearring (i.e., $xRy = 0$ implies $x = 0$ or $y = 0$), then the nonzero elements of $L(R)$ are the left unities of R.*

Proof. Let $0 \neq v \in L(R)$. Then $vx(y - vy) = 0$, for all $x, y \in R$. Hence $vy = y$, for all $y \in R$. □

REFERENCES

[1] J. Angerer and G. Pilz, *The structure of near-rings of small order*, Lecture Notes in Computer Science No. **144** (Computer Algebra, Marseille 1982), Springer-Verlag, 1982, 57–64.

[2] G. M. Bergman, *Boolean rings of projection maps*, J. London Math. Soc. (2) **4** (1972), 593–598.

[3] G. F. Birkenmeier, *Seminearrings and nearrings induced by the circle operation*, Riv. Mat. Pura Appl. **5** (1989), 59–68.

[4] G. F. Birkenmeier and Y. U. Cho, *Additive endomorphisms generated by ring endomorphisms*, East-west J. Math. **1** (1998), 73–84.

[5] G. F. Birkenmeier and H. E. Heatherly, *Operation inducing systems*, Algebra Universalis **24** (1987), 137–148.

[6] G. F. Birkenmeier and H. E. Heatherly, *Medial near-rings*, Monatsh. Math. **107** (1989), 89–110.

[7] G. F. Birkenmeier and H. E. Heatherly, *Polynomial identity properties for near-rings on certain groups*, Near-Ring Newsletter **12** (1989), 5–15.

[8] G. F. Birkenmeier and H. E. Heatherly, *Left self-distributive near-rings*, J. Austral. Math. Soc. (Ser. A) **49** (1990), 273–296.

[9] G. F. Birkenmeier and H. E. Heatherly, *Rings whose additive endomorphisms are ring endomorphisms*, Bull. Austral. Math. Soc. 42 (1990), 145–152.

[10] G. F. Birkenmeier and H. E. Heatherly, *Self-distributively generated algebras*, Contributions to General Algebra 10, Proceedings of the Klagenferrt Conference (eds., D. Dorninger, G. Eigenthaler, H. K. Kaiser, H. Kautschitsch, W. More and W. B. Müller), Verlag Johannes Heyn, Klagenferrt, 1998.

[11] G. F. Birkenmeier and H. E. Heatherly, *Left self-distributively generated algebras*, Comm. Algebra, to appear.

[12] G. F. Birkenmeier, H. E. Heatherly, and T. Kepka, *Rings with left self distributive multiplication*, Acta Math. Hungar. **60** (1992), 107–114.

[13] M. Ćirić and S. Bogdanović, *Rings whose multiplicative semigroups are nil-extensions of a union of groups*, Pure Math. Appl. (Ser. A) **1** (1990), 217–234.

[14] M. Ćirić, S. Bogdanović, and T. Petković, *Rings satisfying some semigroup identities*, Acta Sci. Math. (Szeged) **61** (1995), 123–137.

[15] J. R. Clay, *The near-rings on groups of low order*, Math. Z. **104** (1968), 364–371.

[16] C. Ferrero Cotti, *Sugli stems il cui prodotto è distributivo ríspetto a se stesso*, Riv. Mat. Univ. Parma (3) **1** (1972), 203–220.

[17] P. Dehornoy, *An alternative proof of Laver's results on the algebra generated by an elementary embedding*, Set Theory of the Continuum (Eds. H. Judah, W. Just, H. Woodin), Springer-Verlag, New York, 1992, pp. 27–33.

[18] S. Dhompongsa and J. Sanwong, *Rings in which additive mappings are multiplicative*, Studia Sci. Math. Hungar. **22** (1987), 357–359.

[19] A. Drápal, *Finite left distributive algebras with one generator*, J. Pure Appl. Algebra **121** (1997), 233–251.

[20] M. Dugas, J. Hausen, and J. A. Johnson, *Rings whose additive endomorphisms are multiplicative*, Period. Math. Hungar. **23**(1991), 65–73.

[21] M. Dugas, J. Hausen, J. A. Johnson, *Rings whose additive endomorphisms are ring endomorphisms*, Bull. Austral. Math. Soc. **45** (1992), 91–103.

[22] S. Feigelstock, *Rings whose additive endomorphisms are multiplicative*, Period. Math. Hungar. **19** (1988), 257–260.

[23] S. Feigelstock, *Rings whose additive endomorphisms are N-multiplicative*, Bull. Austral. Math. Soc. **39** (1989), 11–14.

[24] S. Feigelstock, *Rings whose additive endomorphisms are n-multiplicative II*, Period. Math. Hungar. **25** (1992), 21–26.

[25] S. Feigelstock and R. Raphael, *Distributive multiplication rings*, Period. Math. Hungar. **25** (1992), 161–165.

[26] B. Ganter and H. Werner, *Co-ordinatizing Steiner systems*, Topics on Steiner Systems (Eds. C. C. Linder and A. Rosa), North-Holland, Amsterdam, 1980, pp. 3–24.

[27] R. Hill, *The additive group of commutative rings generated by idempotents*, Proc. Amer. Math. Soc. **38** (1973), 499–502.

[28] Y. Hirano, *On rings whose additive endomorphisms are multiplicative*, Period. Math. Hungar. **23** (1991), 87–89.

[29] J. Ježek and T. Kepka, *Distributive groupoids and symmetry-by-mediality*, Algebra Universalis **19** (1984), 208–216.

[30] J. Ježek, T. Kepka, and P. Němec, *Distributive Groupoids*, Rozpravy Ceskoslovenske Acad. Ved. Rada Mat. Prirod. Ved., **91/3**, Prague, 1981.

[31] D. Joyce, *A classifying invariant of knots, the knot quandle*, J. Pure Appl. Algebra **23** (1982), 37–66.

[32] A. V. Kelarev, *On left self distributive rings*, Acta Math. Hungar. **71** (1996), 121–122.

[33] T. Kepka, *Varieties of left distributive semigroups*, Acta Univ. Carolin. Math. Phys. **25** (1984), 3–18.

[34] T. Kepka, *Semirings whose additive endomorphisms are multiplicative*, Comment. Math. Univ. Carolinae **34** (1993), 213–219.

[35] T. Kepka and P. Němec, *Distributive groupoids and the finite basis property*, J. Algebra **70** (1981), 229–237.

[36] K. H. Kim and F. W. Roush, *Additive endomorphisms of rings*, Period. Math. Hungar. **12** (1981), 241–242.

[37] J. J. Malone, *Near-rings with trivial multiplication*, Amer. Math. Monthly **74** (1967), 1111–1112.

[38] J. J. Malone, *D. G. near-rings on the infinite dihedral group*, Proc. Royal Soc. Edinburgh **78A** (1977), 67–70.

[39] S. V. Matveev, *Distributive groupoids in knot theory*, Math. USSR Sbornik. **47** (1984), 73–83.

[40] C. J. Maxson, *Idempotent generated algebras and Boolean pairs*, Fund. Math. **93** (1976), 15–22.

[41] Lee Sin-Min, *Lattice of equational subclasses of distributive semigroups*, Nanta Math. J. **9** (1976), 65–69.

[42] M. Petrich, *Structure des demi-groupes et anneaux distributifs*, C. R. Acad. Sc. Paris Sec. A-B **268** (1969), A849-A852.

[43] G. Pilz, *List of low order near-rings, along with some special properties*, Near-Ring Newsletter **2** (1980), 5–23.

[44] G. Pilz, Near-rings, North-Holland, Amsterdam, 1983.

[45] A. Romanowska and B. Roszkowska, Algebra Universalis **26** (1989), 7–15.

[46] R. Scapellato, *On autodistributive near-rings*, Riv. Mat. Univ. Parma **10** (1984), 303–310.

[47] R. P. Sullivan, Research problem No. 23, Period. Math. Hungar. **8** (1977), 313–314.

[48] M. Willhite, *Distributively generated near-rings on the dihedral group of order eight*, M. S. Thesis, Texas A&M Univ., College Station, 1970.

[49] R. Yearbly and H. Heatherly, *The near-ring multiplications on Alt*(4), Near-Ring Newsletter **6** (1983), 59–73.

DEPARTMENT OF MATHEMATICS, UNIVERSITY OF SOUTHWESTERN LOUISIANA, LAFAYETTE, LA 70504, USA

ON THE DEVELOPMENT OF MATRIX NEARRINGS AND RELATED NEARRINGS OVER THE PAST DECADE

J. H. MEYER

ABSTRACT. Matrix nearrings (in their present form) originated at Stellenbosch in 1984. It is therefore appropriate to present a talk on this subject at this specific conference, and I thank the organizers for the opportunity to do so. In the talk I intend to highlight the origin (by APJ van der Walt and JDP Meldrum [17]), the development (by several people), and the possible future of these structures. Some attention will also be given to certain related structures which arose by using techniques similar to those employed in the definition of matrix nearrings.

1. INTRODUCTION

The object of this paper is to give a survey of what has been done over the past ten years or so in connection with matrix nearrings. How they originated, how they developed, and what next.

Obviously not all results can be included here. We shall therefore focus only on certain key aspects and some interesting examples, many of which will only be referred to — the idea being that the relevant references can be consulted by the reader interested in specific aspects. No radical theory, for example, will be discussed here, although substantial work has been done in this area (see for example [28], [34] and [35]). Every now and then an open question will be posed, for the obvious reason regarding the further development of the theory of matrix nearrings.

All nearrings in our discussion will be right nearrings, and unless otherwise specified, will also have an identity element.

2. FIRST ATTEMPTS

In 1964 Beidleman ([3]) tried to define matrix nearrings by considering the set $M_n(R)$ of $n \times n$ square arrays with entries from a given nearring R. It turned out that with the usual addition and multiplication of matrices, and if R has an identity, $M_n(R)$, $n > 1$, is a nearring if and only if R is a ring.

The only interesting case left was therefore to consider nearrings without identity. In subsequent papers by Heatherly and Ligh ([9], [11]) those nearrings R for which $M_n(R)$ is also a nearring were completely characterized:

Definition 2.1 ([11, p. 383]). A nearring R is called *n-distributive*, n a positive integer, if for all a,b,c,d,r,a_i and b_i in R,

1991 *Mathematics Subject Classification.* 16Y30 .

Y. Fong et al. (eds.), Near-Rings and Near-Fields, 23–34.
© 2001 Kluwer Academic Publishers. Printed in the Netherlands.

(1) $ab + cd = cd + ba$, and

(2) $(\sum_{i=1}^{n} a_i b_i) r = \sum_{i=1}^{n} (a_i b_i r)$

Theorem 2.2 ([11, p. 383]). *Let R be a nearring. Then $M_n(R)$, $n > 1$, is a nearring if and only if R is n-distributive.*

It follows that there are nearrings R (without identity, of course) such that $M_3(R)$ is a nearring, but neither $M_2(R)$ nor $M_4(R)$ is a nearring under the usual addition and multiplication of matrices (see Clay[5, 2.2, no. 22] for an example of a nearring which is 3-distributive, but not 2- or 4-distributive).

3. THE MATRIX NEARRINGS INTRODUCED BY MELDRUM AND VAN DER WALT

The notion of a matrix nearring over a general nearring was coined in 1984 during a visit by John Meldrum to Andries van der Walt in Stellenbosch. They found the hint for their construction in a combination of the facts that in the case of a ring R with identity

(1) an $n \times n$ matrix is a linear mapping $R^n \to R^n$ (where R^n denotes the direct sum of n copies of the abelian group $(R, +)$); and

(2) all $n \times n$ matrices over the ring R can be generated by the elementary matrices E_{ij} (with 1 in position (i, j) and zeros elsewhere).

Let then R denote any nearring with identity and n a positive integer. The $n \times n$ matrices over R are then certain mappings from R^n to R^n (where, in this case, R^n denotes the direct sum of n copies of the not necessarily abelian group $(R, +)$), defined in terms of elementary building-blocks which are in their turn defined as follows: for each $r \in R$ and $1 \leq i, j \leq n$,

$$f_{ij}^r : R^n \to R^n$$

is the mapping given by $f_{ij}^r \langle x_1, x_2, \ldots, x_n \rangle := \langle 0, 0, \ldots, rx_j, \ldots, 0 \rangle$, with rx_j in the i-th position and zeros elsewhere. Note that if R happens to be a ring, then f_{ij}^r corresponds to the $n \times n$ matrix with r in position (i, j) and zeros elsewhere, i.e., rE_{ij}. We are now ready to state the formal

Definition 3.1 ([17]). The *nearring of $n \times n$ matrices over R*, denoted by $M_n(R)$, is the subnearring of $M(R^n)$ generated by the set $\{f_{ij}^r \mid r \in R, 1 \leq i, j \leq n\}$. The elements of $M_n(R)$ will be referred to as *$n \times n$ matrices over R*.

It follows immediately that $M_n(R)$ is a right nearring with identity element $I = f_{11}^1 + f_{22}^1 + \cdots + f_{nn}^1$. Many of the basic properties, such as rules for matrix calculations and some straightforward relationships between R and $M_n(R)$ can be found in [17] and [22]. Interesting to note is the fact that $M_n(M_m(R)) \cong M_{mn}(R)$ for any $m, n \geq 1$ (which is well-known in ring theory) was only proved several years after the definition of a matrix nearring. See [23].

Let us look at two finite examples:

Example 3.2.   (1) Consider the nearring $R = \{0, 1, 2, 3\}$ with addition and multiplication given by:

+	0	1	2	3
0	0	1	2	3
1	1	0	3	2
2	2	3	0	1
3	3	2	1	0

·	0	1	2	3
0	0	0	0	0
1	0	0	0	1
2	0	1	2	2
3	0	1	2	3

This is the nearring listed as E(8) in [25, p. 408]. It follows from [19] that $M_2(R)$ contains 2^{16} elements with one nontrivial proper two-sided ideal I of size 2^{12}, where $I = \mathcal{J}_0(M_2(R)) = \mathcal{J}_2(M_2(R))$. In total, $M_2(R)$ contains 141 left ideals, including $\{0\}$, I and $M_2(R)$.

(2) Use the same R as in 1 with the same addition, but with multiplication given by:

·	0	1	2	3
0	0	0	0	0
1	1	1	1	1
2	0	1	2	3
3	1	0	3	2

This is the nearring listed as E(20) in [25, p. 408]. Here we find that $M_2(R)$ contains only 2^6 elements, less than a 2×2 matrix ring over a ring with 4 elements. As above, it contains exactly one nontrivial proper two-sided ideal I of size $2^2 = 4$, but not equal to the $\mathcal{J}$-radicals, since here we have $\mathcal{J}_0(M_2(R)) = \mathcal{J}_2(M_2(R)) = \{0\}$. In total, $M_2(R)$ contains 16 left ideals, including $\{0\}$, I and $M_2(R)$.

See [19] for a detailed description of the matrix nearrings shown above, as well as some other finite examples.

Open question 1: Suppose the matrix $U : R^n \to R^n$ is a bijection. So the inverse function $U^{-1} : R^n \to R^n$ exists. Is U^{-1} also a matrix?

4. GENERALIZED MATRIX NEARRINGS

In the definition of $M_n(R)$, R is viewed as a faithful, cyclic left module over itself. The action of R on the module $_RR$ is then used to interpret an element of $M_n(R)$ as a member of $M(R^n)$. This viewpoint was generalized by Smith ([27]), as follows: Let M be any faithful left R-module, and let M^n denote the direct sum of n copies of the group $(M, +)$. For each $r \in R$ and $1 \leq i, j \leq n$, define $f_{ij}^r : M^n \to M^n$ by $f_{ij}^r \langle m_1, m_2, \ldots, m_n \rangle :=$ $\langle 0, \ldots, 0, rm_j, 0, \ldots, 0 \rangle$ with rm_j in the i-th position and zeros elsewhere.

Definition 4.1 ([27, section 2]). The *nearring of* $n \times n$ *generalized matrices over* R, denoted by $M_n(R; M)$, is the subnearring of $M(M^n)$ generated by the set $\{f_{ij}^r \mid r \in R, 1 \leq i, j \leq n\}$. The elements of $M_n(R; M)$ will be referred to as *generalized* $n \times n$ *matrices over* R.

It is possible that for a nearring R and two faithful modules $_RM_1$ and $_RM_2$, $M_n(R; M_1)$ is not isomorphic to $M_n(R; M_2)$ ([27, example 2]). It follows trivially that $M_n(R, R) \cong M_n(R)$, and a natural question is: when is $M_n(R, M) \cong M_n(R)$? In this regard, we quote the following special cases:

Theorem 4.2 ([27, corollary 1]). *If M is a faithful R-module which is a homomorphic image of $_RR$, then $M_n(R; M) \cong M_n(R)$.*

Theorem 4.3 ([27, corollary 3]). *Let G be a finite group and $A \leq \text{Aut}(G)$. Then G is a faithful module over $R = M_A(G)$ in the natural way. If there exists a $v \in G$ with $\text{stab}(v) = \{1\}$, then $M_n(R, G) \cong M_n(R)$.*

This leads to:

> **Open question 2:** For a given nearring R, characterize all those faithful R-modules M such that $M_n(R, M) \cong M_n(R)$.

Let N be a zero-symmetric nearring with identity. The *right nearring* of N, denoted by **R**, is the subnearring of $M_0(N)$ generated by the right multiplication maps of N. In the following theorem, **R** is shown to be isomorphic to a suitable generalized matrix nearring for a certain type of nearring N.

Theorem 4.4 ([27, theorem 3]). *If $N = M_A(G)$, where G is finite and $A \leq \text{Aut}(G)$ is fixed point free, then $\mathbf{R} \cong M_n(S; G)$, where n is the number of nonzero A-orbits of G and S is the subnearring of $M_0(G)$ generated by A.*

5. GROUP NEARRINGS

As in the case of matrix nearrings, a group nearring for any nearring R and (multiplicative) group G can be defined by using certain elementary functions from the group $(R^G, +)$ into itself as generators, where R^G denotes the direct sum of $|G|$ copies of the group $(R, +)$. The idea is to define, for each $r \in R$, $g \in G$, a function $\langle r, g \rangle$ which will take the element, say x, indexed by $g^{-1}h$ in R^G, and put rx in position h, for each $h \in G$. More formally (see [10]), $\langle r, g \rangle : R^G \to R^G$ is the function defined by

$$(\langle r, g \rangle(\mu))(h) = r\mu(g^{-1}h)$$

for all $\mu \in R^G$ and $h \in G$. The *group nearring* $R[G]$ constructed from R and G is the subnearring of $M(R^G)$ generated by the set $\{\langle r, g \rangle \mid r \in R, g \in G\}$.

In [13] a definition for group nearrings is given, but specifically for the case where R is dg. Let us denote this group nearring by $R(G)$. Then we have

Theorem 5.1 ([10, theorem 3.2]). *If R is dg then $R[G]$ is a homomorphic image of $R(G)$.*

We also have a significant relationship between the matrix nearring $M_n(R[G])$ and the group nearring $(M_n(R))[G]$:

Theorem 5.2 ([6]). *There is an epimorphism of $(M_n(R))[G]$ onto $M_n(R[G])$.*

This leads to:

> **Open question 3:** Prove or disprove: $(M_n(R))[G] \cong M_n(R[G])$.

Several results on the ideal theory of $R[G]$, and in particular, the augmentation ideal of $R[G]$, can be found in [7], [8] and [10].

6. POLYNOMIAL NEARRINGS

In order to define a polynomial nearring over an arbitrary nearring R, an approach similar to that for matrix nearrings and group nearrings was initiated by van der Walt. This approach has been followed up by Bagley ([2]).

Let $R^{\mathbb{N}}$ denote the set of maps from $\mathbb{N} = \{0,1,2,\ldots\}$ into R. So each element of $R^{\mathbb{N}}$ can be written as $f = (f_0, f_1, f_2, \ldots)$, where $f_i = f(i) \in R$, $i \in \mathbb{N}$. For each $a \in R$, define the mapping $L_a : R^{\mathbb{N}} \to R^{\mathbb{N}}$ by

$$L_a(f_0, f_1, f_2, \ldots) := (af_0, af_1, af_2, \ldots)$$

and for each $i \in \mathbb{N}$, define the mapping $x^i : R^{\mathbb{N}} \to R^{\mathbb{N}}$ by

$$x^i(f_0, f_1, f_2, \ldots) := (\underbrace{0, 0, \ldots, 0}_{i \text{ zeros}}, f_0, f_1, f_2, \ldots).$$

Definition 6.1 ([2, p. 7]). The polynomial nearring $R[x]$ is the subnearring of $M(R^{\mathbb{N}})$ generated by $\{L_a \mid a \in R\} \cup \{x^i \mid i \in \mathbb{N}\}$.

Several elementary properties of $R[x]$ can be found in [2]. Various ideals exist in $R[x]$, and one such ideal is defined by making use of

Definition 6.2 ([2, p. 12]). Let $f \in R[x]$ be nonzero. The *least degree* of f, $\mathrm{ld}(f)$, is given by

$$\mathrm{ld}(f) := \min_{i \in \mathbb{N}, \alpha \in R^{\mathbb{N}}} \{i \mid f(\alpha)(i) \neq 0\}.$$

The *least degree* of the zero polynomial is ∞.

It can be shown that $I_n = \{f \in R[x] \mid \mathrm{ld}(f) \geq n\}$ is an ideal of $R[x]$, for each $n \in \mathbb{N}$. (In the ring case, I_n is simply the ideal $\langle x^n \rangle$ of $R[x]$, generated by x^n.) This gives us an interesting connection between polynomial nearrings and matrix nearrings:

Theorem 6.3 ([2, theorem 8]). $R[x]/I_n \cong L$, where L is the subnearring of $M_n(R)$ generated by $\{f_{i,0}^r + f_{i+1,1}^r + \cdots + f_{n-1,n-i-1}^r \mid r \in R, 0 \leq i \leq n-1\}$.

This is the analogue of the ring result which states that if R is a ring, then

$$R[x]/\langle x^n \rangle \cong L := \left\{ \begin{pmatrix} a_1 & 0 & 0 & \cdots & 0 & 0 & 0 \\ a_2 & a_1 & 0 & \cdots & 0 & 0 & 0 \\ a_3 & a_2 & a_1 & \cdots & 0 & 0 & 0 \\ \vdots & \vdots & \vdots & \ddots & \vdots & \vdots & \vdots \\ a_n & a_{n-1} & a_{n-2} & \cdots & a_3 & a_2 & a_1 \end{pmatrix} \, \middle| \, a_i \in R \right\}.$$

Let $D(R)$ denote the distributor ideal of R, i.e., the ideal of R generated by the set of distributors $\{a(b+c) - ac - ab \mid a,b,c \in R\}$. Furthermore, for any ideal I of R, define I^+ as the ideal of $R[x]$ generated by $\{ax^i \mid a \in I, i \in \mathbb{N}\}$ and I^* as the ideal $\{f \in R[x] \mid f(R^{\mathbb{N}}) \subseteq I^{\mathbb{N}}\}$ of $R[x]$. Then we have

Theorem 6.4 ([2, lemma 28]). $D(R)^+ \subseteq D(R[x]) \subseteq D(R)^*$.

This result should be compared with ([25, theorem 7.84, definition 7.85]) on "enclosing ideals." Similar results regarding ideals in matrix nearrings will be discussed later on (8.5).

Note that the definitions for matrix nearrings, group nearrings and polynomial nearrings all reduce to the standard ring theory situation whenever R happens to be a ring.

7. MODULES OVER MATRIX NEARRINGS AND PRIMITIVITY

Given a module $_RG$, it is natural to ask how we could define an action of $M_n(R)$ on $G^n = \oplus_{i=1}^n (G,+)$ which will turn G^n into a left $M_n(R)$-module. It turns out that, in order to have an action as natural as possible, the module $_RG$ needs to be somewhat special. This special property, originally defined by van der Walt ([33, definition 3.1]), is defined as follows:

Definition 7.1 ([15, definition 1.2]). The module $_RG$ is called *locally monogenic* if for each finite subset $H \subseteq G$, there exists a $g \in G$ such that $H \subseteq Rg$.

For a locally monogenic module $_RG$, we can define an action of $M_n(R)$ on G^n as follows: Let $U \in M_n(R)$ and $\langle g_1, g_2, \ldots, g_n \rangle \in G^n$. So there are $r_1, r_2, \ldots, r_n \in R$ and $g \in G$ such that $r_i g = g_i$, $i = 1, 2, \ldots, n$. Then

$$U \langle g_1, g_2, \ldots, g_n \rangle = U \langle r_1 g, r_2 g, \ldots, r_n g \rangle := \langle s_1 g, s_2 g, \ldots, s_n g \rangle,$$

where

$$U \langle r_1, r_2, \ldots, r_n \rangle = \langle s_1, s_2, \ldots, s_n \rangle.$$

This action of $M_n(R)$ on G^n ensures that G^n is a (left) $M_n(R)$-module. We agree to call it the "natural action." Type 2 modules behave quite well as far as this natural action is concerned:

Theorem 7.2 ([34, proposition 3.2]). *If Γ is an $M_n(R)$-module of type 2, then $\Gamma \cong_{M_n(R)} G^n$, where G is an R-module of type 2, and the action of $M_n(R)$ on G^n is the natural one.*

This result, however, is not true for type 0 modules in general. There do exist type 0 modules $_{M_n(R)}\Gamma$ which cannot be obtained by the natural action ([15, example 2.3]). Because of this "bad" behaviour of type 0 modules, it is possible to construct a nearring R such that $\mathcal{J}_0(R) = \mathcal{J}_2(R)$, but $\mathcal{J}_0(M_n(R)) \subsetneq \mathcal{J}_2(M_n(R))$ for all $n \geq 2$ ([15, example 3.3]).

2-Primitive nearrings are dense subnearrings of suitable centralizer nearrings ([12, theorem 3.35]). The next result shows how this phenomenon carries over to matrix nearrings:

Theorem 7.3 ([22, propositions 2.22 and 2.24]). *If R is not a ring, and 2-primitive on G, then $M_n(R)$ is dense in $M_A(G^n)$, where $A = \text{Aut}_{M_n(R)}(G^n)$. Furthermore, if A has only finitely many orbits on G^n, then $M_n(R) = M_A(G^n)$.*

Corollary 7.4 ([22, corollary 2.25]). *If F is a finite nearfield (not a field), then $M_n(F) = M_F(F^n)$, where the action of F on F^n is considered to be multiplication on the right ($\langle \alpha_1, \alpha_2, \ldots, \alpha_n \rangle \alpha := \langle \alpha_1 \alpha, \alpha_2 \alpha, \ldots, \alpha_n \alpha \rangle$, $\alpha, \alpha_1, \alpha_2, \ldots, \alpha_n \in F$).*

Do we have any flow of primitivity between R and $M_n(R)$? For 2-primitivity, once again, the situation seems to be as expected, namely, $M_n(R)$ is 2-primitive if and only if R is 2-primitive ([33, theorem 3.10]). The situation for 0-primitivity is a bit more involved (as could be expected!). For finite nearrings we do have a result:

Theorem 7.5 ([20, theorem 3.4]). *If R is finite, then $M_n(R)$ is 0-primitive if and only R is 0-primitive.*

This immediately leads to:

> **Open question 4:** Prove or disprove: for an arbitrary nearring R, $M_n(R)$ is 0-primitive if and only if R is 0-primitive.

8. LEFT AND TWO-SIDED IDEALS IN MATRIX NEARRINGS

The left ideal structure of a nearring (especially as far as minimality, DCC and maximality of the left ideals are concerned) tells us a lot about primitivity in the nearring. The next few results give some information on this issue, with the focus on matrix nearrings, of course.

When D is a division ring, then the set of matrices of the form

$$\begin{pmatrix} 0 & \cdots & D & \cdots & 0 \\ 0 & \cdots & D & \cdots & 0 \\ \vdots & \ddots & \vdots & \ddots & \vdots \\ 0 & \cdots & D & \cdots & 0 \end{pmatrix}$$

in $M_n(D)$ with arbitrary elements in the k-th column and zeros elsewhere constitute a minimal left ideal. The nearring analogue of this left ideal is $\text{li}\langle f_{1k}^1 \rangle$, the left ideal of $M_n(R)$ generated by the matrix f_{1k}^1. Surprisingly, it seems to be a very rare occasion for this left ideal to be minimal in the general nearring setting:

Theorem 8.1 ([21, theorems 15 and 16]). *If R contains at least one nondistributive element, then $\text{li}\langle f_{1k}^1 \rangle$ is not a minimal left ideal of $M_n(R)$, $n \geq 2$.*

Closely related to this result, is:

Theorem 8.2 ([22, proposition 2.29]). *Suppose $n \geq 2$ and let F be an infinite nearfield which is not a division ring. Then $M_n(F)$ does not satisfy the DCC for left ideals.*

By considering several examples (see, for instance, example 10.3) and in the light of theorems 8.1 and 8.2, it seems plausible to put forward the following

Open question 5: Prove or disprove: If F is an infinite nearfield (not a division ring), then $M_n(F)$ $(n \geq 2)$ does not contain a minimal left ideal.

This brings us to the matter of maximal left ideals in $M_n(R)$. Stone ([30]) characterized all maximal left ideals of a matrix ring as follows:

Theorem 8.3 ([30, theorem 1.2]). *If L is a maximal left ideal of a ring R and $\alpha \in R^n \setminus L^n$, then $(L^n : \alpha) = \{U \in M_n(R) \mid U\alpha \in L^n\}$ is a maximal left ideal of $M_n(R)$. Moreover, every maximal left ideal of $M_n(R)$ is of this form.*

This result is not true for arbitrary nearrings (see [20, example 2.5]). The best counterpart so far for this theorem in the nearing case, can be found in:

Theorem 8.4 ([15, theorems 4.2 and 4.3]). *Let $\alpha = \langle s_1, s_2, \ldots, s_n \rangle \in R^n$. Let M be the R-subgroup of R generated by $\{s_1, s_2, \ldots, s_n\}$ and let K be a maximal R-ideal of M. (K is in general not an R-ideal of R). Then $(K^n : \alpha)$ is a maximal left ideal of $M_n(R)$.*

Conversely, suppose L is a maximal left ideal of $M_n(R)$. Then L is of the form described above, or L has the following density property: there exists an $m \geq 1$ such that for any m vectors $\alpha_1, \alpha_2, \ldots, \alpha_m \in R^n$, and any $U \in M_n(R)$, there exists $L \in L$ such that $U\alpha_i = L\alpha_i$, $i = 1, 2, \ldots, m$.

It is conjectured that the part on the denseness of L need not be part of the theorem. Confirmation of this would have the implication that the first part of the theorem fully characterizes the maximal left ideals of $M_n(R)$. We formalize this as our next open problem:

Open question 6: Characterize the maximal left ideals of $M_n(R)$.

Let us now turn to (two-sided) ideals in matrix nearrings. For an ideal I of R, there are two natural ways (see [32]) to define a corresponding ideal in $M_n(R)$, namely

$$I^+ := \mathrm{id}\langle f_{ij}^a \,|\, a \in I \text{ and } 1 \le i, j \le n \rangle$$

and

$$I^* := (I^n : R^n) = \{U \in M_n(R) \,|\, U\alpha \in I^n \text{ for all } \alpha \in R^n\}.$$

That $I^+ \subseteq I^*$ follows trivially. When R is a ring, $I^+ = I^*$, but in the general nearring setting it can happen that $I^+ \subsetneqq I^*$ (see , for example, [32] and [15]). It was shown ([33, theorem 4.4] and [22, theorem 2.34]) that $\mathcal{J}_2(M_n(R)) = (\mathcal{J}_2(R))^*$, while $\mathcal{J}_0(M_n(R)) \subseteq (\mathcal{J}_0(R))^*$. In [15] an example was constructed to show that $\mathcal{J}_0(M_n(R)) \subsetneqq (\mathcal{J}_0(R))^*$ is indeed possible.

In trying to measure the "size" of the gap $I^+ \subsetneqq I^*$, the notion of an intermediate ideal was introduced:

Definition 8.5 ([14, definition 2.1]). An ideal I of $M_n(R)$ such that $I^+ \subsetneqq I \subsetneqq I^*$ for some ideal I of R is called an *intermediate ideal* of $M_n(R)$.

A consequence of this definition is that an intermediate ideal can never be of the form I^+ or I^* for any ideal I of R, and any ideal of $M_n(R)$ not of the form I^+ or I^* for all ideals I of R, must be an intermediate ideal of $M_n(R)$ (see [14, lemmas 2.2 and 2.3]).

Several examples of intermediate ideals have been constructed. In [18] examples of arbitrary long chains of intermediate ideals are given where the base nearring R is abelian, and in [14] lattices of intermediate ideals are constructed where R is a d-weakly distributive dg nearring of d-weak distributivity class 2. The question was also raised whether the $\mathcal{J}$-radicals of $M_n(R)$ can be intermediate. Because of the relation $\mathcal{J}_2(M_n(R)) = (\mathcal{J}_2(R))^*$ it follows trivially that $\mathcal{J}_2(M_n(R))$ can never be intermediate. In [16] an example of a finite abelian zero-symmetric nearring R is constructed with

$$(\mathcal{J}_0(R))^+ \subsetneqq \mathcal{J}_0(M_2(R)) \subsetneqq (\mathcal{J}_0(R))^*,$$

which shows that the $\mathcal{J}_0$-radical of a matrix nearring can be intermediate.

9. CENTRALIZER NEARRINGS AND MATRIX NEARRINGS

In [29] an extensive investigation was conducted to determine when a centralizer nearring $M_A(G)$, G a finite group and A a group of automorphisms of G, is isomorphic to a (nontrivial) matrix nearring. The main result in this regard is:

Theorem 9.1 ([29, theorem 3.2]). *If A is a group of fixed point free automorphisms of the finite group G, then, if $M_A(G)$ is not a ring, $M_A(G^n) \cong M_n(M_A(G))$ for every $n \ge 1$.*

Examples are also given in [29] to show that if A is not fixed point free, then it does not necessarily follow that $M_A(G^n)$ is isomorphic to any matrix nearring $M_n(R)$, $n \ge 2$.

Open question 7: Characterize those centralizer nearrings which are isomorphic to nontrivial matrix nearrings.

To conclude this section, we state the main result from [26], namely:

Theorem 9.2 ([26, corollary 17]). *If G is a finite group with $|G| \ge 3$, then*

$$M_0(G^n) \cong M_n(M_0(G)) \quad \text{for every } n \ge 1.$$

10. 2-PRIMITIVE MATRIX NEARRINGS AND NEAR-VECTOR SPACES

This section is mainly devoted to the work done by van der Walt in [31]. The main purpose of this paper was to show that every 2-primitive nearring with suitable finiteness conditions has a certain kind of matrix nearring associated with it, almost like the well-known Artin-Wedderburn theorem for left Artinian primitive rings.

First of all, let us quote two results (from [12]) by Betsch ([4]) on 2-primitive nearrings with minimal left ideals:

Theorem 10.1 ([12, theorem 4.5]). *Let R be a 2-primitive nearring with minimal left ideal K. If G is an R-module of type 2, then*

- (1) $G \cong_R K$
- (2) *There is an idempotent $e \in K \setminus \{0\}$ such that $K = Re = Ke$*
- (3) *Every nonzero element of K has rank 1.*

As a partial converse, we have:

Theorem 10.2 ([12, theorem 4.12]). *Suppose R is 2-primitive on the R-module G. If R contains an idempotent e of rank 1, then R has a minimal left ideal, or else we have*

$$(*) \qquad e \neq 1 \text{ and } \operatorname{Ann}_R(G \setminus eG) = \{0\}.$$

These two theorems were proved in 1971, and it was not known whether the exceptional case $(*)$ in theorem 10.2 can actually occur. In 1986, however, an example, using the theory of matrix nearrings, was constructed in [24] to show that $(*)$ is indeed possible. We provide this example explicitly:

Example 10.3 ([24]). Consider the infinite Dickson nearfield $F = (\mathbb{Q}(x), +, \circ)$ arising from the field of rational functions over the rationals, by defining multiplication as follows:

$$\frac{g(x)}{h(x)} \circ \frac{p(x)}{q(x)} := \begin{cases} 0 & \text{if } p(x) = 0 \\ \frac{g(x+d)}{h(x+d)} \cdot \frac{p(x)}{q(x)} & \text{if } p(x) \neq 0 \end{cases}$$

where $d = \operatorname{degree}(p(x)) - \operatorname{degree}(q(x))$.

Then $R := M_2(F)$ is 2-primitive on $G := F^2$ by [33, corollary 3.9]. Furthermore, $e = f_{11}^1$ is an idempotent in $M_2(F)$ and not equal to the identity $f_{11}^1 + f_{22}^1$, while $\operatorname{rank}(e) = 1$, since $f_{11}^1 G = \{\langle a, 0 \rangle \mid a \in F\}$ (see [24, proposition 1]). It then follows from [24, proposition 2] that

$$\operatorname{Ann}_R(G \setminus eG) = \{0\}.$$

It can also be shown directly that R does not contain any minimal left ideals ([24, proposition 3]).

The idea in [31] is to focus on 2-primitive nearrings with minimal R-subgroups rather that minimal left ideals. This gives us:

Theorem 10.4 ([31, theorem 2.1]). *Suppose R is 2-primitive on the R-module G. Then R has a minimal left R-subgroup if and only if R contains an idempotent e of rank 1.*

This result led to the concept of a CDI-nearring ([31, section 3]), which in turn links up with the theory of near-vector spaces. So in order to formulate the main results of this section, it is necessary to introduce the concept of a near-vector space:

Definition 10.5 ([1]). A *near-vector space* is a pair (G,A) where G is an abelian group and A is a group (with zero) of automorphisms of G such that:

   (1) Each nonzero element of A is fixed point free.
   (2) The automorphism -1 is in A.
   (3) $\mathrm{gp}\langle Q(G,A)\rangle = G$, where

$$Q(G,A) = \{g \in G \mid (\forall \alpha, \beta \in A)(\exists \gamma \in A) \text{ with } g\alpha + g\beta = g\gamma\}.$$

The elements of $M_A(G)$ are (naturally) called the *near-linear transformations* of the near-vector space (G,A). We have the following characterization of finite dimensional near-vector spaces:

Theorem 10.6 ([31, theorem 3.4]). *Let G be a group and let $A = D \cup \{0\}$, where D is a fixed point free group of automorphisms of G. Then (G,A) is a finite dimensional near-vector space if and only if there exist a finite number of nearfields $F_1, F_2, \ldots, F_n$, semigroup isomorphisms $\psi_i : A \to F_i$ ($1 \le i \le n$), and a group isomorphism $\Phi : G \to F_1 \oplus F_2 \oplus \cdots \oplus F_n$ such that if*

$$\Phi(g) = x_1 + x_2 + \cdots + x_n,$$

then

$$\Phi(g\alpha) = x_1 \psi_1(\alpha) + x_2 \psi_2(\alpha) + \cdots + x_n \psi_n(\alpha),$$

for all $g \in G$ and $\alpha \in A$.

Thus a finite dimensional near-vector space can be specified by taking n nearfields $F_1, F_2, \ldots, F_n$ such that there are semigroup isomorphisms $\theta_{ij} : (F_j, \cdot) \to (F_i, \cdot)$ with $\theta_{ij}\theta_{jk} = \theta_{ik}$ for $1 \le i, j, k \le n$. Then take $G = F_1 \oplus F_2 \oplus \cdots \oplus F_n$ as the additive group of the near-vector space and any one of the semigroups $(F_i, \cdot)$, say $(F_1, \cdot)$, as the semigroup of endomorphisms by defining

$$(x_1 + x_2 + \cdots + x_n)\alpha = x_1 \theta_{11}(\alpha) + x_2 \theta_{21}(\alpha) + \cdots + x_n \theta_{n1}(\alpha)$$

for all $x_j \in F_j$ and all $\alpha \in F_1$. The object is to study the nearring $R := M_{F_1}(G)$, namely, to determine the smallest subnearring S of R which contains the complete set of distributive idempotents $\{e_1, e_2, \ldots, e_n\}$ (where e_j is defined by $e_j(x_1 + x_2 + \cdots + x_n) = x_j$), and which is 2-primitive on G. It turns out that every square matrix $C = (c_{ij})$, where $c_{ij} \in F_i$, represents an element of S, with the action on G defined by

$$\begin{pmatrix} c_{11} & \cdots & c_{1n} \\ \vdots & \ddots & \vdots \\ c_{n1} & \cdots & c_{nn} \end{pmatrix} \begin{pmatrix} x_1 \\ \vdots \\ x_n \end{pmatrix} := \begin{pmatrix} c_{11}\theta_{11}(x_1) + \cdots + c_{1n}\theta_{1n}(x_n) \\ \vdots \\ c_{n1}\theta_{n1}(x_1) + \cdots + c_{nn}\theta_{nn}(x_n) \end{pmatrix}.$$

Moreover, it follows that the subnearring of R generated by all these elements is indeed equal to S, which is thence called the *nearring of matrices determined by the nearfields F_1, $F_2, \ldots, F_n$ and the matrix of isomorphisms* (θ_{ij}), and is denoted by $M_n(\{F_i\}, (\theta_{ij}))$.

Let us summarize this in:

Theorem 10.7 ([31, theorem 3.5]). *Suppose (G,A) is an n-dimensional near-vector space. Then $M_A(G)$ contains a subnearring S isomorphic to a nearring of matrices determined by n nearfields with isomorphic multiplicative semigroups. If S is not a ring, then S is dense in $M_A(G)$.*

11. CONCLUSION

It is clear that matrix nearrings play an important role in the general theory of nearrings: there are several connections between them and other nearrings (section 9); they are used in fundamental structure theorems (section 10); they seem to provide us with a rich and useful source of examples (10.3, for instance) — and in the theory of matrix nearrings itself, there are still many important open questions to be answered.

Although group nearrings and polynomial nearrings have up till now not yet been studied to the extent matrix nearrings have, it does not imply that these structures will not evolve in the near future as a useful source of algebraic knowledge.

ACKNOWLEDGEMENT

Financial assistance by the FRD and the Research Fund of the UOFS is gratefully acknowledged.

REFERENCES

[1] J. André, Algebra über fastkörpern, *Math. Z.* **136** (1974), 295-313.

[2] S. W. Bagley, *Polynomial near-rings, distributor and J_2 ideals of generalized centralizer near-rings.* Dissertation (Texas A&M University, 1993).

[3] J. C. Beidleman, *On near-rings and near-ring modules.* Dissertation (Pennsylvania State University, 1964).

[4] G. Betsch, *Some structure theorems on 2-primitive near-rings.* (In "Rings, Modules and Radicals, Hungary, 1971," Colloq. Math. Soc. János Bolyai, North-Holland, **6** (1973), 73-102.)

[5] J. R. Clay, The nearrings on groups of low order, *Math. Z.* **104** (1968), 364-371.

[6] R. L. Fray, Connections between group near-rings and matrix near-rings. Manuscript.

[7] R. L. Fray, On group distributively generated near-rings, *J. Austral. Math. Soc. (Series A)* **52** (1992), 40-56.

[8] R. L. Fray, On ideals in group near-rings, *Acta Math. Hungar.* **74** (1997), 155-165.

[9] H. E. Heatherly, Matrix near-rings, *J. London Math. Soc.* **7** (1973), 355-356.

[10] L. R. le Riche, JDP Meldrum and APJ van der Walt, On group near-rings, *Arch. Math.* **52** (1989), 132-139.

[11] S. Ligh, A note on matrix near-rings, *J. London Math. Soc.* **11** (1975), 383-384.

[12] J. D. P. Meldrum, *Near-rings and their links with groups*, (Pitman, 1985).

[13] J. D. P. Meldrum, The group distributively generated near-ring, *Proc. London Math. Soc.* **32** (1976), 323-346.

[14] J. D. P. Meldrum and JH Meyer, Intermediate ideals in matrix near-rings, *Comm. Algebra* **24**(5) (1996), 1601-1619.

[15] J. D. P. Meldrum and JH Meyer, Modules over matrix near-rings and the J_0-radical, *Monatsh. Math.* **112** (1991), 125-139.

[16] J. D. P. Meldrum and JH Meyer, The J_0-radical of a matrix nearring can be intermediate, *Canad. Math. Bull.* **40**(2) (1997), 198-203.

[17] J. D. P. Meldrum and APJ van der Walt, Matrix near-rings, *Arch. Math.* **47** (1986), 312-319.

[18] J. H. Meyer, Chains of intermediate ideals in matrix near-rings, *Arch. Math.* **63** (1994), 311-315.

[19] J. H. Meyer, *Computer generated matrix near-rings over some small near-rings with identity.* (Research report no.1, Dept. of Mathematics, University of the Orange Free State, 1993).

[20] J. H. Meyer, Left ideals and 0-primitivity in matrix near-rings, *Proc. Edinburgh Math. Soc.* **35** (1992), 173-187.

[21] J. H. Meyer, Left ideals in matrix near-rings, *Comm. Algebra* **17**(6) (1989), 1315-1335.

[22] J. H. Meyer, *Matrix near-rings.* Dissertation (Stellenbosch University, 1986).

[23] J. H. Meyer, On the near-ring counterpart of the matrix ring isomorphism $M_{mn}(R) \cong M_n(M_m(R))$, *Rocky Mountain J. Math.* **27**(1) (1997), 231-240.

[24] J. H. Meyer and APJ van der Walt, *Solution of an open problem concerning* 2-*primitive near-rings*. (In Near-rings and Near-fields, G. Betsch, ed, North-Holland, 1987, 185-191.)

[25] G. Pilz, *Near-rings,* revised edition (North-Holland, 1983).

[26] R. S. Rao, On near-rings with matrix units, *Quaestiones Math.* **17** (1994), 321-332.

[27] K. C. Smith, Generalized matrix near-rings, *Comm. Algebra* **24**(6) (1996), 2065-2077.

[28] K. C. Smith and L van Wyk, Solution of the J_2-radical problem in structural matrix near-rings. Submitted.

[29] K. C. Smith and L van Wyk, When is a centralizer near-ring isomorphic to a matrix near-ring?, *Comm. Algebra* **24**(14) (1996), 4549-4562.

[30] D. R. Stone, Maximal left ideals and idealizers in matrix rings, *Canad. J. Math.* **32** (1980), 1397-1410.

[31] A. P. J. van der Walt, Matrix near-rings contained in 2-primitive near-rings with minimal subgroups, *J. Algebra* **148** (1992), 296-304.

[32] A. P. J. van der Walt, *On two-sided ideals in matrix near-rings.* (In Near-rings and Near-fields, G Betsch, ed, North-Holland, 1987, 267-272.)

[33] A. P. J. van der Walt, Primitivity in matrix near-rings, *Quaestiones Math.* **9** (1986), 459-469.

[34] A. P. J. van der Walt and L van Wyk, The J_2-radical in structural matrix near-rings, *J. Algebra* **123** (1989), 248-261.

[35] S. Veldsman, Special radicals and matrix near-rings, *J. Austr. Math. Soc. (Series A)* **52** (1992), 356-367.

DEPARTMENT OF MATH. AND APPLIED MATH., UNIVERSITY OF THE ORANGE FREE STATE, PO BOX 339, 9300 BLOEMFONTEIN, SOUTH AFRICA. E-MAIL: MeyerJH@wis.nw.uovs.ac.za

THE USE OF COMPUTERS IN NEAR-RING THEORY

ERHARD AICHINGER, JÜRGEN ECKER, AND CHRISTOF NÖBAUER

ABSTRACT. We give a survey of our use of computers in near-ring theory.

1. THE PROBLEMS

In this section we want to state some problems in near-ring theory, and how computer algebra can be used to attack those. Our notation concerning near-rings follows [13].

(1) *Given a set of mappings on a group, compute the near-ring generated by these mappings.* Finding the near-rings $I(G), E(G), A(G)$ falls into this category of questions.

(2) *Given a group, compute its (unary) polynomial functions.* Polynomial functions on groups have received special interest from the viewpoint of near-ring theory [11], universal algebra [9], and at Vienna [10]; recent advances include [8, 3].

(3) *Given a set S of endomorphisms of a group G, compute the centralizer near-ring $M_S(G)$.* The centralizer near-ring is the near-ring of all mappings commuting with each single endomorphism in S; these near-rings play a central role in the structure theory of near-rings [5].

In all of these problems it is asked to compute a certain near-ring. But what does it mean to "compute" an algebraic structure? It definitely does not mean to compute a list of all of its elements, because although everybody likes the near-ring $E(S_4)$ and the large sporadic simple groups, no one wants to have all their elements on his desk. What we need instead is a representation of the algebraic structure that allows us to answer typical questions about it quickly. Returning to a near-ring of mappings on a group G, such typical questions might be:

(1) Is a given mapping $f : G \to G$ in the near-ring?
(2) What is the size of the near-ring?
(3) What are the ideals of the near-ring?
(4) Is it possible to produce "random elements" in the near-ring such that each element is chosen with equal probability?

We will now explain how we treat each of these problems using computer algebra. But before giving some aspects of the theory that stands behind the solution, we show what is possible using computers.

Supported by the project P 11486-TEC of the Austrian "Fonds zur Förderung der wissenschaftlichen Forschung"

The first author is also supported by a "Doktorandenstipendium" of the Austrian Academy of Sciences

35

Y. Fong et al. (eds.), Near-Rings and Near-Fields, 35–41.
© 2001 *Kluwer Academic Publishers. Printed in the Netherlands.*

2. A FEW PRACTICAL EXAMPLES OF THE USE OF COMPUTERS IN NEAR-RING THEORY

During the last two years, we have tried to implement all available methods for computing with near-rings into the group theory system GAP [15]. GAP is a computer algebra system with emphasis on groups that contains most advanced algorithms such as the Reidemeister-Schreier algorithm, the Todd-Coxeter algorithm, Knuth-Bendix methods, Tietze transformations, subgroup lattice algorithms and the Schreier-Todd-Coxeter algorithm (described e.g. in [17, 16]). GAP allows to compute and draw lattices of subgroups, to compute the orbits of group actions, to find the size of the permutation group generated by some permutations, etc. It also contains a programming language suitable for dealing with algebraic structures. In the most recent version of GAP, namely GAP 4, this language has the features of a real object-oriented programming language, and most of the group algorithms in GAP have been written in it; only a small kernel is in C. Furthermore, the user of GAP can read – and check – each single line of its source code, which is in contrast to other computer algebra systems.

The package we have written is called SONATA. A technical description of this product is given in [2].

2.1. Polynomial functions and distributively generable near-rings.
We examine the polynomial near-ring on the symmetric group S_4 (cf. [7]). In the following section we compute the near-rings $I(S_4)$, and then we find all the mappings i in $I(S_4)$ that satisfy $i(S_4) \subseteq A_4$. The lines starting with **gap>** show our input, the other lines the answers the system gives; all computations of this example can be done in a few seconds.

```
gap> S4 := SymmetricGroup ( 4 );
Sym[ 1 .. 4 ]

gap> I := InnerAutomorphismNearring ( S4 );
InnerAutomorphismNearring( Sym[ 1 .. 4 ] )

gap> Size (I);
927712935936

gap> A4 := CommutatorSubgroup (S4, S4);
Group( [ (1,2,3), (2,4,3) ], ... )

gap> N := NoetherianQuotient (I, A4, S4);
NoetherianQuotient( Group( [ (1,2,3), (2,4,3) ], ... ),
  Sym[ 1 .. 4 ] )

gap> Size ( N );
463856467968
```

2.2. Finding nice PM near-rings.
Prime ideals need not necessarily be maximal, not even in finite d.g. near-rings with identity. The following computations show how the

near-ring library can be used to find a near-ring with a non maximal prime ideal. We
search among all near-rings whose additive group is elementary abelian of order 8.

```
gap> Filtered( AllLibraryNearringsWithIdentity( GTW8_3 ), n ->
>          IsZeroSymmetricNearring( n ) and
>          not IsPMNearring( n ) );

[ LibraryNearring(8/3, 674) ]
```

2.3. Analyzing the ideals of a near-ring. We analyse a certain near-ring on the fourth
group of order 27 in Thomas and Wood's book [18]. This is the non-abelian group of
order 27 in which every element has order 3. The near-ring we are going to examine is
the 13th of all near-rings with identity on this group in SONATA's near-ring library. We
compute all its left ideals and its lattice of two-sided ideals. Finally, we find the finite field
GF(3) as a quotient of the near-ring.

```
gap> n := LibraryNearringWithIdentity(GTW27_4,13);
LibraryNearringWithIdentity(27/4, 13)

gap> i := NearringLeftIdeals(n);
[ NearringLeftIdeal(...), NearringLeftIdeal(...),
  NearringLeftIdeal(...), NearringLeftIdeal(...) ]

gap> List(i, Size);
[ 1, 3, 9, 27 ]

gap> Filtered(i, IsNearringRightIdealInParent);
[ NearringIdeal(...), NearringIdeal(...),
  NearringIdeal(...), NearringIdeal(...) ]

gap> Size( Intersection(i[2],i[3]) );
3

gap> f := n / i[3];
FactorNearring( LibraryNearringWithIdentity(27/4, 13),
  NearringIdeal(...) )

gap> IsNearfield(f) and IsCommutative(f);
true
```

3. HOW TO ATTACK THE PROBLEMS

The way of computing the polynomial functions on a group G reveals one of the typical
approaches for computing near-rings: The key observation is that it is a very easy task to
compute the subgroup generated by some elements of a big group $(M, +)$. Again, computing the subgroup does not mean computing a list of all its elements, but rather means
computing a representation from which we can quickly answer the typical questions on
this subgroup. For subgroups of permutation groups, all information lies in their stabilizer
chains [16].

Since one can therefore compute with groups very efficiently [16, 17], it is a good idea to translate a problem about near-rings into a problem about groups. For computing the polynomial functions on a group, this is an easy task: The group G^G can be represented as the $G^{|G|}$; this power can easily be computed from G. The polynomial functions are now just (the universe of) the subgroup of G^G generated by the constant mappings and the identity function. This allows to compute, e.g., the size of $P(S_4)$, which is 22 265 110 462 464, in a few seconds. In a similar way, one computes the near-rings generated by all endomorphisms and automorphisms of a group G. However, what one has to do first is to compute the set of all endomorphisms and automorphisms on this group. For computing the endomorphisms, we do not use any non-trivial mathematical insight, so we compute the endomorphisms by finding paths to the leaves of a certain search tree: our favourite books on these topics are [12, 1]. Finding the automorphisms is a problem that falls into the scope of computational group theory. Concluding, we see that for all the near-rings whose additive generators are known we obtain a nice representation of their additive groups. This allows us to compute e.g. random elements of a near-ring and its size, and to perform membership tests very quickly. Results of this approach have appeared in [4, 14].

For centralizer near-rings, it is not known in general what the generators of their additive groups are. But fortunately, it is easy to compute the intersection of two subgroups of a group. We can write the additive group of the centralizer near-ring $M_S(G)$ as the subgroup of G^G defined by

$$M_S(G) = \bigcap_{s \in S} \bigcap_{x \in G} \{f : G \to G \,|\, f(s(x)) = s(f(x))\}.$$

In both cases $s(x) = x$ and $s(x) \neq x$, it is easy to compute the generators of the subgroup $\{f : G \to G \,|\, f(s(x)) = s(f(x))\}$ of G^G from the generators and the elements of G, and we find the centralizer near-ring as the intersection of all those subgroups of G^G.

Now let A be a subset of G and let I be a subgroup of G. Intersecting any near-ring F of functions on G with the group $\{f : G \to G \,|\, f(A) \subseteq I\}$ yields the Noetherian Quotient $(I : A)_F$. Putting certain conditions on I and A, one can guarantee that $(I : A)_F$ is a left ideal or a subnear-ring of F. We know that in studying the polynomial functions on a group, it is important to describe those polynomial functions that map every group element into some normal subgroup I of G. These functions can be computed using the above technique.

In special cases, we do not have to do computations in G^G when computing the polynomial functions on a group, but only in G. This is the case for abelian groups because there polynomial functions have a nice normal form. A generalization to groups of nilpotency class 2 has been done in [6].

Nevertheless, the problem of finding the near-ring generated by some functions on a group G becomes hard if we have no idea how to find generators of the additive group of this near-ring. Even for one function and for special groups, the problem is completely unsettled. As we have mentioned, computing the additive closure is a purely group theoretic problem and can easily be solved with the help of computer algebra systems that contain the algorithms for working with groups, such as GAP or MAGMA. So knowing the additive generators of near-ring of transformations is the most advantageous situation.

In the case that the transformations are distributive generators of the near-ring, the closure problem can be solved in two steps: First, we compute the multiplicative semigroup generated by the transformations, and then its additive closure. For both closure operations GAP has built-in algorithms.

In all remaining cases no fast algorithm (i.e., an algorithm of average complexity less than linear in the order of the resulting near-ring) is known. It shall be remarked that in the case that we have additive generators, near-rings with 10^{50} elements can still be handled. Algorithms that compute a near-ring by enumerating its elements may work with at most 10^6 elements at the moment.

What is the straight-forward method to compute the near-ring generated by some transformations on a group? We could procede as follows: Starting with a set of generators, first we compute the additive group generated by those. Then we test whether this group is already multiplicatively closed. This test can only be performed in time nm in a near-ring of size n with m generators: each product of a generator with a near-ring element has to be tested. Now if such a product happens to be outside the group, we have to add it as a new generator and start again. This shows that the computing time is at least linear in the size of the resulting near-ring; therefore this approach allows to compute near-rings up to size 10^6.

In order to make this procedure feasible for near-rings of size up to 10^{50}, we avoid testing all possible products and test only a few instead. The hope is that if the group is not multiplicatively closed, then the probability that a randomly chosen product stays inside the group is extremely low. So testing enough products assures that the probability that the group passes all our tests, but is not multiplicatively closed is smaller than a certain ε. Yet, examples show that it can well be that a group is "nearly multiplicatively closed", i.e., only very few products do not lie in the group. To describe and somehow eliminate these cases would be a way of putting the random closure algorithm on solid theoretical ground. In practice a heuristic algorithm which uses a fixed number of a few hundred elements (instead of all) seems to work well, but we do not see why. It yields correct results much more often than one would expect. For those with strong nerves, the system SONATA offers the possibility to use this heuristic algorithm.

The SONATA package contains a library of small near-rings consisting of all near-rings up to order 15 and all near-rings with identity up to order 31 with the exception of the near-rings on the elementary abelian groups of order 16 and 27. Compiling this library near-rings is not as easy as it may seem at first sight. The algorithm constructing all near-rings with given additive group finds all possible near-ring multiplication tables by systematically filling all columns with group endomorphisms, keeping in mind that the resulting multiplication needs to be associative. This leads to the problem of traversing a tree. In our case, the leaves of the tree are completely filled multiplication tables such that the resulting multiplications are associative. In order to prune the search tree as much as possible, we use an extension procedure, which – based on the requirement of associativity – fills any partially filled multiplication table as much as possible. This seems to be similar to Yearby's approach [19]. Once all possible near-ring multiplications on a given group are computed, they are classified according to isomorphism.

4. Open Problems in Computing with Near-Rings

There are two problems concerning subnear-rings of the near-ring of all functions on a group that we would like to have answered. Solutions would mean a great speed-up in computing with certain near-rings

(1) *Given one function f on a group G, what is the near-ring generated by f?* This question has already been asked by G. Ferrero. Of course, to make the problem more difficult, one can increase the number of functions, but we think that an answer for one function, and for special groups, would already be nice. We recall that Sim's stabilizer chains solve similar problems in the theory of permutation groups.

(2) *Given a subgroup $(H, +)$ of $(G^G, +)$ that is not closed under functional composition, give a lower bound for the probability that a product $h_1 \circ h_2$ does not lie in H.* In the cases where such a bound exists, Jürgen Ecker's probabilistic closure algorithm would stand on solid ground.

5. SONATA - A System of Near-Rings and their Applications

All the methods mentioned in this note (and many more) have been implemented into the system SONATA, which can be obtained from its authors. For technical questions, the reader is either referred to [2] or cordially invited to contact the authors.

References

1. A. Aho, J. Hopcroft, and J. Ullman, *The design and analysis of computer algorithms*, Addison Wesley – Reading, Massachusetts, 1974.
2. E. Aichinger, F. Binder, J. Ecker, R. Eggetsberger, P. Mayr, and C. Nöbauer, *Sonata – a system of nearrings and their applications*, Near-ring newsletter - number 17 (Y. Fong, G. Pilz, A. Oswald, and K.C. Smith, eds.), Johannes Kepler University Linz, Austria, October 1997, pp. 65 – 86.
3. E. Aichinger and P.M. Idziak, *Polynomial interpolation in expanded groups*, in preparation, 1998.
4. E. Aichinger and C. Nöbauer, *The cardinalities of the endomorphism near-rings I(G), A(G), and E(G) for all groups G with $|G| \leq 31$*, Near-rings, near-fields and K-loops (G. Saad and M.J. Thomsen, eds.), Kluwer Acad. Publisher, 1997, pp. 175–178.
5. G. Betsch, *Some structure theorems on 2-primitive near-rings*, Coll. Math. Soc. J. Bolyai 6, North-Holland, Amsterdam, 1973.
6. J. Ecker, *On the number of polynomial functions on nilpotent groups of class 2*, Contributions to General Algebra, vol. 10, Verlag Johannes Heyn, Klagenfurt, 1998.
7. Y. Fong, *The endomorphism near-rings of the symmetric groups*, Ph.D. thesis, University of Edinburgh, 1979.
8. Y. Fong and K Kaarli, *Unary polynomials on a class of groups*, Acta Sci. Math. (Hungary) **61** (1995), 137–152.
9. G. Grätzer, *Universal algebra*, 2nd ed., Springer-Verlag, 1979.
10. H. Lausch and W. Nöbauer, *Algebra of polynomials*, North-Holland, Amsterdam, London; American Elsevier Publishing Company, New York, 1973.
11. J.D.P. Meldrum, *Near-rings and their links with groups*, Research Notes in Mathematics, vol. 134, Pitman Publishing Ltd., 1985.
12. N.J. Nilsson, *Principles of artificial intelligence*, Springer Verlag. Berlin-Heidelberg-New York, 1982.
13. G.F. Pilz, *Near-rings*, 2nd ed., North-Holland Publishing Company – Amsterdam, New York, Oxford, 1983.
14. G. Saad, S.A. Syskin, and M.J. Thomsen, *The inner automorphism nearrings I(G) on all nonabelian groups of order $|G| \leq 100$*, Near-rings, near-fields and K-loops (G. Saad and M.J. Thomsen, eds.), Kluwer Acad. Publisher, 1997, pp. 377–402.

15. M. Schönert et al., *GAP - Groups, Algorithms and Programming*, Lehrstuhl D für Mathematik, RWTH Aachen, 1994.
16. C. Sims, *Computational methods in the study of permutation groups*, Computational problems in abstract algebra, Conf. Oxford (J. Leech, ed.), 1970, pp. 169–183.
17. _____, *Computation with finitely presented groups*, Encyclopedia of Mathematics and Its Applications, vol. 48, Cambridge University Press, 1994.
18. A.D. Thomas and G.V. Wood, *Group tables*, Shiva Publishing Limited, 1980.
19. R.L. Yearby, *A computer-aided investigation of near-rings on low order groups*, Ph.D. thesis, University of Southwestern Louisiana, May 1973.

INSTITUT FÜR ALGEBRA, STOCHASTIK UND WISSENSBASIERTE MATHEMATISCHE SYSTEME, JOHANNES KEPLER UNIVERSITÄT LINZ, ALTENBERGERSTR. 69, A-4040 LINZ

SOME RESULTS ON DERIVATIONS IN NEARRINGS

NURCAN ARGAÇ* AND HOWARD E. BELL*

ABSTRACT. Let N denote a 3-prime near-ring. We prove that if $2N \neq \{0\}$ and d_1 and d_2 are nonzero derivations on N, then $d_1 d_2$ cannot act as a derivation on a nonzero additively-closed semigroup ideal. We then establish some results involving conditions of form $d(x)f(x) = 0$, where d is a derivation on N and f is an endomorphism of N.

There is an increasing body of evidence that prime nearrings with derivations have ring-like behavior; indeed, there are several results asserting that the existence of a suitably-constrained derivation on a 3-prime nearring forces the nearring to be a ring. It is our purpose to explore further this ring-like behavior. We first give a result on composition of derivations, which extends a recent result of Beidar, Fong, and Wang [1]; and then we extend some results of Brešar on derivations satisfying conditions of the form $d(x)f(x) = 0$ or $f(x)d(x) = 0$, where f is an appropriately-chosen map from a prime ring to itself [4, 5].

1. PRELIMINARIES

Throughout the paper, N will denote a zero-symmetric left nearring. An additive map $d : N \rightarrow N$ will be called a derivation if

$$(1:1) \qquad d(ab) = ad(b) + d(a)b$$

for all $a, b \in N$. More generally, an additive map d will be said to act as a derivation on the subset S of N if (1:1) holds for all $a, b \in S$.

As in [2], a nonempty subset U of N will be called a semigroup right ideal (resp. semigroup left ideal) if $UN \subseteq U$ (resp. $NU \subseteq U$); and if U is both a semigroup right ideal and a semigroup left ideal, it will be called a semigroup ideal. The symbol Z will denote the multiplicative center of N. The nearring N will be called 3-prime if $aNb = \{0\}$ implies $a = 0$ or $b = 0$.

We recall some elementary results from earlier papers.

Lemma 1.1 ([7, Proposition 1]). *If d is a derivation on N, then*

$$(1:2) \qquad d(ab) = d(a)b + ad(b) \qquad \text{for all } a, b \in N.$$

1991 *Mathematics Subject Classification.* 16Y30, 16W25 .

*Supported by the Natural Sciences and Engineering Research Council of Canada, Grant 3961.

Y. Fong et al. (eds.), Near-Rings and Near-Fields, 42–46.
© 2001 Kluwer Academic Publishers. Printed in the Netherlands.

Lemma 1.2 ([3, Lemma 1]). *If d is a derivation on N, then*

$$(ad(b) + d(a)b)c \ = \ ad(b)c + d(a)bc \qquad \text{for all } a,b,c \in N$$

$$\text{and}$$

$$(d(a)b + ad(b))c \ = \ d(a)bc + ad(b)c \qquad \text{for all } a,b,c \in N.$$

Lemma 1.3 ([3, Lemma 3]). *Let N be 3-prime. Then*

(i) *Z contains no nonzero zero divisors of N;*
(ii) *for any derivation d on N, $d(Z) \subseteq Z$.*

Lemma 1.4 ([2, Lemma 1.3]). *Let N be 3-prime, and let d be a nonzero derivation on N.*

(i) *If U is a nonzero semigroup right ideal (resp. semigroup left ideal) and x is an element of N such that $Ux = \{0\}$ (resp. $xU = \{0\}$), then $x = 0$.*
(ii) *If U is a nonzero semigroup right ideal or semigroup left ideal, then $d(U) \neq \{0\}$.*

Lemma 1.5 ([2, Lemma 1.4]). *Let N be 3-prime, and U a nonzero semigroup ideal of N. Let d be a nonzero derivation on N.*

(i) *If $x, y \in N$ and $xUy = \{0\}$, then $x = 0$ or $y = 0$.*
(ii) *if $x \in N$ and $d(U)x = \{0\}$, then $x = 0$.*

Our final lemma does not appear in earlier papers.

Lemma 1.6. *Let N be 3-prime, and U a nonzero semigroup ideal of N. If $2N \neq \{0\}$, then $2U \neq \{0\}$; and if d is any nonzero derivation on N, then $d(2U) \neq \{0\}$.*

Proof. Let $x \in N$ with $x + x \neq 0$. Then for every $u \in U$, $u(x + x) = ux + ux \in 2U$; and by Lemma 1.4(i), we get $\{0\} \neq U(x + x) \subseteq 2U$. Since $2U$ is a semigroup left ideal, it follows by Lemma 1.4(ii) that $d(2U) \neq \{0\}$. $\qquad\square$

2. A THEOREM ON COMPOSITION OF DERIVATIONS

In [1], Beidar, Fong, and Wang proved the following theorem, which extends a classic result of Posner [6] on prime rings.

Proposition 2.1. *Let N be 3-prime with $2N \neq \{0\}$, and let d_1 and d_2 be derivations on N. If $d_1 d_2$ is a derivation, then $d_1 = 0$ or $d_2 = 0$.*

We now provide an extension of this result.

Theorem 2.2. *Let N be 3-prime with $2N \neq \{0\}$, and let U be a nonzero semigroup ideal which is closed under addition. If d_1 and d_2 are derivations on N such that $d_1 d_2$ acts as a derivation on U, then $d_1 = 0$ or $d_2 = 0$.*

Proof. Assume that $d_1 \neq 0$ and $d_2 \neq 0$. Since $d_1 d_2$ acts as a derivation on U, we have

$$(2{:}1) \qquad d_1 d_2(xy) = x d_1 d_2(y) + d_1 d_2(x)y \text{ for all } x, y \in U.$$

On the other hand, we can write

$$d_1 d_2(xy) = d_1(d_2(xy)) = d_1(x d_2(y) + d_2(x)y) = d_1(x d_2(y)) + d_1(d_2(x)y);$$

therefore

$$(2{:}2) \qquad d_1 d_2(xy) = x d_1 d_2(y) + d_1(x)d_2(y) + d_2(x)d_1(y) + d_1 d_2(x)y \text{ for all } x, y \in U.$$

Comparing (2:1) and (2:2), we obtain

$$(2:3) \qquad d_1(x)d_2(y) + d_2(x)d_1(y) = 0 \qquad \text{for all } x, y \in U.$$

Replacing x by xz, $z \in U$, and making use of (1:2) and Lemma 1.2, we have

$$
\begin{aligned}
0 &= d_1(xz)d_2(y) + d_2(xz)d_1(y) \\
&= (d_1(x)z + xd_1(z))d_2(y) + (xd_2(z) + d_2(x)z)d_1(y) \\
&= d_1(x)zd_2(y) + xd_1(z))d_2(y) + xd_2(z)d_1(y) + d_2(x)zd_1(y) \\
&= d_1(x)zd_2(y) + x(d_1(z)d_2(y) + d_2(z)d_1(y)) + d_2(x)zd_1(y)
\end{aligned}
$$

for all $x, y, z \in U$; and since the second summand here is 0 by (2:3), we conclude that

$$(2:4) \qquad d_1(x)zd_2(y) + d_2(x)zd_1(y) = 0, \qquad \text{for all } x, y, z \in U.$$

Now substitute yt for y, where $t \in U$, obtaining

$$
\begin{aligned}
0 &= d_1(x)z(d_2(y)t + yd_2(t)) + d_2(x)z(yd_1(t) + d_1(y)t) \\
&= d_1(x)zd_2(y)t + (d_1(x)zyd_2(t) + d_2(x)zyd_1(t)) + d_2(x)zd_1(y)t.
\end{aligned}
$$

Since the middle summand is 0 by (2:4), we have shown that

$$(2:5) \qquad d_1(x)zd_2(y)t + d_2(x)zd_1(y)t = 0 \text{ for all } x, y, z, t \in U.$$

Now assume that t has been chosen so that $d_1(t) \in U$ also. Then

$$(2:6) \qquad d_1(x)zd_2(y)d_1(t) + d_2(x)zd_1(y)d_1(t) = 0.$$

In (2:4) we substitute $zd_1(y)$ for z and t for y, obtaining

$$d_1(x)zd_1(y)d_2(t) + d_2(x)zd_1(y)d_1(t) = 0,$$

that is,

$$d_2(x)zd_1(y)d_1(t) = -d_1(x)zd_1(y)d_2(t).$$

Substituting in (2:6), we now get

$$d_1(x)zd_2(y)d_1(t) - d_1(x)zd_1(y)d_2(t) = 0,$$

which can be written as

$$d_1(x)z(d_2(y)d_1(t) - d_1(y)d_2(t)) = 0.$$

By Lemmas 1.4(ii) and 1.5(i), we conclude that

$$(2:7) \qquad d_2(y)d_1(t) - d_1(y)d_2(t) = 0.$$

But by (2:3), we have $d_1(y)d_2(t) + d_2(y)d_1(t) = 0$; and by adding this to (2:7), we get

$$d_2(y)d_1(t) + d_2(y)d_1(t) = 0,$$

so that

$$d_2(U)(2d_1(t)) = \{0\}.$$

Since d_2 was assumed to be nonzero, by Lemma 1.5(ii) we see that

$$(2:8) \qquad d_1(2t) = 0 \text{ for all } t \in U \text{ for which } d_1(t) \in U.$$

Note that the semigroup ideal $U^2 = \{uv \mid u, v \in U\}$ is nonzero by Lemma 1.4(i), and observe that $d_1(U^2) = \{ud_1(v) + d_1(u)v \mid u, v \in U\} \subseteq U$. It follows from (2:8) that

$$(2:9) \qquad\qquad d_1(2U^2) = \{0\};$$

and by Lemma 1.6, $d_1 = 0$, in contradiction of our original assumption. This completes the proof. $\qquad\qquad\square$

Corollary 2.3. *Let N be 3-prime with $2N \neq \{0\}$, and let U be a nonzero semigroup ideal such that $U \cap Z \neq \{0\}$. If d_1 and d_2 are derivations on N such that $d_1 d_2$ acts as a derivation on U, then $d_1 = 0$ or $d_2 = 0$.*

Proof. In the proof of Theorem 2.2, closure of U under addition was used only in the final paragraph, to show that U contains a nonzero semigroup ideal V such that $d_1(V) \subseteq U$. If $U \cap Z \neq \{0\}$, choose a nonzero element $z \in U \cap Z$, and let $V = z^2 N$. Clearly V is a semigroup ideal contained in U, and it is nonzero by Lemma 1.3(i). Moreover, for each $x \in N$, $d_1(z^2 x) = z^2 d_1(x) + d_1(z^2)x = z^2 d_1(x) + 2z d_1(z)x = z(z d_1(x) + 2 d_1(z)x) \in U$. Therefore $d_1(V) \subseteq U$, and our argument proceeds as before. $\qquad\square$

3. EXTENSIONS OF A THEOREM OF BREŠAR

In [4] Brešar proved the following theorem.

Proposition 3.1. *Let R be a prime ring of characteristic not 2. If d is a nonzero derivation on R and f is an additive mapping on R such that $f(x)d(x) = d(x)f(x) = 0$ for all $x \in R$, then $f = 0$.*

It was observed in [5] that it is not sufficient, in general, to assume only $f(x)d(x) = 0$ or $d(x)f(x) = 0$.

For suitably-restricted f, we have obtained various extensions of Bresar's theorem. We present a sample of our results.

Theorem 3.2. *Let N be 3-prime with 1, let d be a nonzero derivation on N, and let f be an endomorphism of N. If $d(x)f(x) = 0$ for all $x \in N$, then $f = 0$.*

Proof. Since $d(1) = 0$, the condition $d(x+1)f(x+1) = 0$ yields $d(x)(f(x) + f(1)) = 0$, so $d(N)f(1) = \{0\}$. It follows by Lemma 1.5(ii) that $f(1) = 0$; hence $f(x) = f(1x) = f(1)f(x) = 0$ for all $x \in N$. $\qquad\square$

In our final two theorems we confine our attention to the case of f being the identity map.

Theorem 3.3. *Let N be 3-prime and d a derivation on N such that $xd(x) = d(x)x = 0$ for all $x \in N$. Then $d = 0$.*

Proof. Suppose $d \neq 0$. Let c be any constant, i.e., $d(c) = 0$. Then $d(x+c)(x+c) = 0 = d(x)(x+c) = d(x)x + d(x)c = d(x)c$ for all $x \in N$, so that $d(N)c = \{0\}$ and by Lemma 1.5(ii), $c = 0$.

Now $d(x^2) = xd(x) + d(x)x = 0$, so $x^2 = 0$ for all $x \in N$. Thus $x(x+y)^2 = 0 = (x(x+y))(x+y) = xy(x+y) = xyx$ for all $x, y \in N$; hence $xNx = \{0\}$ and therefore $N = \{0\}$, a contradiction. $\qquad\square$

Theorem 3.4. *Let N be 3-prime and let U be a nonzero semigroup ideal of N. If d is a derivation such that $d(x)x = 0$ for all $x \in U$ or $xd(x) = 0$ for all $x \in U$, then $d(Z) = \{0\}$.*

Proof. We consider the case $d(x)x = 0$ for all $x \in U$. Suppose $d(Z) \neq 0$, and let $z \in Z$ such that $d(z) \neq 0$. For each $x \in U$ we have $d(xz)xz = 0 = (xd(z) + d(x)z)xz$, so that by Lemma 1.2 we have $xd(z)xz + d(x)zxz = 0 = x^2 d(z)z + d(x)xz^2 = x^2 d(z)z$. It follows by Lemma 1.3 that $x^2 = 0$ for all $x \in U$. In particular, if $x, y \in U$ and $x + y \in U$ as well, we have $x(x+y)^2 = 0 = xyx$ (as in the proof of Theorem 3.3).

If we take $w \in U \setminus \{0\}$ and $x, y \in wU$, then $x = wu$ and $y = wv$ for $u, v \in U$, and $x + y = w(u + v) \in U$; therefore, $wuwvwu = 0 = wuwvwuw$. Thus $wuwUwuw = \{0\}$, so by Lemma 1.5(i), $wUw = \{0\}$ and hence $w = 0$, a contradiction.

The argument for the case $xd(x) = 0$ for all $x \in U$ is a trivial modification of the one just given. □

REFERENCES

[1] K. I. Beidar, Y. Fong, and X. K. Wang, *Posner and Herstein theorems for derivations of 3-prime near-rings*, Comm. Algebra **24** (1996), 1581–1589.

[2] H. E. Bell, *On derivations in near-rings, II*, Nearings, Nearfields, and K-Loops (G. Saad and M. J. Thomsen, eds.), Kluwer 1997, 191–197.

[3] H. E. Bell and G. Mason, *On derivations in near-rings*, Near-rings and Near-fields (G. Betsch, ed.), North-Holland 1987, 31–35.

[4] M. Brešar, *A note on derivations*, Math. J. Okayama Univ. **32** (1990), 83–88.

[5] M. Brešar, J. Škarabot, and J. Vukman, *On a certain identity satisfied by a derivation and an arbitrary additive mapping*, Aequationes Math. **45** (1993), 219–231.

[6] E. C. Posner, *Derivations in prime rings*, Proc. Amer. Math. Soc. **8** (1957), 1093–1100.

[7] X. K. Wang, *Derivations in prime near-rings*, Proc. Amer. Math. Soc. **121** (1994), 361–366.

DEPARTMENT OF MATHEMATICS, FACULTY OF SCIENCE, EGE UNIVERSITY, BORNOVA, IZMIR, TURKEY

DEPARTMENT OF MATHEMATICS, BROCK UNIVERSITY, ST. CATHARINES, ONTARIO, CANADA L2S 3A1

WEAKLY DIVISIBLE NEARRINGS: GENESIS, CONSTRUCTION AND THEIR LINKS WITH DESIGNS

A. BENINI, F. MORINI, AND S. PELLEGRINI

ABSTRACT. A nearring N is weakly divisible (wd-nearring) if, for each $x, y \in N$, there exists an element $z \in N$ such that $xz = y$ or $yz = x$. In the first part the structure of wd-nearrings with ascending or descending chain conditions is studied and wd-nearrings with ascending chain condition on the right annihilators are completely described. In the second part all zerosymmetric wd-nearrings on the group G of integers (mod p^n), p prime, in which pG is the set of all the nilpotent elements are characterised and constructed. At the end the possibility to construct partially balanced block designs from finite wd-nearrings is investigated. Non balanced designs and their related association schemes are derived from wd-nearrings having cyclic additive group G of order 2^n. For wd-nearrings on a cyclic additive group of order p^n, p prime and $p \neq 2$, an example is given.

1. INTRODUCTION

The first part of the present survey contains the main results on weakly- divisible nearrings obtained in [6, 4, 5]. In the last part we present some new partially balanced block designs and their respective association schemes derived from weakly divisible nearrings having additive group $(\mathbb{Z}_{p^n}, +)$, p prime.

1.1. **Wd-nearrings.** A weakly divisible nearring (wd-nearring in the following) is a nearring N satisfying the following: $\forall a, b \in N \exists x \in N \mid ax = b$ or $bx = a$.

The previous condition is fulfilled, for example, by the integral planar nearrings and it is not trivial even in ring-theory: $(\mathbb{Z}_n, +, \cdot)$ is a wd-ring if, and only if, n is a power of a prime number.

Wd-nearrings have been first introduced and studied in [6]: in the presence of chain conditions, a wd-nearring assumes a significant configuration and, in particular, under ascending chain condition on the right annihilators, it results to be shared in two distinct parts: the set Q of the nilpotent elements and the rest C, disjoint union of isomorphic monoids, consisting of the cancellable elements. This configuration immediately shows a certain resemblance to planar nearrings, but it is in the finite case that wd-nearrings reach a structure very similar to the planar ones [1]. While the wd-nearrings on $(\mathbb{Z}_{2^n}, +)$ are obtained in [6] in a natural way, the construction on $(\mathbb{Z}_{p^n}, +)$, p prime, is more complex (see [4, 5]).

1991 *Mathematics Subject Classification.* 16Y30.
Work carried out on behalf of Italian M.U.R.S.T.

47

Y. Fong et al. (eds.), Near-Rings and Near-Fields, 47–71.
© 2001 *Kluwer Academic Publishers. Printed in the Netherlands.*

In some papers written from 1964 to 1970 (see [7, 8, 10, 11]), James Clay began to work on the construction of nearrings on given additive groups. The problem, which was later developed by various authors, (see [2, 3, 13, 12, 19]), remains substantially open. In fact, except for some general theorems, a method explicitly describing the construction of nearrings on given additive groups is available only for certain specific classes of groups (see [13, 12, 17]). In the same period Giovanni Ferrero characterized finite planar nearrings as pairs (G, Φ), where G is an additive group and Φ is a subgroup of fixed point free automorphisms of $Aut(G)$. Applying the results of J.R. Clay concerning cyclic groups [7] and keeping in mind the construction developed by G. Ferrero for planar nearrings ([12, 14]), in [4] we give a construction method for the class $\mathcal{M}$ of all zerosymmetric wd-nearrings on the group $(\mathbb{Z}_{p^n}, +)$ of integers (mod p^n), p prime, in which $p\mathbb{Z}_{p^n}$ is the ideal Q of all the nilpotent elements. Precisely, if G is a cyclic group of prime power order p^n and Φ is an arbitrary subgroup of $Aut(G)$, all the wd-nearrings of $\mathcal{M}$ are constructible starting from the pair (G, Φ) and from the representatives of orbits of Φ selected in the following way: *if any two elements of two distinct orbits are congruent mod p^j, $j < n$, then also the respective representatives have to be congruent mod p^j.*

Too many computations are necessary to verify whether the above condition holds, even with the use of a computer. Therefore, in [5], after a detailed account of the orbits of an automorphism group of $(\mathbb{Z}_{p^n}, +)$, we define when two orbits are *p-equivalent*, i.e., their elements belong to the same residue classes (mod p), and we prove that the previous condition is automatically guaranteed if, and only if, *the representatives of p-equivalent orbits are chosen congruent mod p, for $p \neq 2$ and for $p = 2$ with Φ generated by $\alpha_{1+2^{n-h}}$: $x \longrightarrow x(1 + 2^{n-h})$, otherwise congruent mod 4.* Clearly, it is not complicated to verify the last condition and this makes the method suitable in view of an implementation of this construction in a computer program.

1.2. Designs. The possibility of constructing balanced and partially balanced block designs [13, 14, 15], beginning with planar nearrings is well-known. Here we show that also wd-nearrings turn out to be a good starting point for new tactical configurations. In [6] we have obtained non balanced designs (and the related association schemes) derived from wd-nearrings having additive group $(\mathbb{Z}_{2^n}, +)$. We now observe that the construction presented in [6] does not always work, in fact this method does not generate significant designs when applied to wd-nearrings constructed as in θ of Section 5 or to those on $(\mathbb{Z}_{p^n}, +)$, p prime and $p \neq 2$. Nevertheless, in this last case we are able to construct partially balanced designs (and the related association schemes) following a method different from the previous one.

2. PRELIMINARIES

In the following we refer to zerosymmetric left nearrings, without any explicit recall. For the notations and their basic elements, we refer to [20] and [9]. With regards to the number theory we refer to [16].

Definition 2.1. A nearring N is called weakly divisible if the following condition is satisfied:

$$(\alpha) \quad \forall a, b \in N \exists x \in N \mid ax = b \text{ or } bx = a.$$

From condition (α) some trivial properties, not expressly recalled in the following, are derived.

Proposition 2.2. *Let N be a wd-nearring.*

    (a) *Each non zero element $n \in N$ has a right identity and this one is non nilpotent.*

    (b) *If N is zero-symmetric, the non zero idempotent elements of N are left identities.*

    (c) *N has the identity if, and only if, its idempotent elements are central.*

Proof. (a) From condition (α) the equation $ax = a$ must have a solution x_0. If $ax_0 = a$, also $a(x_0)^n = a$ and x_0 is not a nilpotent element.

    (b) Let e be a non zero idempotent element of N. We can observe that e is not a left zero-divisor. In fact, suppose $et = 0$ with $t \neq 0$. From condition (α) there is an element x of N such that $tx = e$ or $ex = t$. If $e = tx$, then $e = ee = etx = 0$, but now $e \neq 0$; if $ex = t$, then $et = 0 = eex = ex = t$, but $t \neq 0$. So, $\forall h \in N, e(eh - h) = 0$ implies $eh = h$.

    (c) If the idempotent elements are central, there is only one idempotent element in N and this one is the identity of N. The converse is trivial. $\square$

Now we can explain the connection between the planar nearrings and the wd-nearrings.

Proposition 2.3. *A planar nearring is a wd-nearring if, and only if, it is integral.*

Proof. If N is an integral planar nearring, condition (α) is obviously verified. Viceversa, if N is a planar nearring and a is a zero-divisor, then it follows that $aN = 0$ and this is excluded in a wd-nearring. $\square$

In ring theory also, condition (α) is not trivially verified, in fact:

Proposition 2.4. *The rings $(\mathbb{Z}_n, +, \cdot)$ are wd-rings if, and only if, n is a power of a prime number.*

Proof. Let $n = p^h$ and a, b be elements of $\mathbb{Z}_n$. If at least one of them is relatively prime to n, it turns out to be invertible and (α) is obviously verified. If both a and b are not relatively prime to n, they are both multiples of powers of p, where the factors of multiplicity are invertible elements, and condition (α) also holds.

    Viceversa, if n is not a power of a prime number, it is in any case a product of prime numbers. Let p_1 and p_2 be two different prime divisors of n. Obviously it does not exist any $x \in \mathbb{Z}_n$ such that $p_1 = p_2 x$ or $p_2 = p_1 x$, hence (α) does not hold. $\square$

3. $ACCA_R$ IN WD-NEARRINGS

From [6], Prop. 9, a wd-nearring with the $ACCA_R$ is the disjoint union of Q, the set of the nilpotent elements, and C, the set of the cancellable elements. From Proposition 2.2(a), we know that C is a non empty set, so, for $a \in C$, we can define $B_a = \{x \in C \mid x1_a = x\}$, where 1_a is the right identity of a. Obviously $a \in B_a$ and $C = \bigcup_{a \in C} B_a$. The following theorem describes the structure of such a class of nearrings.

Theorem 3.1. *Let N be a wd-nearring with the $ACCA_R$.*

    (a) *If $a \in C$ then B_a is a multiplicative monoid and 1_a is its identity.*

    (b) *$\{Q, B_a, (a \in C)\}$ is a partition of N.*

(c) *Every* $1_a (a \in C)$ *is a left identity of N and the unique idempotent element of* B_a.

(d) *For every* $a \in C$ *we have* $B_a = CB_a$, *in particular* $B_a = C1_a$.

(e) *The map* $\Theta : B_a \to B_b$, *defined by* $\Theta(x) = x1_b$ $(a, b \in C)$, *is a monoid isomorphism.*

Proof. (a) If $a_1, a_2 \in B_a$, then $(a_1 a_2)1_a = a_1(a_2 1_a) = a_1 a_2$ and therefore $a_1 a_2 \in B_a$.

(b) If $x \in B_a \cap B_b$, then $x1_a = x1_b = x$, from which $x(1_a - 1_b) = 0$, and, due to $x \in C$, $1_a = 1_b$. Therefore C is the disjoint union of the B_as $(a \in C)$ and N is the disjoint union of Q and C.

(c) For every $n \in N$ and $a \in B_a$ we have $a(1_a n) = (a1_a)n = an$ and, by cancelling a, $1_a n = n$. Obviously, 1_a is idempotent and if h is another idempotent of B_a, they coincide.

(d) In fact $(c1_a)1_a = c(1_a 1_a) = c1_a$ implies $CB_a \subseteq B_a$, hence $CB_a = B_a$. In particular $C1_a = B_a$.

(e) Applying (d), it is a standard routine to verify that Θ is an isomorphism. $\square$

4. FINITE WD-NEARRINGS

In a finite wd-nearring N the set Q of the nilpotent elements is an ideal (see [6], Th. 6) and $N \setminus Q = C$ is the set of the cancellable elements (see [6], Prop. 9). Since the integral nearrings are well-known (see [13], Oss. 5), the following theorem describes finite non integral (and hence non-planar) wd-nearrings. From the previous Theorem 3.1, we get:

Theorem 4.1. *Let N be a finite wd-nearring.*

   (a) *The set C of the cancellable elements is the disjoint union of isomorphic groups.*

   (b) *For each* $a \in C$ *there exists a positive integer t such that* a^t *is a left identity of N.*

   (c) *The set Q of the nilpotent elements is a prime ideal and contains every right ideal of N.*

   (d) *The Jacobson radicals and the prime radical coincide with Q.*

Proof. (a) A finite monoid in which all the elements are left cancellable turns out to be a group.

(b) Obvious, from (a).

(c) From Th. 4 and Th. 6 of [6] we know that Q is a prime ideal. Moreover, from Prop. 8 of [6] the N-subgroups (and therefore the right ideals) are contained in Q.

(d) From (c) and Th. 5.61 of [20]. $\square$

Proposition 4.2. *Let N be a finite wd-nearring and q a nilpotent element of N. The set of the right identities of q is a multiplicative subsemigroup of C which contains at least one idempotent element.*

Proof. Let q be a non trivial nilpotent element of N. From Proposition 2.2 the set $R(q)$ of the right identities of q is a subset of C. Furthermore, $R(q)$ is closed with respect to the multiplication, hence it is a multiplicative semigroup of cancellable elements. Since each cancellable element has a power which is a left identity of N (see Theorem 4.1(b)), $R(q)$ obviously contains some idempotent elements. $\square$

We recall here some notations. Let $(N, +, \cdot)$ be a finite nearring with additive group N^+ and Φ a subgroup of $Aut(N^+)$. Let e be a selected representative of any orbit of Φ. For every h belonging to $\Phi(e)$, ϕ_h will denote an automorphism of Φ such that $\phi_h(e) = h$.

Obviously ϕ_h exists for every non zero element h in N, and, if the automorphisms of Φ are fixed point free, it is the only one. The identity of $Aut(N^+)$ will be denoted by id_N. The left translation defined by a, for $a \in N$, will be denoted by γ_a, that is $\gamma_a(x) = ax$, for every $x \in N$. If H is a subset of N, $\Gamma(H)$ denotes the set of the left translations identified by the elements of H. Recall that γ_a is an endomorphism of N^+ and it turns out to be an automorphism if, and only if, a is a cancellable element of N.

Proposition 4.3. *Let N be a finite wd-nearring.*

    (1) *$\Gamma(C)$ is a group of automorphisms of N^+.*
    (2) *$\Gamma(C) = \Gamma(B_a), \forall a \in C$.*
    (3) *Let $c \in C$ with $\gamma_c \neq id_N$. The fixed points of γ_c are nilpotent and form an N-subgroup of N.*
    (4) *The set $A = \{x \in N \mid \exists \gamma_c \in \Gamma(C)^*, \gamma_c(x) = x\}$ is an N-subgroup of N.*

Proof. (1) Obviously, $\Gamma(C)$ is a semigroup of automorphisms of N^+. Furthermore, from Theorem 4.1(b), for each $c \in C$ there is a power c^t which is a left identity of N. Thus $id_N = \gamma_{c^t}$ belongs to $\Gamma(C)$ and $\gamma_c^{-1} = \gamma_{c^{t-1}}$ is the inverse of γ_c.

(2) For all $a, b \in C$, $\Gamma(B_a) = \Gamma(B_b)$. In fact, for every $h \in B_a$, $\gamma_h(x) = hx = h(1_b x) = (h1_b)x = \gamma_{h1_b}(x)$. From $h1_b \in B_b$ it follows that $\gamma_h \in \Gamma(B_b) \, \forall h \in B_a$. In the same way we obtain $\gamma_k \in \Gamma(B_a) \, \forall k \in B_b$.

(3) Let $c \in C$ and $\gamma_c \neq id_N$. Let h be a fixed point of γ_c, that is $ch = h$. If h is cancellable, there is a power h^t which is a left identity of N thus, from $ch^t = h^t$, we obtain $ch^t x = h^t x$ and this implies $cx = x$, now excluded. Therefore h is nilpotent. It is routine to verify that $S(c) = \{x \in N \mid \gamma_c(x) = x\}$ is an N-subgroup of N.

(4) From (3) and Prop. 5 of [6], we know that A is the union of a finite number of linearly ordered N-subgroups, hence A is an N-subgroup. □

5. WD-NEARRINGS ON $(\mathbb{Z}_{2^n}, +)$

We are now able to present a class of examples by the following construction.

Firstly, to define a product over the additive group $(\mathbb{Z}_{2^n}, +)$ of integers (mod 2^n) in such a way that $(\mathbb{Z}_{2^n}, +, \cdot)$ becomes a nearring, it is sufficient to define the product $[x] \cdot [1]$, in fact $[x] \cdot [y] = y([x] \cdot [1])$.

Moreover, we can write each element of $\mathbb{Z}_{2^n}$ in the form $[x] = [2^t(\pm 1 + 4h)]$, where t is a non negative integer and h belongs to $\mathbb{Z}$. We obtain finite wd-nearrings by:

(θ) $[x] \cdot [1] = [\pm 2^t]$,   if $[x] = [k2^t]$ with k odd.

$$[x] \cdot [1] = [1], \quad \text{if} \begin{cases} 0 \leq h < 2^{n-3} & \text{and } [x] = [1 + 4h]; \\ 2^{n-3} \leq h < 2^{n-2} & \text{and } [x] = [3 + 4h], \end{cases}$$

$$[x] \cdot [1] = [2^{n-1} + 1], \quad \text{if} \begin{cases} 2^{n-3} \leq h < 2^{n-2} & \text{and } [x] = [1 + 4h]; \\ 0 \leq h < 2^{n-3} & \text{and } [x] = [3 + 4h]. \end{cases}$$

(δ) $[x] \cdot [1] = [1]$ if $t = 0$

      $[x] \cdot [1] = [2^t]$ if $t > 0$

(ζ) $[x] \cdot [1] = [\pm 2^t]$ if $t > 0$

      $[x] \cdot [1] = [1]$ if $t = 0$ and $[x] = [1 + 4h]$

      $[x] \cdot [1] = [2^{n-1} - 1]$ if $t = 0$ and $[x] = [-1 + 4h]$

In all these cases we obtain wd-nearrings in which the ideal of all the nilpotent elements $Q = \{[2^t(\pm 1 + 4h)], t > 0\}$ consists of the even numbers, while their cancellable part $C = \{[\pm 1 + 4h], h \in \mathbb{Z}\}$ is composed of the odd ones, but these wd-nearrings are not isomorphic.

In case (δ), every cancellable element is a left identity of N and forms a trivial group, so C is the disjoint union of such trivial groups.

In case (ζ) we obtain: $C = \bigcup_{h \in \mathbb{Z}} \{[1 + 4h], [-1 + 4(2^{n-3} - h)]\}$ disjoint union of groups of order two, whose identities are $[1 + 4h]$.

In case (θ), C is equal to the union of the following sets:

$$\bigcup_{0 \leq h < 2^{n-3}} \{[\underline{1 + 4h}], [1 + 4(2^{n-3} + h)]\} \text{and} \bigcup_{2^{n-3} \leq h < 2^{n-2}} \{[\underline{3 + 4h}], [3 + 4(2^{n-3} + h)]\}$$

again disjoint union of groups of order two, whose identities are the underlined elements.

The previous construction has suggested we investigate the wd-nearrings on $(\mathbb{Z}_m, +)$. In Section 6 we will study the case with $m = p^n$, p prime.

6. WD-NEARRINGS ON $(\mathbb{Z}_{p^n}, +)$

The particular additive structure of a nearring N on the group of integers (mod p^n) acts very strongly to determine the weaker multiplicative structure. For instance, we know that, for any x and y in N, $x \circ y = y \cdot (x \circ 1)$, where "$\circ$" and "$\cdot$" denote the multiplications in N and in the ring of integers (mod p^n) respectively (see [7]). In the following $p^{(t)}$ and p^t will denote the powers of p with respect to "$\circ$"and "$\cdot$". As usual, "$\cdot$" will be omitted.

We recall here that every automorphism α_k of $(\mathbb{Z}_{p^n}, +)$, where k is relatively prime to p, is defined by $\alpha_k : x \to kx$. The automorphism group of $(\mathbb{Z}_{p^n}, +)$ is a well-known group of order $p^{n-1}(p-1)$ whose subgroups containing only fixed point free automorphisms have order t which divides $p - 1$ (see [9], Chapter 2). We also recall that a nearring on a cyclic group admits the identity if, and only if, it is a commutative ring with identity (see [11]).

The following propositions describe some further properties of wd-nearrings with the additive group $G = (\mathbb{Z}_{p^n}, +)$. Firstly, we observe that $Q \subseteq p\mathbb{Z}_{p^n}$, because Q^+ is a subgroup of G.

Proposition 6.1. *Let N be a wd-nearring on $G = (\mathbb{Z}_{p^n}, +)$. If p divides the order of $\Gamma(C)$ then p is nilpotent.*

Proof. From Proposition 4.3(1) we know that $\Gamma(C)$ is an (abelian) subgroup of $Aut(G)$ hence, if p divides the order of $\Gamma(C)$, then there is an element of order p in $\Gamma(C)$: let γ_c be, for some $c \in C$.

Let $p \neq 2$. The elements of $Aut(G)$ of order p are those automorphisms of G defined by an element of the form $hp^{n-1} + 1$, with $1 \leq h \leq p - 1$, so $\gamma_c(p) = (hp^{n-1} + 1)p = p$. From Proposition 4.3(3) it follows that p is nilpotent.

Let $p = 2$. It is well-known that the elements of $Aut(G)$ of order 2 are the automorphisms $\alpha_{a_i} (i = 1, 2, 3)$ defined by the elements $a_1 = 1 + 2^{n-1}$, $a_2 = -1$, $a_3 = -1 + 2^{n-1}$ of $\mathbb{Z}_{2^n}$. When $|\Gamma(C)| = 2$, it results either $\Gamma(C) = \{id_N, \alpha_{a_1}\}$ or $\Gamma(C) = \{id_N, \alpha_{a_2}\}$ or $\Gamma(C) = \{id_N, \alpha_{a_3}\}$, thus we have to examine the following complementary cases: (1) $|\Gamma(C)| > 2$; (2) $\Gamma(C) = \{id_N, \alpha_{a_1}\}$; (3) $\Gamma(C) = \{id_N, \alpha_{a_2}\}$; (4) $\Gamma(C) = \{id_N, \alpha_{a_3}\}$.

Cases (1) and (2). Now α_{a_1} belongs to $\Gamma(C)$, hence 2 is nilpotent because it is fixed by α_{a_1}.

Case (3). If $\Gamma(C) = \{id_N, \alpha_{a_2}\}$ and we suppose 2 is cancellable, then $\gamma_2 \in \Gamma(C)$ and, hence, it must be $2 \circ 1 = \gamma_2(1) = \pm 1$. In both cases, it cannot be $2^{n-1} \circ 1 = 2^{n-1}$, otherwise

$$2^{n-1} = \pm 2^{n-1} = 2^{n-1} \circ (\pm 1) = 2^{n-1} \circ (2 \circ 1) = (2^{n-1} \circ 2) \circ 1 = [2(2^{n-1} \circ 1)] \circ 1 = 0,$$

and this is absurd. So $2^{n-1} \circ 1 \neq 2^{n-1}$.

Nevertheless, 2^{n-1} is always nilpotent because it is a fixed point of each element of $Aut(G)$, hence $2^{n-1} \circ 1$ is nilpotent too. Since $Q \subseteq 2\mathbb{Z}_{2^n}$, $2^{n-1} \circ 1 = 2^k b$ with (b,2)=1 and $1 < k < n - 1$. A direct verification shows that 2^{n-1-k} is a right identity of 2^{n-1} and, therefore, it is a cancellable element of N (see Proposition 2.2), hence $\gamma_{2^{n-1-k}} \in \Gamma(C)$ and thus $2^{n-1-k} \circ 1 = \gamma_{2^{n-1-k}}(1) = \pm 1$. We examine the two possibilities separately.

Suppose $2^{n-1-k} \circ 1 = 1$. Since $B_{2^{n-1-k}} = \{2^{n-1-k} - 2^{n-1-k}\}$, we have

$$(-2^{n-1-k}) \circ 1 = -1.$$

Thus

$$2^k b = 2^{n-1} \circ 1 = (-2^{n-1}) \circ 1 = [-(2^{n-1} \circ 2^{n-1-k})] \circ 1$$
$$= [2^{n-1} \circ (-2^{n-1-k})] \circ 1 = 2^{n-1} \circ [(-2^{n-1-k} \circ 1)]$$
$$= 2^{n-1} \circ -1 = -(2^{n-1} \circ 1) = -2^k b,$$

that is $2^k b = -2^k b$, but now this is excluded because of $k < n - 1$. Thus 2 is nilpotent.

Suppose $2^{n-1-k} \circ 1 = -1$. We have again

$$B_{2^{n-1-k}} = \{2^{n-1-k}, -2^{n-1-k}\},$$

but now $(-2^{n-1-k}) \circ 1 = 1$. As above, it results $-2^k b = 2^k b$ which is absurd.

Case (4). If $\Gamma(C) = \{id_N, \alpha_{a_3}\}$, the statement arises analogously to case (3). $\square$

Proposition 6.2. *Let N be a wd-nearring on $G = (\mathbb{Z}_{p^n}, +)$. The following statements are equivalent:*

(1) *p is a nilpotent element;*
(2) *$Q = p\mathbb{Z}_{p^n}$;*
(3) *the right identities of p are relatively prime to p.*

Proof. (1)$\Rightarrow$ (2) If p belongs to the subnearring Q, obviously $p\mathbb{Z}_{p^n}$ is included in Q. But $p\mathbb{Z}_{p^n}$ is a maximal subgroup of $\mathbb{Z}_{p^n}$, so $Q = p\mathbb{Z}_{p^n}$.

$(2) \Rightarrow (3)$ The right identities of p are cancellable (see Proposition 4.2) and if $Q = p\mathbb{Z}_{p^n}$, the cancellable elements of N are those relatively prime to p.

$(3) \Rightarrow (1)$ Let g be a right identity of p. Since g is relatively prime to p, then g is one of the generators of $(\mathbb{Z}_{p^n}, +)$, hence, for some k in $\mathbb{Z}_{p^n}$, it follows that $p = kg$, where p divides k because p and g are relatively prime. By induction, we can show that $p^{(t)} = k^{t-1} p$. In particular, we obtain $p^{(n)} = k^{n-1} p = 0$, because k is a multiple of p, hence p is nilpotent. $\square$

The previous Propositions 6.1 and 6.2 imply the following:

Theorem 6.3. *If N is a wd-nearring on $G = (\mathbb{Z}_{p^n}, +)$ and p divides the order of $\Gamma(C)$, then the set Q of the nilpotent elements coincides with $p\mathbb{Z}_{p^n}$.*

Thus all wd-nearrings on $(\mathbb{Z}_{2^n}, +)$ have $Q = 2\mathbb{Z}_{2^n}$, while, if $p \neq 2$, there exist wd-nearrings on $(\mathbb{Z}_{p^n}, +)$ with $Q = p\mathbb{Z}_{p^n}$ and also with $Q \neq p\mathbb{Z}_{p^n}$, when p does not divide the order of $\Gamma(C)$. That is shown by the following example.

Example 6.4. Let $G = (\mathbb{Z}_{81}, +)$ and define on $\mathbb{Z}_{81}$ the following multiplications:

$$a \circ b = \begin{cases} 0, & \text{if } a=0; \\ b, & \text{if } a \equiv_3 1 \text{ or } a = 3k \text{ with } k \equiv_3 1; \\ 80b, & \text{if } a \equiv_3 2 \text{ or } a = 3k \text{ with } k \equiv_3 2; \\ 9b, & \text{if } a = 27 \text{ or } a = 9k \text{ with } k \equiv_3 1; \\ 72b, & \text{if } a = 54 \text{ or } a = 9k \text{ with } k \equiv_3 2, \end{cases}$$

$$a \circ' b = \begin{cases} 0, & \text{if } a=0; \\ b, & \text{if } a \equiv_3 1; \\ 80b, & \text{if } a \equiv_3 2; \\ 3b, & \text{if } a = 3k \text{ with } k \equiv_3 1; \\ 78b, & \text{if } a = 3k \text{ with } k \equiv_3 2; \\ 9b, & \text{if } a = 9k \text{ with } k \equiv_3 1; \\ 72b, & \text{if } a = 9k \text{ with } k \equiv_3 2; \\ 27b, & \text{if } a = 27; \\ 54b, & \text{if } a = 54, \end{cases}$$

then $(\mathbb{Z}_{81}, +, \circ')$ turns out to be a wd-nearring with $Q = 3\mathbb{Z}_{81}$, while $(\mathbb{Z}_{81}, +, \circ)$ results a wd-nearring with $Q \neq 3\mathbb{Z}_{81}$. Both these constructions are possible, because $p = 3$ does not divide the order of $\Gamma(C) = \{id_G, -id_G\}$, in according to Theorem 6.3.

6.1. Construction.

In [12] Giovanni Ferrero shows how to construct, in the finite case, strongly monogenic nearrings, starting from an additive group G and a subgroup Φ of $Aut(G)$. The construction essentially consists in considering some, or even all, principal orbits of Φ and defining a non trivial product on them, so that they become mutually isomorphic groups, while a zero product is defined on the other orbits. With a suitable choice of Φ, in [13], the author can build a particular class of strongly monogenic nearrings, the planar and, specifically, integral planar nearrings. It is exactly in [13] that the (G, Φ) pair is introduced, where G is an additive group and Φ is a subgroup of $Aut(G)$ which only includes fixed point free automorphisms. This pair (G, Φ) is known in literature as *Ferrero pair*.

Even if in accordance with Ferrero's work, the construction described in this paper starts from a pair (G, Φ) which is not necessarily a Ferrero pair, in fact G equals $(\mathbb{Z}_{p^n}, +)$ and Φ is any subgroup of not necessarily fixed point free automorphisms of G. Thus, we are able to build non integral nearrings but with a trivial left annihilator, therefore, in particular, non integral planar nearrings and not even strongly monogenic.

In [7] Clay proved that every nearring on $(\mathbb{Z}_m, +)$ arises from a function π of $\mathbb{Z}_m$ in itself such that $\pi(a)\pi(b) = \pi(a\pi(b))$, for all $a, b \in \mathbb{Z}_m$, hereinafter called *Clay function*. Therefore, using this result, our purpose is *to determine a Clay function defining wd-nearrings on $(\mathbb{Z}_{p^n}, +)$ with $Q = p\mathbb{Z}_{p^n}$*.

In short, with K we will indicate the set of the elements relatively prime to p. Since K is fixed by any automorphism of G, we will denote B the set of the orbits of Φ in K, so $B = \{\Phi(a) \mid a \in K\}$, where, for each $a \in K$, $|\Phi(a)| = |\Phi|$.

Definition 6.5. Let $G = (\mathbb{Z}_{p^n}, +)$ and let Φ be a subgroup of $Aut(G)$. Select a representative for each orbit of B and let e be a choice element among the fixed representatives. For every $a \in \mathbb{Z}_{p^n}$ define[1]:

$$\pi(a) = \begin{cases} 0, & \text{if } a = 0; \\ p^r \phi_{ke^r}(e^{-r}), & \text{if } a = kp^r \text{ with } k \in \mathbb{Z}, (k, p) = 1 \text{ and } 0 \le r < n. \end{cases}$$

This definition turns out to be right because when π is a function, it is a Clay function, as proved by the following:

Lemma 6.6. *Let G, Φ and π be as in Definition 6.5. When π is a function, it fulfils the following:*

$$\pi(a)\pi(b) = \pi(a\pi(b)), \quad \text{for all } a, b \in G.$$

Proof. Take $a, b \in \mathbb{Z}_{p^n}$ with $a = hp^r$, $b = kp^s$, where h, k are relatively prime to p and $0 \le r, s < n$. We have:

$$\begin{aligned} \pi(a)\pi(b) &= p^r \phi_{he^r}(e^{-r}) p^s \phi_{ke^s}(e^{-s}) = \\ &= p^{r+s} e^{-(r+s)} \phi_{he^r}(1) \phi_{ke^s}(1) = p^{r+s} e^{-(r+s)} \phi_{ke^s}(\phi_{he^r}(1)), \end{aligned}$$

$$\begin{aligned} \pi(a\pi(b)) &= \pi(hp^{r+s} \phi_{ke^s}(e^{-s})) = p^{r+s} \phi_{h\phi_{ke^s}(e^{-s})e^{r+s}}(e^{-(r+s)}) = \\ &= p^{r+s} e^{-(r+s)} \phi_{he^r \phi_{ke^s}(1)}(1) = p^{r+s} e^{-(r+s)} \phi_{\phi_{ke^s}(he^r)}(1). \end{aligned}$$

Because $\phi_{\phi_x(y)} = \phi_x \circ \phi_y$, for each $x, y \in K$, then the proof is complete. $\qquad \square$

Unfortunately, π is not always a function, as shown in the following:

Example 6.7. Let $G = (\mathbb{Z}_{16}, +)$ and $\Phi = \{id_G, \alpha_7, \alpha_9, \alpha_{15}\}$. Since $|\Phi| = 4$, there are exactly two orbits in K: $\Phi(1) = \{1, 7, 9, 15\}$ and $\Phi(3) = \{3, 5, 11, 13\}$.

---

[1] We recall that ϕ_x denotes the automorphism of Φ such that $\phi_x(e_x) = x$, where e_x is the selected representative of $\Phi(x)$.

(a) Let 7 and 5 be the selected representatives of $\Phi(1)$ and $\Phi(3)$, respectively. Choose $e = 7$. In this case, for instance,

$$\pi(4) = 4\phi_{72}(7^{-2}) = 4\phi_1(1) = 12$$

while

$$\pi(5\cdot 4) = 4\phi_{5\cdot72}(7^{-2}) = 4\phi_5(1) = 4,$$

hence π is not a function.

(b) Let 7 and 11 be the selected representatives of $\Phi(1)$ and $\Phi(3)$, respectively. Choose $e = 7$. Now we are able to prove that π turns out to be a function. For every $a \in G$, $\pi(a)$ exists, so we now have to prove that $a = b$ implies $\pi(a) = \pi(b)$. If a belongs to K the statement is obvious. If a belongs to $2\mathbb{Z}_{16}$, keeping in mind that $\phi_x(1) = e_x^{-1}x$ $(x \in K)$, for $h, k \in \mathbb{Z}$ and $(k, 2) = 1$, we consider:

(i) $a = k2$ and $b = k2 + h16$. By Definition 6.5:

$$\pi(k2) = k2e_{k7}^{-1} \quad \text{and} \quad \pi(k2 + h16) = \pi((k + h8)2) = k2e_{(k+h8)7}^{-1}$$

Since $k7$ and $(k + h8)7$ belong to the same orbit, they have the same representative, so $\pi(a) = \pi(b)$;

(ii) $a = k4$ and $b = k4 + h16$. By Definition 6.5:

$$\pi(k4) = k4e_{k\cdot1}^{-1} \quad \text{and} \quad \pi(k4 + h16) = \pi((k + h4)4) = k4e_{(k+h4)1}^{-1}$$

Since the selected representatives are congruent to each other (mod 4), clearly, it follows that $4e_{k\cdot1}^{-1} = 4e_{(k+h4)1}^{-1}$, thus $\pi(a) = \pi(b)$;

(iii) $a = k8$ and $b = k8 + h16$. Similarly:

$$\pi(k8) = k8e_{k7}^{-1} \quad \text{and} \quad \pi(k8 + h16) = \pi((k + h2)8) = k8e_{(k+h2)7}^{-1}$$

Because of the same previous reason we have $8e_{k7}^{-1} = 8e_{(k+h2)7}^{-1}$, thus $\pi(a) = \pi(b)$ in any case.

In the previous example we can see that the choice of the representatives of the orbits of B is essential in order to make π a function. The following lemma provides a necessary and sufficient condition so that this always holds.

Lemma 6.8. *Let G, Φ and π be as in Definition 6.5. Then π is a function from $\mathbb{Z}_{p^n}$ in itself if, and only if, the following condition holds:*

(ω)   *if there are $x \in \Phi(a)$ and $y \in \Phi(b)$ such that $x \equiv_{p^j} y$, then the selected representatives of $\Phi(a)$ and $\Phi(b)$ are congruent to each other (mod p^j).*

Proof. Let π, defined as in the hypothesis, be a function. Let a, b be two elements of K congruent (mod p^j) and belonging to distinct orbits. Let e_a and e_b be the choice representatives of $\Phi(a)$ and $\Phi(b)$ respectively.

From $a \equiv_{p^j} b$ it follows $p^{n-j}e^{-(n-j)}a \equiv_{p^n} p^{n-j}e^{-(n-j)}b$. Keeping in mind that π is a function, then $\pi(p^{n-j}e^{-(n-j)}a) = \pi(p^{n-j}e^{-(n-j)}b)$, hence $\phi_a(1) \equiv_{p^j} \phi_b(1)$. Since $\phi_a(1) = e_a^{-1}a$ and $\phi_b(1) = e_b^{-1}b$, then $e_a^{-1}a \equiv_{p^j} e_b^{-1}b$ that is $e_b \equiv_{p^j} e_a$.

Conversely, let π be defined as in the hypothesis and let condition (ω) hold. Clearly, for every $a \in \mathbb{Z}_{p^n}$, $\pi(a)$ exists. Hence it is sufficient to show that $a = b$ implies $\pi(a) = \pi(b)$. If $a, b \in K$ the statement is clear.

If $a \notin K$ then $b \notin K$ too. Denote $a = kp^r$ and $b = (k + tp^{n-r})p^r$, for some $t \in \mathbb{Z}$, with $(k, p) = 1$ and $0 \leq r < n$. It follows:

$$(\beta) \qquad \pi(a) = p^r \phi_{ke^r}(e^{-r}) = e^{-r} p^r \phi_{ke^r}(1),$$

$$(\gamma) \qquad \pi(b) = p^r \phi_{(k+tp^{n-r})e^r}(e^{-r}) = e^{-r} p^r \phi_{(k+tp^{n-r})e^r}(1).$$

Comparing (β) and (γ), we can see that our statement is true if $\phi_{ke^r}(1) \equiv_{p^{n-r}} \phi_{(k+tp^{n-r})e^r}(1)$.

Let e_1 and e_2 denote the selected representatives of $\Phi(ke^r)$ and $\Phi((k+tp^{n-r})e^r)$ respectively, from (ω) it follows $e_1 \equiv_{p^{n-r}} e_2$, hence

$$\phi_{(k+tp^{n-r})e^r}(e_1) \equiv_{p^{n-r}} \phi_{(k+tp^{n-r})e^r}(e_2) = (k+tp^{n-r})e^r \equiv_{p^{n-r}} ke^r = \phi_{ke^r}(e_1).$$

Thus, since $e_1 \in K$, $\phi_{(k+tp^{n-r})e^r}(1) \equiv_{p^{n-r}} \phi_{ke^r}(1)$. $\qquad\qquad\square$

Therefore, if condition (ω) of Lemma 6.8 holds, then π is a Clay function and we can give the following:

Definition 6.9. Let G, Φ and π as in Definition 6.5. When π is a function, we define: $a * b = b\pi(a)$, for all $a, b \in G$.

Example 6.10. Definition 6.9 provides the following multiplication "$*$" on $\mathbb{Z}_{16}$:

*	0	1	2	3	4	5	6	7	8	9	10	11	12	13	14	15
0	0	0	0	0	0	0	0	0	0	0	0	0	0	0	0	0
1	0	7	14	5	12	3	10	1	8	15	6	13	4	11	2	9
2	0	14	12	10	8	6	4	2	0	14	12	10	8	6	4	2
3	0	9	2	11	4	13	6	15	8	1	10	3	12	5	14	7
4	0	12	8	4	0	12	8	4	0	12	8	4	0	12	8	4
5	0	15	14	13	12	11	10	9	8	7	6	5	4	3	2	1
6	0	2	4	6	8	10	12	14	0	2	4	6	8	10	12	14
7	0	1	2	3	4	5	6	7	8	9	10	11	12	13	14	15
8	0	8	0	8	0	8	0	8	0	8	0	8	0	8	0	8
9	0	15	14	13	12	11	10	9	8	7	6	5	4	3	2	1
10	0	14	12	10	8	6	4	2	0	14	12	10	8	6	4	2
11	0	1	2	3	4	5	6	7	8	9	10	11	12	13	14	15
12	0	4	8	12	0	4	8	12	0	4	8	12	0	4	8	12
13	0	7	14	5	12	3	10	1	8	15	6	13	4	11	2	9
14	0	2	4	6	8	10	12	14	0	2	4	6	8	10	12	14
15	0	9	2	11	4	13	6	15	8	1	10	3	12	5	14	7

where π is the function

$$\begin{pmatrix} 0 & 1 & 2 & 3 & 4 & 5 & 6 & 7 & 8 & 9 & 10 & 11 & 12 & 13 & 14 & 15 \\ 0 & 7 & 14 & 9 & 12 & 15 & 2 & 1 & 8 & 15 & 14 & 1 & 4 & 7 & 2 & 9 \end{pmatrix}$$

described in Example 6.7(b). Now $N = (\mathbb{Z}_{16}, +, *)$ turns out to be a nearring (see [7]) and, in particular, a wd-nearring with $Q = 2\mathbb{Z}_{16}$. Thus N is non integral, without non trivial left annihilators, and, therefore, non planar and not even strongly monogenic.

In general:

Theorem 6.11. *The structure* $N=(\mathbb{Z}_{p^n},+,*)$, *where "$*$" is defined as in Definition 6.9, is a wd-nearring whose ideal Q of the nilpotent elements coincides with $p\mathbb{Z}_{p^n}$.*

Proof. From Th. II of [7], Lemma 6.6 and Lemma 6.8, N is a (left) nearring. Now we have to verify that $(\mathbb{Z}_{p^n},+,*)$ is weakly divisible.

Assume $x = hp^r$ and $y = kp^s$ and suppose $s \leq r$. Take $a = hp^{r-s}(\phi_{ke^s}(e^{-s}))^{-1}$, it results $y * a = x$. In the same way we can proceed when $r \leq s$.

Finally, from Proposition 6.2, to prove $Q = p\mathbb{Z}_{p^n}$ can be reduced to show that p is nilpotent. Applying the induction principle we can show that $p^{(t)} = p^t [\phi_e(e^{-1})]^{t-1}$. From this it follows $p^{(n)} = 0$, hence p is nilpotent. $\qquad\square$

Conversely, we now prove that *every wd-nearring on* $(\mathbb{Z}_{p^n},+)$ *with* $Q = p\mathbb{Z}_{p^n}$ *is constructible as in Theorem* 6.11. For our purpose the following will be useful.

Lemma 6.12. *Let N be a wd-nearring on $G = (\mathbb{Z}_{p^n},+)$, with $Q = p\mathbb{Z}_{p^n}$. For every $k \in \mathbb{Z}$, it results $kp^t \circ 1 = p^t e^{-t}(ke^t \circ 1)$, where $1 \leq t < n$ and e is an idempotent right identity of p.*

Proof. From the hypothesis we have $p \circ e = p$ where e is relatively prime to p and hence a unit of the ring $\mathbb{Z}_{p^n}$, so $p \circ 1 = e^{-1}p$.

Consequently, $kp = \underbrace{p \circ e + p \circ e + \cdots + p \circ e}_{k\ \text{terms}} = p \circ ke$ and $p^{(2)} = p^2 e^{-1}$. By induction we can prove $kp^t = p^{(t)} \circ ke^t$ and $p^{(t)} \circ 1 = p^t e^{-t}$. Thus

$$kp^t \circ 1 = p^{(t)} \circ ke^t \circ 1 = (ke^t \circ 1)(p^{(t)} \circ 1) = p^t e^{-t}(ke^t \circ 1).$$

$\qquad\square$

Theorem 6.13. *Every wd-nearring N on $G = (\mathbb{Z}_{p^n},+)$ with $Q = p\mathbb{Z}_{p^n}$ is constructible as in Theorem 6.11 when:*

(1) $\Phi = \Gamma(C)$;

(2) *the representatives of the orbits of Φ equal the idempotent elements of N;*

(3) *e equals an idempotent right identity of p.*

Proof. Let $N = (\mathbb{Z}_{p^n},+,\circ)$ be as in hypothesis. According to (1), (2) and (3), assume "$*$" as in Definition 6.5. We need to verify that, for every $a,b \in N$, we have $a \circ b = a * b$, but, since $a * b = b(a * 1)$ and $a \circ b = b(a \circ 1)$, it will be sufficient to show $a \circ 1 = a * 1$.

Let $a \in C$. From the hypothesis, now we have $\Phi = \Gamma(C)$ hence $\Phi(a) = B_a$, consequently $a \circ 1_a = a$ and also $\phi_a(1_a) = a$. Then $a \circ 1 = 1_a^{-1}[1_a(a \circ 1)] = 1_a^{-1}(a \circ 1_a) = 1_a^{-1}a = 1_a^{-1}\phi_a(1_a) = \phi_a(1_a 1_a^{-1}) = \phi_a(1) = a * 1$.

Let $a \in Q$, $a = kp^t$ with $(k,p) = 1$. Using Lemma 6.12 and since $ke^t \in C$, we have $a \circ 1 = kp^t \circ 1 = p^t e^{-t}(ke^t \circ 1) = p^t e^{-t}(ke^t * 1) = p^t e^{-t}\phi_{ke^t}(1) = p^t \phi_{ke^t}(e^{-t}) = a * 1$ and this completes the proof. $\qquad\square$

Now we proceed to describe a method for choosing the representatives of the orbits of K so that condition (ω) is automatically guaranteed.

To our purpose we need to show some properties of the orbits of Φ distinguishing two cases: $p \neq 2$ and $p = 2$.

6.2. **Case** $p \neq 2$. In this section G denotes the additive group of integers (mod p^n) with $p \neq 2$ and Φ an arbitrary subgroup of $Aut(G)$.

It is well-known that $|Aut(G)| = (p-1)p^{n-1}$ and if the order of Φ is tp^h, with $(p,t) = 1$, then $\Phi = T \otimes \Phi_h$, where $T \leq Aut(G)$ is a group of fixed point free automorphisms of order t and $\Phi_h = \{\alpha_k : a \to ka \mid k = bp^{n-h} + 1, 0 \leq b \leq p^h - 1\} \leq Aut(G)$ has order p^h (see [9], Chapter 2).

Proposition 6.14. Let $G = (\mathbb{Z}_{p^n}, +)$ with $p \neq 2$.

    (1) If β_1 and β_2 are distinct automorphisms of G whose orders divide $p-1$, then $\beta_1(a) \not\equiv_p \beta_2(a)$, for every a relatively prime to p.

    (2) If φ_1 and φ_2 are automorphisms of G of orders p^r and p^h, $r \leq h$, respectively, then $\varphi_1(a) \equiv_{p^{n-h}} \varphi_2(a)$, for every a relatively prime to p.

Proof. (1) Suppose $\beta_1(a) \equiv_p \beta_2(a)$, for some $a \in K$. Then $p^{n-1}\beta_1(a) = p^{n-1}\beta_2(a)$, so $(\beta_2^{-1}\beta_1)(p^{n-1}a) = p^{n-1}a$, but this is excluded because otherwise $p^{n-1}a$ should be a fixed point of $\beta_2^{-1}\beta_1$.

(2) It is well known that φ_1 and φ_2 are determined by elements of the form $bp^{n-h} + 1$, $0 \leq b \leq p^h - 1$. Thus, for all $a \in K$, we have $\varphi_1(a) = (b_1p^{n-h} + 1)a$ and $\varphi_2(a) = (b_2p^{n-h} + 1)a$ for suitable b_1 and b_2, hence $\varphi_1(a) - \varphi_2(a) \equiv_{p^{n-h}} 0$. $\square$

The previous statements allow us to describe the elements of the orbits of Φ included in K.

Corollary 6.15. Let $G = (\mathbb{Z}_{p^n}, +)$, with $p \neq 2$, and $\Phi = T \otimes \Phi_h \leq Aut(G)$ of order tp^h, where t divides $p-1$. Then:

    (1) every orbit $\Phi(a)$, $a \in K$, can be partitioned into t distinct residue classes (mod p) whose representatives are $\beta_i(a)$'s, $\beta_i \in T$ and $i = 1, \ldots, t$, and each of these classes contains exactly p^h elements congruent (mod p^{n-h}) to $\beta_i(a)$;

    (2) for every $a \in K$, two elements of $\Phi(a)$ are congruent (mod p) if, and only if, they are congruent (mod p^{n-h}).

Hereinafter, for each orbit $\Phi(a)$, $a \in K$, $[\Phi(a)]_m$ will denote the set of the residue classes (mod m) in which the elements of $\Phi(a)$ are gathered and, in short, such a set will be called *m-class* of $\Phi(a)$.

At this point, it results natural to give the following definition, which can be obviously extended also to the case $p = 2$.

Definition 6.16. Let $G = (\mathbb{Z}_{p^n}, +)$, with any prime p, and $\Phi \leq Aut(G)$. Two orbits $\Phi(a)$ and $\Phi(b)$, with $a, b \in K$, are called *p-equivalent* if $[\Phi(a)]_p = [\Phi(b)]_p$, that is $\Phi(a)$ and $\Phi(b)$ have the same *p-class*.

Example 6.17. Take $G = (\mathbb{Z}_{49}, +)$ and $\Phi \leq Aut(G)$ generated by the automorphism α_4 : $x \to 4x$ of order 21. Using the notations of Corollary 6.15, Φ equals $T \otimes \Phi_1$, where $T = \langle \alpha_{18} \rangle = \{id_G, \alpha_{18}, \alpha_{30}\}$ and $\Phi_1 = \langle \alpha_{22} \rangle = \{id_G, \alpha_{22}, \alpha_{43}, \alpha_{15}, \alpha_{36}, \alpha_8, \alpha_{29}\}$. Hence, in this case, $n = 2$, $h = 1$, $t = 3$ and the orbits of K are:

$$\Phi(1) = \{1,4,16,15,11,44,29,18,23,43,25,2,8,32,30,22,39,9,36,46,37\},$$

$$\Phi(3) = \{3,12,48,45,33,34,38,5,20,31,26,6,24,47,41,17,19,27,10,40,13\}.$$

We can observe that in these orbits the elements can be gathered in three disjoint subsets, each of which contains seven elements congruent (mod 7) to each other. Precisely, in $\Phi(1)$ we find elements congruent (mod 7) to 1, 2 and 4 and in $\Phi(3)$ we find elements congruent (mod 7) to 3, 5 and 6 respectively. So the 7-class of $\Phi(1)$ is $\{1,2,4\} \subseteq \mathbb{Z}_7$ and the one of $\Phi(3)$ is $\{3,5,6\} \subseteq \mathbb{Z}_7$. Thus $[\Phi(1)]_7 \neq [\Phi(3)]_7$ that is $\Phi(1)$ and $\Phi(3)$ are not 7-equivalent.

Now we show that the p-classes of two arbitrary orbits either coincide or are disjoint and, moreover, we note that if $|\Phi| = t p^h$, there are exactly $s = \frac{p-1}{t}$ distinct p-classes under Φ. In other words we can prove the following:

Proposition 6.18. Let $G = (\mathbb{Z}_{p^n}, +)$, with $p \neq 2$, and $\Phi = T \otimes \Phi_h \leq Aut(G)$ of order $t p^h$, where t divides $p - 1$. The set $\{\ [\Phi(a)]_p \mid a \in K\}$ of p-classes under Φ determines a partition of $\mathbb{Z}_p^*$ of order $s = \frac{p-1}{t}$.

Proof. Obviously, $[\Phi(a)]_p$ is non empty for every $a \in K$ and $\bigcup_{a \in K} [\Phi(a)]_p = \mathbb{Z}_p^*$. To complete we have to show that distinct blocks are disjoint. Suppose $x \in [\Phi(a)]_p \cap [\Phi(b)]_p$. Then, from Corollary 6.15(1) there exist $\beta_k, \beta_j \in T$ such that $\beta_k(a) \equiv_p x \equiv_p \beta_j(b)$. Consequently, $\beta_i(\beta_k(a)) \equiv_p \beta_i(\beta_j(b))$, for any $\beta_i \in T$, thus $[\Phi(a)]_p = [\Phi(b)]_p$. Again from Corollary 6.15(1), $[\Phi(a)]_p$ contains exactly t different elements, hence the partition determined by $[\Phi(a)]_p$ contains exactly $\frac{p-1}{t} = s$ blocks. $\qquad\square$

In the hypothesis of the previous proposition, in K there are obviously $s p^{n-h-1}$ distinct orbits which can be divided into s subsets, each of them contains p^{n-h-1} orbits p-equivalent to each other. Precisely, we have the following:

Corollary 6.19. Let $G = (\mathbb{Z}_{p^n}, +)$, with $p \neq 2$, and $\Phi = T \otimes \Phi_h \leq Aut(G)$ of order $t p^h$, where t divides $p - 1$. In K:

(1) there are $\frac{p-1}{t} = s$ orbits $\Phi(e_i)$, $i \in \{1, 2, \dots, s\}$, non p-equivalent pairwise;

(2) there are exactly p^{n-h-1} orbits p-equivalent to $\Phi(e_i)$, for all $i \in \{1, 2, \dots, s\}$.

Proof. (1) From Proposition 6.18, in K there are s distinct orbits $\Phi(e_i)$, $i \in \{1, 2, \dots, s\}$, such that $[\Phi(a)]_p \neq [\Phi(b)]_p$, for all $i, j \in \{1, 2, \dots, s\}$ with $i \neq j$.

(2) Let $i \in \{1, 2, \dots, s\}$ and $a \in \Phi(e_i)$. Since in $\mathbb{Z}_{p^n}$ there are p^{n-1} elements congruent to a (mod p) and, by Corollary 6.15, each orbit $\Phi(e_i)$ contains p^h elements congruent to a (mod p^{n-h}), then there are p^{n-h-1} orbits p-equivalent to $\Phi(e_i)$. $\qquad\square$

Finally, we prove the next proposition used in the following to derive a condition equivalent to condition (ω).

Proposition 6.20. Let $G = (\mathbb{Z}_{p^n}, +)$, with $p \neq 2$, and $\Phi = T \otimes \Phi_h \leq Aut(G)$ of order $t p^h$, where t divides $p - 1$. Let $\Phi(a)$, $\Phi(b)$ be distinct p-equivalent orbits of Φ such that $a \equiv_{p^j} b$. For all $x \in \Phi(a)$, $y \in \Phi(b)$, if x is congruent (mod p) to y, then x is congruent (mod p^j) to y.

Proof. In $\mathbb{Z}_{p^n}$ there are exactly p^h elements congruent to each other (mod p^{n-h}), thus they all belong to the same orbit (Corollary 6.15). Hence, if $a \equiv_{p^j} b$ then $j < n - h$. Since x is an element of $\Phi(a)$, there exists an automorphism of Φ, say α, such that $\alpha(x) = a$.

From $x \equiv_p y$ it follows $\alpha(x) \equiv_p \alpha(y)$, hence $b \equiv_{p^j} a = \alpha(x) \equiv_p \alpha(y)$. Therefore, from Corollary 6.15(2), we have $b \equiv_{p^j} \alpha(y)$, and, hence, $\alpha(x) \equiv_{p^j} \alpha(y)$. $\qquad\square$

The next example explains all the notations and the results presented in this section.

Example 6.21. Take $G = (\mathbb{Z}_{125}, +)$.

(a) Take $T = \langle \alpha_{124} \rangle = \{id_G, \alpha_{124}\}$, $\Phi_1 = \langle \alpha_{26} \rangle = \{id_G, \alpha_{26}, \alpha_{51}, \alpha_{76}, \alpha_{101}\}$, so $\Phi = T \otimes \Phi_1 = \langle \alpha_{49} \rangle = \{id_G, \alpha_{49}, \alpha_{26}, \alpha_{24}, \alpha_{51}, \alpha_{124}, \alpha_{76}, \alpha_{99}, \alpha_{101}, \alpha_{74}\}$. Now $n = 3$, $h = 1$, $t = 2$, $s = 2$. List the orbits of K:

$$\begin{aligned}
\Phi(1) &= \{1, 49, 26, 24, 51, 124, 76, 99, 101, 74\}, \\
\Phi(2) &= \{2, 98, 52, 48, 102, 123, 27, 73, 77, 23\}, \\
\Phi(3) &= \{3, 22, 78, 72, 28, 122, 103, 47, 53, 97\}, \\
\Phi(4) &= \{4, 71, 104, 96, 79, 121, 54, 21, 29, 46\}, \\
\Phi(6) &= \{6, 44, 31, 19, 56, 119, 81, 94, 106, 69\}, \\
\Phi(7) &= \{7, 93, 57, 43, 107, 118, 32, 68, 82, 18\}, \\
\Phi(8) &= \{8, 17, 83, 67, 33, 117, 108, 42, 58, 92\}, \\
\Phi(9) &= \{9, 66, 109, 91, 84, 116, 59, 16, 34, 41\}, \\
\Phi(11) &= \{11, 39, 36, 14, 61, 114, 86, 89, 111, 64\}, \\
\Phi(12) &= \{12, 88, 62, 38, 112, 113, 37, 63, 87, 13\}.
\end{aligned}$$

We can see the 5-class of $\Phi(1)$, $\Phi(4)$, $\Phi(6)$, $\Phi(9)$ and $\Phi(11)$ is $\{1, 4\}$, while the 5-class of $\Phi(2)$, $\Phi(3)$, $\Phi(7)$, $\Phi(8)$ and $\Phi(12)$ is $\{2, 3\}$.

So there are $s = 2$ orbits non 5-equivalent (Corollary 6.19(1)), for instance $\Phi(1)$ and $\Phi(2)$, and there are exactly $5^{3-1-1} = 5$ orbits 5-equivalent to $\Phi(1)$ and 5 orbits 5-equivalent to $\Phi(2)$ (Corollary 6.19(2)). Here no elements congruent to each other (mod 25) lie in different orbits (see Proposition 6.20), because 5 divides $|\Phi|$ and so every orbit contains the five elements of $\mathbb{Z}_{125}$ congruent (mod 25) to each other (see Corollary 6.15(1)).

(b) Take $\Phi = T = \langle \alpha_{57} \rangle = \{id_G, \alpha_{57}, \alpha_{124}, \alpha_{68}\}$. Now $n = 3$, $h = 0$, $t = 4$, $s = 1$, hence all the orbits are 5-equivalent. In this case distinct orbits exist containing elements congruent (mod 25) to each other respectively—for instance $\Phi(3) = \{3, 46, 122, 79\}$ and $\Phi(103) = \{103, 121, 22, 4\}$ - since 5 is here relatively prime to $|\Phi|$. Thus, for all elements of these orbits, congruence (mod 5) implies congruence (mod 25) (see Proposition 6.20).

6.3. **Case $p = 2$.** Let now $G = (\mathbb{Z}_{2^n}, +)$ and $\Phi < Aut(G)$ of order 2^h. The following cases are possible (see [16], Chap. 4)[2]:

(A) $\Phi = \langle \alpha_{1+2^{n-h}} \rangle = \{\alpha_k : x \to kx \mid k = 1 + b2^{n-h}, \ 0 \le b \le 2^h - 1\}$

(B) $\Phi = \langle \alpha_{-1+2^{n-h}} \rangle = \{\alpha_k : x \to kx \mid k = (-1)^b + b2^{n-h}, \ 0 \le b \le 2^h - 1\}$

(C) $\Phi = \langle \alpha_{1+2^{n-h+1}}, -id_G \rangle = \{\alpha_k : x \to kx \mid k = \pm(1 + b2^{n-h+1}), \ 0 \le b \le 2^{h-1} - 1\}$.

To obtain a useful description of the orbits in K, it is better to examine case (A) and cases (B) and (C) separately.

Case (A). The orbits in K are described by the following:

---

[2] Here id_G denotes the identity map of G and $-id_G$ is defined by $x \to -x$

Proposition 6.22. *Let* $G = (\mathbb{Z}_{2^n}, +)$ *and let* $\Phi = \langle \alpha_{1+2^{n-h}} \rangle$ *be a subgroup of* $Aut(G)$. *In* K:

    (1) *all the orbits are 2-equivalent pairwise;*
    (2) *every orbit contains* 2^h *elements congruent to each other* $(\bmod\ 2^{n-h})$;
    (3) *if* $\Phi(a)$, $\Phi(b)$ *are distinct orbits of* Φ *such that* $a \equiv_{2^j} b$, *then each element of* $\Phi(a)$ *is congruent to each element of* $\Phi(b)$ $(\bmod\ 2^j)$.

Proof. Immediately (1) follows by the definition of 2-equivalent orbits, while the proof of (2) and (3) is analogous to the case $p \neq 2$, because of the form of the elements of Φ.   □

Cases (B) and (C). The orbits in K are now described by the following:

Proposition 6.23. *Let* $G = (\mathbb{Z}_{2^n}, +)$ *and let* Φ *be a subgroup of* $Aut(G)$ *having form* (B) *or* (C). *In* K:

    (1) *all the orbits have the same 4-class* $[\Phi(a)]_4 = \{a, -a\} = \{1, -1\} \subseteq \mathbb{Z}_4$, *for each* $a \in K$;
    (2) *let* $\Phi(a)$, $\Phi(b)$ *be distinct orbits such that* $a \equiv_{2^j} b$. *For all* $x \in \Phi(a)$, $y \in \Phi(b)$, *if* x *is congruent* $(\bmod\ 4)$ *to* y, *then* x *is congruent* $(\bmod\ 2^j)$ *to* y.

Proof. (1) It is clear, because of the form of Φ's elements.

    (2) Let $|\Phi| = 2^h$ and let $x \in \Phi(a)$ and $y \in \Phi(b)$ such that $x \equiv_4 y$.

If $j = 2$, the statement is clear. Furthermore, since the elements congruent to a modular 2^{n-h+1} are 2^{h-1}, all of them belong to the orbit of a, so we have $j < n - h + 1$. Thus we consider: $2 < j < n - h + 1$.

From (1) we have only two possibilities. The first one is $a \equiv_{2^{n-h+1}} \pm x$ and $b \equiv_{2^{n-h+1}} \pm y$, then:

$$\pm x \equiv_{2^{n-h+1}} a \equiv_{2^j} b \equiv_{2^{n-h+1}} \pm y,$$

that is $x \equiv_{2^j} y$.

Otherwise $a \equiv_{2^{n-h+1}} \pm x$ and $b \equiv_{2^{n-h+1}} \mp y$, then:

$$\pm x \equiv_{2^{n-h+1}} a \equiv_{2^j} b \equiv_{2^{n-h+1}} \mp y \equiv_4 \mp x,$$

that is $\pm x \equiv_4 \mp x$, but this is false.   □

6.4. Conclusion. We are now able to prove a necessary and sufficient condition about the choice of the representatives of the orbits so that π of Definition 6.5 can be a function and hence, in particular, a Clay function.

Theorem 6.24. *Let* $G = (\mathbb{Z}_{p^n}, +)$, *p any prime, let* Φ *be a subgroup of* $Aut(G)$ *and* π *as in Definition 6.5. Condition* (ω) *is fulfilled if, and only if, the representatives of p-equivalent orbits in K are chosen:*

$$\begin{cases} congruent(\bmod\ p) & if\ p \neq 2\ or\ p = 2\ and\ \Phi = \langle \alpha_{1+2^{n-h}} \rangle, \\ congruent(\bmod\ 4) & otherwise. \end{cases}$$

Proof. Suppose that condition (ω) is satisfied, that is π of Definition 6.5 is a function by Lemma 6.8.

Assume $p \neq 2$ or $p = 2$ and $\Phi = \langle \alpha_{1+2^{n-h}} \rangle$.

Let e_k and $e_{k'}$ be the selected representatives of two p-equivalent orbits in K and let $k \in \Phi(e_k)$, $k' \in \Phi(e_{k'})$ such that $k \equiv_p k'$. Consider the elements $a = e^{-(n-1)}kp^{n-1}$ and $a' = e^{-(n-1)}k'p^{n-1}$. They are obviously congruent (mod p^n), hence, since π is a function, we have $\pi(a) = \pi(a')$. Therefore $e^{-(n-1)}p^{n-1}\phi_k(1) = e^{-(n-1)}p^{n-1}\phi_{k'}(1)$ and thus $\phi_k(1) \equiv_p \phi_{k'}(1)$. Consequently $e_k\phi_k(1) = \phi_k(e_k) = k \equiv_p k' = \phi_{k'}(e_{k'}) = e_{k'}\phi_{k'}(1) \equiv_p e_{k'}\phi_k(1)$, from which, since $\phi_k(1)$ is relatively prime to p, we obtain $e_k \equiv_p e_{k'}$.

Assume $p = 2$ and $\Phi = \langle \alpha_{-1+2^{n-h}} \rangle$ or $\Phi = \langle \alpha_{1+2^{n-h+1}}, -id_G \rangle$.

Since all the orbits have the same 4-class, any two of them contain respectively elements which are congruent (mod 4) to each other, hence condition (ω) implies that the representatives of the orbits are at least congruent (mod 4) pairwise.

The converse arises directly from Propositions 6.18, 6.22(3) and 2.3(2). $\square$

Remark. The wd-nearrings constructed in Section 5 can be also obtained by the previous method. Take $(\mathbb{Z}_{2^n}, +)$. The case (δ) derives from choosing $\Phi = \{id\}$, the set of the selected representatives $R = \{[\pm 1 + 4h], h \in \mathbb{Z}\}$, and $e = 1$. The case (ζ) derives from choosing $\Phi = \{id, \alpha_{-1+2^{n-1}}\}$, $R = \{[1 + 4h], h \in \mathbb{Z}\}$ and $e = 1$. The case (θ) derives from choosing $\Phi = \{id, \alpha_{1+2^{n-1}}\}$, $R = \{[1 + 4h], 0 \leq h < 2^{n-3}\} \cup \{[3 + 4h], 2^{n-3} \leq h < 2^{n-2}\}$ and $e = 1$.

An application of the above theorem is shown in the following:

Example 6.25. Take $G = (\mathbb{Z}_{49}, +)$ and $\Phi = \langle \alpha_{18} \rangle = \{id_G, \alpha_{18}, \alpha_{30}\}$. The 7-class of $\Phi(1)$, $\Phi(2)$, $\Phi(4)$, $\Phi(8)$, $\Phi(9)$, $\Phi(16)$ and $\Phi(29)$ is $\{1, 2, 4\}$. The 7-class of $\Phi(3)$, $\Phi(6)$, $\Phi(12)$, $\Phi(13)$, $\Phi(19)$, $\Phi(24)$ and $\Phi(26)$ is $\{3, 5, 6\}$. So, in K there are $s = 2$ orbits non 7-equivalent, for instance $\Phi(1)$ and $\Phi(3)$. There are exactly 7 orbits 7-equivalent to $\Phi(1)$ and by Theorem 6.24 their representatives must be selected congruent (mod 7) to each other: choose 18, 11, 4, 46, 25, 39, 28. There are 7 orbits 7-equivalent to $\Phi(3)$ and, for the same reason, their representatives have to be selected congruent (mod 7) to each other: choose 3, 10, 17, 18, 37, 24, 45. Fix arbitrarily $e = 46$ among the selected representatives and define:

$$\pi(a) = \begin{cases} 0, & \text{if } a = 0; \\ 7^r \phi_{k46^r}(46^{-r}), & \text{if } a = k7^r \text{ with } (k, 7) = 1 \text{ and } 0 \leq r < n. \end{cases}$$

Because of the choice of the representatives, Theorem 6.24 guarantees that π is a function. Specifically, π is a Clay function, so the structure $(\mathbb{Z}_{49}, +, *)$, where "$*$" is defined by $a * b = b\pi(a)$, turns out a wd-nearring with $Q = 7\mathbb{Z}_{49}$.

7. TACTICAL CONFIGURATIONS AND PBIBDs

An incidence structure $(P, \mathcal{B})$ is a *tactical configuration* (or *design*) if all the blocks contain the same number k of elements, and all the elements occur in the same number r of blocks (see [18]). The parameters of the configuration are (v, b, r, k), where v is the cardinality of P and b is the number of blocks.

7.1. Case (ζ). Beginning with the nearring described in (ζ) of Section 5, we are able to construct exactly 2^{n-3} isomorphic designs of parameters $(2^{n-1}, 2^{n-1}, 2, 2)$.

Consider a group $B_x = \{[x_1], [x_2]\}$. We can generate a design with the set of blocks $B_x + Q = \{\{[x_1 + q], [x_2 + q]\} \mid q \in Q\}$. The points of the design are the elements of C, therefore 2^{n-1}; the blocks are 2^{n-1}, because 2^{n-1} are the elements of Q; each block contains two elements and each element belongs exactly to two blocks.

Finally, the groups of C are 2^{n-2} and we can show that there are exactly two different groups which generate the same design. In fact, if we consider two different groups $B_x = \{[x_1], [x_2]\}$ and $B_y = \{[y_1], [y_2]\}$, we can easily show, with a direct calculation, that $B_x + Q$ and $B_y + Q$ coincide if, and only if, $[x_2 - x_1] = [y_2 - y_1]$. Now $[x_2 - x_1] = [2^{n-1} - 2] - 8[h_1]$ and $[y_2 - y_1] = [2^{n-1} - 2] - 8[h_2]$, because of $[x_1] = [1 + 4h_1]$, $[x_2] = [-1 + 4(2^{n-3} - h_1)]$, $[y_1] = [1 + 4h_2]$ and $[y_2] = [-1 + 4(2^{n-3} - h_2)]$. Therefore $[x_2 - x_1] = [y_2 - y_1]$ if, and only if, $[h_1] - [h_2] = [2^{n-3}]$. Since $[h_1]$ and $[h_2]$ belong to $\mathbb{Z}_{2^{n-2}}$, for each fixed $[h_1]$ there is a unique $[h_2]$ such that this occurs, hence each design derives from exactly two distinct groups of C. Thus the different and obviously isomorphic designs are 2^{n-3}.

The following example shows the design and the related association scheme induced by the wd-nearring $(\mathbb{Z}_{2^4}, +, \cdot)$. The following table shows the product "$\cdot$", defined as in (ζ) of Section 5:

·	0	1	2	3	4	5	6	7	8	9	10	11	12	13	14	15
0	0	0	0	0	0	0	0	0	0	0	0	0	0	0	0	0
1	0	1	2	3	4	5	6	7	8	9	10	11	12	13	14	15
2	0	2	4	6	8	10	12	14	0	2	4	6	8	10	12	14
3	0	7	14	5	12	3	10	1	8	15	6	13	4	11	2	9
4	0	4	8	12	0	4	8	12	0	4	8	12	0	4	8	12
5	0	1	2	3	4	5	6	7	8	9	10	11	12	13	14	15
6	0	14	12	10	8	6	4	2	0	14	12	10	8	6	4	2
7	0	7	14	5	12	3	10	1	8	15	6	13	4	11	2	9
8	0	8	0	8	0	8	0	8	0	8	0	8	0	8	0	8
9	0	1	2	3	4	5	6	7	8	9	10	11	12	13	14	15
10	0	2	4	6	8	10	12	14	0	2	4	6	8	10	12	14
11	0	7	14	5	12	3	10	1	8	15	6	13	4	11	2	9
12	0	12	8	4	0	12	8	4	0	12	8	4	0	12	8	4
13	0	1	2	3	4	5	6	7	8	9	10	11	12	13	14	15
14	0	14	12	10	8	6	4	2	0	14	12	10	8	6	4	2
15	0	7	14	5	12	3	10	1	8	15	6	13	4	11	2	9

The wd-nearring $(\mathbb{Z}_{2^4}, +, \cdot)$ has the cancellable part which is a disjoint union of the groups $B_1 = (\underline{1}, 7)$; $B_2 = (3, \underline{5})$; $B_3 = (\underline{9}, 15)$ and $B_4 = (11, \underline{13})$, where the underlined elements are the identities of the respective groups and the left identities of the nearring. If we consider the following:

$$B_1 + Q = \{(1,7), (3,9), (5,11), (7,13), (9,15), (11,1), (13,3), (15,5)\} = B_3 + Q,$$
$$B_2 + Q = \{(3,5), (5,7), (7,9), (9,11), (11,13), (13,15), (15,1), (1,3)\} = B_4 + Q,$$

we obtain 2 isomorphic designs of parameters $(8, 8, 2, 2)$.

The previous designs are non-balanced, but we are able to construct an association scheme that makes them PBIBDs[3].

---

[3]For the definitions we refer to [9, 18]

Firstly, we can observe that the design $B_1 + Q$, for instance, can be represented, in a natural way, by the following graph:

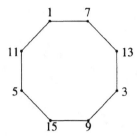

where the blocks and the related points are represented by the edges and the related vertices, respectively.

Consequently, we can define the relations $R_{d'}$ on C in this way: for $x, y \in C$, $xR_{d'}y$ if, and only if, $d(x, y) = d'$, where $d(x, y)$ means the usual distance between the vertices of a graph. These $R_{d'}$ are four symmetric anti-reflexive binary relations, suitable to define the following association scheme on C:

	Associates			
Element	First	Second	Third	Fourth
1	7, 11	5, 13	3, 15	9
3	9, 13	7, 15	1, 5	11
5	11, 15	1, 9	7, 3	13
7	1, 13	11, 3	5, 9	15
9	15, 3	5, 13	7, 11	1
11	5, 1	7, 15	9, 13	3
13	7, 3	1, 9	11, 15	5
15	5, 9	3, 11	1, 13	7

Now, the numbers n_i of ith associates of each element are: $n_1 = n_2 = n_3 = 2$ and $n_4 = 1$, while the numbers λ_i of blocks in which any couple of ith associates occur together are $\lambda_1 = 1, \lambda_2 = \lambda_3 = \lambda_4 = 0$.

Denoted by P_i the matrices of the p^i_{jl} (the number of elements which are both jth associates of x and lth associates of y, for any ith associates $x, y \in C$), we have:

$$P_1 = \begin{pmatrix} 0 & 1 & 1 & 0 \\ 1 & 0 & 1 & 0 \\ 1 & 1 & 0 & 1 \\ 0 & 0 & 1 & 0 \end{pmatrix}, \qquad P_2 = \begin{pmatrix} 1 & 0 & 1 & 0 \\ 0 & 0 & 0 & 1 \\ 1 & 0 & 1 & 0 \\ 0 & 1 & 0 & 0 \end{pmatrix},$$

$$P_3 = \begin{pmatrix} 0 & 1 & 0 & 1 \\ 1 & 0 & 1 & 0 \\ 0 & 1 & 0 & 0 \\ 1 & 0 & 0 & 0 \end{pmatrix}, \qquad P_4 = \begin{pmatrix} 0 & 0 & 2 & 0 \\ 0 & 2 & 0 & 0 \\ 2 & 0 & 0 & 0 \\ 0 & 0 & 0 & 0 \end{pmatrix}.$$

7.2. Case (θ). If we consider the wd-nearring described in (θ) of Section 5 with $n = 4$, the product "$\cdot$" is the following:

·	0	1	2	3	4	5	6	7	8	9	10	11	12	13	14	15
0	0	0	0	0	0	0	0	0	0	0	0	0	0	0	0	0
1	0	1	2	3	4	5	6	7	8	9	10	11	12	13	14	15
2	0	2	4	6	8	10	12	14	0	2	4	6	8	10	12	14
3	0	9	2	11	4	13	6	15	8	1	10	3	12	5	14	7
4	0	4	8	12	0	4	8	12	0	4	8	12	0	4	8	12
5	0	1	2	3	4	5	6	7	8	9	10	11	12	13	14	15
6	0	2	4	6	8	10	12	14	0	2	4	6	8	10	12	14
7	0	9	2	11	4	13	6	15	8	1	10	3	12	5	14	7
8	0	8	0	8	0	8	0	8	0	8	0	8	0	8	0	8
9	0	9	2	11	4	13	6	15	8	1	10	11	12	5	14	7
10	0	2	4	6	8	10	12	14	0	2	4	6	8	10	12	14
11	0	1	2	3	4	5	6	7	8	9	10	11	12	13	14	15
12	0	4	8	12	0	4	8	12	0	4	8	12	0	4	8	12
13	0	9	2	11	4	13	6	15	8	1	10	3	12	5	14	7
14	0	2	4	6	8	10	12	14	0	2	4	6	8	10	12	14
15	0	1	2	3	4	5	6	7	8	9	10	11	12	13	14	15

The wd-nearring $(\mathbb{Z}_{2^4}, +, \cdot)$ has the cancellable part C which is the disjoint union of the groups $B_1 = (\underline{1}, 9)$; $B_2 = (3, \underline{11})$; $B_3 = (\underline{5}, 13)$ and $B_4 = (7, \underline{15})$, where the underlined elements are the identities of the respective groups and left identities of the nearring. If we try to construct designs as before we obtain only the original partition of C:

$$B_1 + Q = \{(1,9), (3,11), (5,13), (7,15)\} = B_3 + Q = B_2 + Q = B_4 + Q.$$

The previous different cases exemplify a more general situation. In fact, we can show the following:

Proposition 7.1. *If N is a wd-nearring of type (ζ), $B_a + Q$ is a non trivial design for all $a \in C$. If N is a wd-nearring of type (θ), $B_a + Q$ is always the original partition $\{B_a, a \in C\}$ of C.*

The proof, not difficult but with laborious calculations, is here omitted.

Now our aim is to establish when it is possible to derive PBIB-designs starting from any wd-nearring on $(\mathbb{Z}_{2^n}, +)$. Following the line drawn in the previous sections, a wd-nearring on $(\mathbb{Z}_{2^n}, +)$ can be thought as a pair (G, Φ), where $G = (\mathbb{Z}_{2^n}, +)$ and Φ is a group of automorphisms of G. $\mathrm{Aut}(\mathbb{Z}_{2^n}, +)$ is well-known (see [16]): it is isomorphic to $\langle -id_G, \alpha_5 \rangle$, has order 2^{n-1}, has 3 subgroups of order 2: $A_2 = \langle \alpha_{5^{2^{n-3}}} \rangle$, $B_2 = \langle \alpha_{-5^{2^{n-3}}} \rangle$ and $C_2 = \langle -id_G \rangle$, has 3 subgroups of order 4: $A_4 = \langle \alpha_{5^{2^{n-4}}} \rangle$, $B_4 = \langle \alpha_{-5^{2^{n-4}}} \rangle$ and C_4 isomorphic to the Klein group, and, among its subgroups of order 2^k ($2 < k < n - 1$) there are $A_{2^k} = \langle \alpha_{5^{2^{n-k-2}}} \rangle$ and $B_{2^k} = \langle \alpha_{-5^{2^{n-k-2}}} \rangle$.

We note that the wd-nearrings of type (ζ), described above, are pairs (G, Φ) where Φ equals B_2, while those of type (θ) have $\Phi = A_2$. Therefore, if we examine the problem connected to the possibility of constructing designs following the previous method, we can observe that the choice of A_2, A_4 or A_8, as the group of automorphisms defining the wd-nearring, does not imply the construction of designs, whereas the choice of B_2, B_4 or B_8 allows the construction of wd-nearrings from which we are able to obtain designs, as the following examples show.

From (G, Φ), where $G = (\mathbb{Z}_{2^5}, +)$ and $\Phi = B_2$, we obtain a design of parameters $(2^4, 2^4, 2, 2)$, with $P = \mathbb{Z}_{2^5} \setminus 2\mathbb{Z}_{2^5}$ and $\mathcal{B} = \{\{1, 15\} + q \mid q \in 2\mathbb{Z}_{2^5}\}$, represented by the following graph:

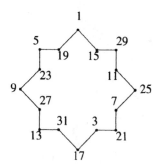

where the blocks and the related points are represented by the edges and the related vertices, respectively.

If we consider $\Phi = B_{2^2}$ and $G = (\mathbb{Z}_{2^5}, +)$, we obtain a design of parameters $(2^4, 2^3, 2, 2^2)$, with $P = \mathbb{Z}_{2^5} \setminus 2\mathbb{Z}_{2^5}$ and $\mathcal{B} = \{\{1, 15, 17, 31\} + q \mid q \in 2\mathbb{Z}_{2^5}\}$, which can be represented by the following:

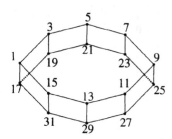

where the blocks and the related points are represented by the quadrangles and the related vertices, respectively.

Last example. The pair $\Phi = B_{2^3}$ and $G = (\mathbb{Z}_{2^6}, +)$ determine a design of parameters $(2^5, 2^3, 2, 2^3)$, with $P = \mathbb{Z}_{2^6} \setminus 2\mathbb{Z}_{2^6}$ and $\mathcal{B} = \{\{1, 39, 49, 55, 33, 17, 23\} + q \mid q \in 2\mathbb{Z}_{2^6}\}$. Analogously to the previous cases, this design can be represented by the following graph:

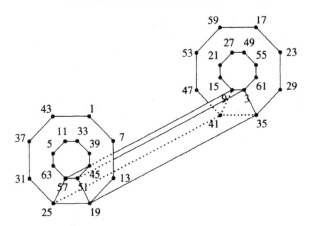

We observe that this method, applied to wd-nearrings on $\mathbb{Z}_{p^n}$ $(p \neq 2)$, does not determine significant designs but the following example shows that it is possible to construct PBIB-designs with another technique.

Here we present an example but our aim is to generalize this method in future research.

As usual let (G, Φ) be a nearring where $G = (\mathbb{Z}_9, +)$ and $\Phi = \{id_G, -id_G\}$. Clearly, the non trivial orbits of Φ are:

$$\Delta_1 = \{\underline{1}, 8\}, \Delta_2 = \{\underline{2}, \underline{7}\}, \Delta_3 = \{3, 6\}, \Delta_4 = \{\underline{4}, 5\},$$

where the underlined elements are the fixed representatives and the choice representative is 1. Thus the multiplicative table turns out to be:

$\circ$	0	1	2	3	4	5	6	7	8
0	0	0	0	0	0	0	0	0	0
1	0	1	2	3	4	5	6	7	8
2	0	8	7	6	5	4	3	2	1
3	0	3	6	0	3	6	0	3	6
4	0	1	2	3	4	5	6	7	8
5	0	8	7	6	5	4	3	2	1
6	0	6	3	0	6	3	0	6	3
7	0	1	2	3	4	5	6	7	8
8	0	8	7	6	5	4	3	2	1

We are now able to describe a tactical configuration $(P, \mathcal{B})$ where $P = \mathbb{Z}_9$ and $\mathcal{B} = \{N \circ a + b \mid a \in C, b \in N\}$. Precisely the blocks are:

(01836), (12047), (23158), (34260), (45371), (56482), (67503), (78614), (80725),
(02736), (13847), (24058), (35160), (46271), (57382), (68403), (70514), (81625),
(04536), (15647), (26758), (37860), (48071), (50182), (61203), (72314), (83425),

and its parameters are $(v, b, r, k) = (9, 27, 15, 5)$.

Obviously $(P, \mathcal{B})$ is non-balanced, but on $\mathbb{Z}_9$ it is possible to define the suitable relations $\mathcal{R}_i = \{(d, x + d) \mid x \in \Delta_i, d \in \mathbb{Z}_9\}$, $i = 1, 2, 3, 4$, which describe the following association scheme:

Element	Associates			
	First	Second	Third	Fourth
0	8, 1	2, 7	3, 6	4, 5
1	0, 2	3, 8	4, 7	5, 6
2	1, 3	4, 0	5, 8	6, 7
3	2, 4	5, 1	6, 0	7, 8
4	3, 5	6, 2	7, 1	8, 0
5	4, 6	7, 3	8, 2	0, 1
6	5, 7	8, 4	0, 3	1, 2
7	6, 8	0, 5	1, 4	2, 3
8	7, 0	1, 6	2, 5	3, 4

Hence $(\mathbb{Z}_9, \mathcal{B})$ turns out to be a PBIB-design with the following parameters $n_1 = n_2 = n_3 = n_4 = 2$, $\lambda_1 = 7$ $\lambda_2 = 7$ $\lambda_3 = 9$ $\lambda_4 = 7$ and the matrices P_i of p^i_{jl} are:

$$
P_1 = \begin{pmatrix} 0 & 1 & 0 & 0 \\ 1 & 0 & 1 & 0 \\ 0 & 1 & 0 & 1 \\ 0 & 0 & 1 & 1 \end{pmatrix}, \qquad
P_2 = \begin{pmatrix} 1 & 0 & 1 & 0 \\ 0 & 0 & 0 & 1 \\ 1 & 0 & 0 & 1 \\ 0 & 1 & 1 & 0 \end{pmatrix},
$$

$$
P_3 = \begin{pmatrix} 0 & 1 & 0 & 1 \\ 1 & 0 & 0 & 1 \\ 0 & 0 & 1 & 0 \\ 1 & 1 & 0 & 0 \end{pmatrix}, \qquad
P_4 = \begin{pmatrix} 0 & 0 & 1 & 1 \\ 0 & 1 & 1 & 0 \\ 1 & 1 & 0 & 0 \\ 1 & 0 & 0 & 0 \end{pmatrix}.
$$

Consider now the wd-nearring $N' = (\mathbb{Z}_9, +, *)$ where the multiplication is defined by the following table:

$*$	0	1	2	3	4	5	6	7	8
0	0	0	0	0	0	0	0	0	0
1	0	7	5	3	1	8	6	4	2
2	0	4	8	3	7	2	6	1	5
3	0	3	6	0	3	6	0	3	6
4	0	1	2	3	4	5	6	7	8
5	0	1	2	3	4	5	6	7	8
6	0	3	6	0	3	6	0	3	6
7	0	4	8	3	7	2	6	1	5
8	0	7	5	3	1	8	6	4	2

In this example $\Phi = \{id_{N'}, \alpha_4, \alpha_7\}$ and its non trivial orbits are:

$$\Delta_1 = \{1, 4, 7\}, \Delta_2 = \{2, 8, 5\}, \Delta_3 = \{3\}, \Delta_4 = \{6\}.$$

As in the previous case the sets of the form $N' * a + b$, where a is a cancellable element of N' and $b \in \mathbb{Z}_9$, are

(01347), (12458), (23560), (34671), (45782), (56803), (67013), (78125), (80236),

(02568), (13670), (24781), (35802), (46013), (57124), (68235), (70346), (81457),

and form the set $\mathcal{B}^*$ of the blocks of a tactical configuration on $\mathbb{Z}_9$ of parameters

$$(v, b, r, k) = (9, 18, 10, 5).$$

Thus $(\mathbb{Z}_9, \mathcal{B}^*)$ results to be a PBIB-design of parameters $n_1 = 2$, $n_2 = 6$, $\lambda_1 = 8$, $\lambda_2 = 4$ with the following association scheme:

Element	Associates	
	First	Second
0	3, 6	1, 2, 4, 5, 7, 8
1	4, 7	2, 3, 5, 6, 8, 0
2	5, 8	3, 4, 6, 7, 0, 1
3	6, 0	4, 5, 7, 8, 1, 2
4	7, 1	5, 6, 8, 0, 2, 3
5	8, 2	6, 7, 0, 1, 3, 4
6	0, 3	7, 8, 1, 2, 4, 5
7	1, 4	8, 0, 2, 3, 5, 6
8	2, 5	0, 1, 3, 4, 6, 7

derived from the following relations on $\mathbb{Z}_9$:

$$\mathcal{R}_1 = \{(d, x+d) \mid x \in \Delta_3 \cup \Delta_4, d \in \mathbb{Z}_9\},$$
$$\mathcal{R}_2 = \{(d, x+d) \mid x \in \Delta_1 \cup \Delta_2, d \in \mathbb{Z}_9\}.$$

REFERENCES

[1] M. Anshel, J.R. Clay, *Planar algebraic systems, some geometric interpretations*, J. Algebra **10** (1968), 166–173.

[2] A. Benini, *Sums of near-rings*, Riv. Mat. Univ. Parma (4) **14** (1988), 135–141.

[3] A. Benini, *Near-rings on certain groups*, Riv. Mat. Univ. Parma (4) **15** (1989), 149–158.

[4] A. Benini, F. Morini, *Weakly divisible nearrings on the group of integers* (mod p^n), Riv. Mat. Univ. Parma, (6) **1** (1998), 1–11.

[5] A. Benini, F. Morini, *On the construction of a class of weakly divisible nearrings*, Riv. Mat. Univ. Parma, (6) **1** (1998), 103–111.

[6] A. Benini, S. Pellegrini, *Weakly Divisible Nearrings*, Discrete Math., (to appear).

[7] J.R. Clay, *The near-rings on a finite cyclic group*, Amer. Math. Monthly **71** (1964), 47–50.

[8] J.R. Clay, *The near-rings on groups of low order*, Math. Z. **104** (1968), 364–371.

[9] J.R. Clay, Nearrings: Geneses and Applications (Oxford University Press, New York, 1992).

[10] J.R. Clay, D.K. Doi, *Near-rings with identity on alternating groups*, Math. Scand. **23** (1968), 54–56.

[11] J.R. Clay, J.J. Malone, Jr., *The near-rings with identities on certain finite groups*, Math. Scand. **19** (1966), 146–150.

[12] G. Ferrero, *Classificazione e costruzione degli stems p-singolari*, Istit. Lombardo Accad. Sci. Lett. Rend. A **102** (1968), 597–613.

[13] G. Ferrero, *Stems planari e BIB-Disegni*, Riv. Mat. Univ. Parma (2) **11** (1970), 79–96.

[14] G. Ferrero, *Su certe geometrie gruppali naturali*, Riv. Mat. Univ. Parma (3), **1** (1972), 97–111.

[15] M. Hall, *Designs with transitive authomorphism group*, Proc. of Symposia in Pure Math. AMS, T. L. Motzkin, ed., (1971), 109–113.

[16] W.J. LeVeque, Fundamentals of Number Theory (Addison-Wesley Publishing Company, Philippines, 1977).

[17] S. Pellegrini, *Φ-sums: medial, permutable and LRD-near-rings*, Near-rings and Near-fields: Proc. of a Conference held at the Math. Forschungsinstitut (Oberwolfach, 1989) (G. Betsch et al. eds., 1995), 152–169.

[18] A. Penfold Street, D. J. Street, *Combinatorics of experimental design*, Oxford Science Publication, Clarendon Press 1987.

[19] G. Pilz, *A construction method for near-rings*, Acta Math. Acad. Sci. Hungar. **24** (1973), 97–105.

[20] G. Pilz, *Near-rings* **23**, (Revised edition) (North Holland Math. Studies, Amsterdam, 1983).

FACOLTÀ DI INGEGNERIA, DIPARTIMENTO DI ELETTRONICA, UNIVERSITÀ DI BRESCIA, VIA BRANZE 38, I-25123 BRESCIA, ITALY

ON SEMI-ENDOMORPHISMS OF ABELIAN GROUPS

Y. FONG AND W.-F. KE, F.-K. HUANG, AND Y.-N. YEH

ABSTRACT. In this paper we investigate the properties of the semi-endomorphism near-rings of abelian groups, and describe the semi-endomorphisms of finitely generated abelian groups. Specifically, characterizations of groups G are given when the semi-endomorphism nearring of G is the endomorphism rings of G, and when the semi-endomorphism nearring is the full transformation nearring $M(G)$; and an explicit description of the semi-endomorphisms of finitely generated abelian groups is given.

1. INTRODUCTION

A nonempty set N with two binary operations $+$ and $\cdot$ is called a *(left) nearring* if (1) $(N,+)$ is a group, (2) $(N,\cdot)$ is a semigroup, and (3) $x(y+z) = xy + xz$ for all $x,y,z \in N$. For an arbitrary group $(G,+)$, the set $M(G)$ of all mappings on G is a nearring under pointwise addition and composition of mappings. $M(G)$ is usually referred to as the *full transformation nearring of G*.

For a general introduction to nearring theory, please refer to Pilz [9], Meldrum [8] and Clay [2].

In this paper, we consider the subnearring $\mathscr{SE}(G) \subseteq M(G)$, G a group, generated by the semi-endomorphisms of G, i.e., $\mathscr{SE}(G) = \langle \{f \in M(G) \mid f \text{ is a semi-endomorphism of } G\}, +, \circ \rangle$. Here a mapping $f \in M(G)$ is called a *semi-endomorphism* if $f(x+y+x) = f(x) + f(y) + f(x)$ for all $x,y \in G$. (See Herstein [6].) The nearring $\mathscr{SE}(G)$ is referred as the *semi-endomorphism nearring* of G.

The study of the semi-endomorphism nearrings on various groups can be found in [3] and [1]. Explicitly, the cases of $\mathscr{SE}(\mathbb{Z})$, $\mathscr{SE}(\mathbb{Z}_n)$, $\mathscr{SE}(\overset{n}{\oplus}\mathbb{Z}_2)$, $\mathscr{SE}(\overset{n}{\oplus}\mathbb{Z}_p)$ and $\mathscr{SE}(G)$ with simple group G, and $\mathscr{SE}(S_n)$ have been studied.

The main goal of the present paper is to exploit $\mathscr{SE}(G)$ for abelian groups G. In section 2, we discuss the general properties of semi-endomorphisms and the semi-endomorphism nearrings of abelian groups. In section 3, the semi-endomorphisms of finitely generated abelian group is considered.

2. SEMI-ENDOMORPHISM NEARRINGS OF ABELIAN GROUPS

Let $(G,+)$ be a group. A nonempty subset K of G is said to be a *semi-subgroup* if $h + k + h \in K$ for all $h,k \in K$. Thus, it is easy to see that if G is abelian, then for a given semi-endomorphism $\phi : G \to G$ the subset $G_\phi = \{(x+y)\phi - x\phi - y\phi \mid x,y \in G\} \subseteq G$ is a semi-subgroup of G.

1991 *Mathematics Subject Classification.* 20K30, 16Y30.

Y. Fong et al. (eds.), Near-Rings and Near-Fields, 72–78.
© 2001 *Kluwer Academic Publishers. Printed in the Netherlands.*

For any group G, let $\mathscr{S}End\,G$ be the set of all semi-endomorphisms from G into G itself. Thus, $\mathscr{S}(G) = \langle \mathscr{S}End\,G, +, \circ \rangle$. Note that when G is abelian we have $\mathscr{S}(G) = \mathscr{S}End\,G$ (see Fong and Pilz [3, Corollary 2.3]).

Remark 2.1. If we call a (left) nearring N satisfying $(x + y + x)z = xz + yz + xz$ for all $x, y, z \in N$ a *(right) semi-distributive nearring*, then we have immediately that for an abelian group the nearring $\mathscr{S}(G)$ is semi-distributive.

For an arbitrary element x of a group G, we denote the order of x by $|x|$. Moreover, let $G[2] = \{x \in G \mid 2x = 0\}$. For any $a \in G$, let θ_a denote the constant map determined by a, i.e., $\theta_a(x) = $ for all $x \in G$.

Lemma 2.2. *Let G be an abelian group. Then the following statements are equivalent:*

   (i) $\theta_a \in \mathscr{S}End\,G$;

   (ii) $|a| = 2$;

   (iii) $\{a\}$ *is a semi-subgroup of G.*

Proof. (i) implies (ii). Since $a = (a + a + a)\theta_a = a\theta_a + a\theta_a + a\theta_a = a + a + a$, we see that $|a| = 2$.

(ii) implies (iii). From $|a| = 2$ we have $a + a + a = a \in \{a\}$, so $\{a\}$ is a semi-subgroup of G.

(iii) implies (i). Since $\{a\}$ is a semi-subgroup of G, we have $a + a + a = a$. Thus, for any $x, y \in G$, we have

$$(x + y + x)\theta_a = a = a + a + a = x\theta_a + y\theta_a + x\theta_a.$$

Hence $\theta \in \mathscr{S}End\,G$. $\qquad\qquad\qquad\qquad\qquad\qquad\qquad\qquad\qquad\qquad\qquad\qquad\qquad\quad\Box$

Example 2.3. Let $V = \{0, a, b, c\}$ be the Klein 4-group with $2a = 2b = 2c = 0$. Then θ_a and θ_b are semi-endomorphisms of V by (2.2). Since

$$x(\theta_a + \theta_a)\theta_b = b$$

and

$$x(\theta_a\theta_b + \theta_a\theta_b) = b + b = 0,$$

we see that $\mathscr{S}(V)$ is not a ring. This disproves the claim made in [7] that the set of semi-endomorphisms of an abelian group form a ring.

Lemma 2.4 (Huq [7]). *Let $\phi \in \mathscr{S}End\,G$, where G is abelian. Then $2((x + y)\phi - x\phi - y\phi) = 0$ for all $x, y \in G$. Consequently, $2(0\phi) = 0$.*

We conclude immediately from this result that if G is an abelian group without elements of order 2, then every semi-endomorphism of G is an endomorphism. Thus $\mathscr{S}End\,G = $ End G, the ring of endomorphisms of G. The converse is also valid as stated in the following result. (We note that it was shown in Fong and Pilz [3] that if n is even, then $\mathscr{S}End\,\mathbb{Z}_n$ is isomorphic to $\mathbb{Z}_n \oplus \mathbb{Z}_2$.)

Theorem 2.5. *Let G be an abelian group. Then the following statements are equivalent:*

   (i) $\mathscr{S}End\,G = $ End G;

   (ii) G *contains no elements of order 2;*

   (iii) G_ϕ *contains no elements of order 2 for any $\phi \in \mathscr{S}End\,G$;*

   (iv) $0\phi = 0$ *for all $\phi \in \mathscr{S}End\,G$.*

Proof. (i) implies (ii). If $x \in G$ and $|x| = 2$, then $\theta_x \in \mathscr{SE}nd\, G$ but θ_x is not an endomorphism of G; hence $\mathscr{SE}nd\, G \neq \text{End}\, G$.

(ii) implies (i). Let $f \in \mathscr{SE}nd\, G$. Then, for any $x, y \in G$, $2((x+y)f - xf - yf) = 0$ by (2.4). Since there are no elements of order 2 in G by hypothesis, we see that $(x+y)f - xf - yf = 0$ for all $x, y \in G$, and so $f \in \text{End}\, G$. Therefore, $\mathscr{SE}nd\, G = \text{End}\, G$.

(ii) implies (iii). Trivial.

(iii) implies (ii). Suppose $a \in G$ and $|a| = 2$. Then $\theta_a \in \mathscr{SE}nd\, G$ by (2.2). But then

$$a = a + a + a = (a+a)\theta_a - a\theta_a - a\theta_a \in G_{\theta_a},$$

a contradiction. Hence G contains no elements of order 2.

(ii) implies (iv). Since that $2(0\phi) = 0$ by (2.4) and that G contains no elements of order 2, we have $0\phi = 0$.

(iv) implies (ii). If $a \in G$ is of order 2, then $\theta_a \in \mathscr{SE}nd\, G$ and $0\theta_a = a \neq 0$ which contradicts to (iv). $\qquad\square$

It also follows directly from (2.4) that a semi-endomorphism ϕ of G induces an endomorphism ϕ' on the quotient group $G/G[2]$ via

$$(x + G[2])\phi' = x\phi + G[2]$$

for all $x \in G$.

Lemma 2.6. *Let $\phi \in \mathscr{SE}nd\, G$, where G is abelian. Then $G[2]$ is a semi-subgroup of G containing G_ϕ.*

Proof. For any $x, y \in G[2]$, we have $2(x + y + x) = 2(2x + y) = 2y = 0$. Hence $x + y + x \in G[2]$ and $G[2]$ is a semi-subgroup of G. Now $G_\phi \subset G[2]$ is a result of (2.4). $\qquad\square$

Theorem 2.7. *Let G be an abelian group. Then $\mathscr{SE}(G) = M(G)$ if and only if $2x = 0$ for all $x \in G$.*

Proof. Suppose $\mathscr{SE}(G) = M(G)$. Then for any $x \in G \setminus \{0\}$, we have $\theta_x \in \mathscr{SE}(G) = \mathscr{SE}nd\, G$; hence, by (2.2), $|x| = 2$.

Conversely, suppose $2x = 0$ for all $x \in G$ and let $f \in M(G)$. For any $x, y \in G$, we have

$$(x + y + x)f = (2x + y)f = yf = 2(xf) + yf = xf + yf + xf.$$

Therefore, $f \in \mathscr{SE}nd\, G$. This completes the proof. $\qquad\square$

As a nearring, $\mathscr{SE}(G)$ is the sum of its zero-symmetric part $\mathscr{SE}(G)_0$ and constant part $\mathscr{SE}(G)_c$ (see Pilz [9]). From (2.2), we have $\mathscr{SE}(G)_c = \{\theta_a \mid a \in G[2]\}$. As is pointed out in Fong and Pilz [3] that $\mathscr{SE}(\mathbb{Z}_n) \cong \mathbb{Z}_n \oplus \mathbb{Z}_2$, so $\mathscr{SE}(\mathbb{Z}_n) \neq E(\mathbb{Z}_n) + M_c(\mathbb{Z}_n)$ for $n \geq 3$. The following example shows that, in general, $\mathscr{SE}(G)_0 \neq \text{End}\,(G)$.

Example 2.8. Let G be an abelian group satisfying $2G \neq G$. Thus there is some element $a \in G$ with $|a| = 2$. Define $f : G \to G$ by

$$xf = \begin{cases} 0, & \text{if } x \in 2G; \\ a, & \text{otherwise.} \end{cases}$$

We claim that $f \in \mathscr{SE}nd\, G$. Let $x, y \in G$. Then

Case 1. $x,y \in 2G$. Then $x+y+x \in 2G$, and so $(x+y+x)f = 0$ and $xf+yf+xf = 0+0+0 = 0$. Thus $(x+y+x)f = xf+yf+xf$ in this case.

Case 2. $x \in 2G$ and $y \notin 2G$. Then $x+y+x \notin 2G$, since otherwise, $y = -2x+z$ for some $z \in 2G$, and so $y \in 2G$, a contradiction. Therefore, $(x+y+x)f = a$. On the other hand, $xf+yf+xf = 0+a+0 = a$, also. Hence $(x+y+x)f = xf+yf+xf$ in this case.

Case 3. $x \notin 2G$ and $y \in 2G$. This is similar to Case 2.

Case 4. $x,y \notin 2G$. Again, we cannot have $x+y+x \in 2G$. Otherwise, $y = -2x+z \in 2G$ for some $z \in 2G$, which cannot be. Therefore, $(x+y+x)f = a$ and $xf+yf+xf = a+a+a = a$; hence $(x+y+x)f = xf+yf+xf$ in this case, also.

In all cases, we have $(x+y+x)f = xf+yf+xf$ for any $x,y \in G$. Therefore $f \in \mathcal{S}End\,G$ as claimed. Moreover, $0f = 0$ since $0 \in 2G$. Thus, $f \in \mathcal{SE}(G)_0$. One easily sees that $f \notin End\,G$. Hence $\mathcal{SE}(G)_0 \neq End\,G$.

3. SEMI-ENDOMORPHISMS OF FINITELY GENERATED ABELIAN GROUPS

To study $\mathcal{S}End\,G$ for arbitrary finitely generated abelian groups G, we collect here some basic facts about semi-endomorphisms. We assume that G is abelian in this section.

Lemma 3.1 (Fong-Pilz [3]). *Let* $\phi \in \mathcal{S}End\,G$. *If* $k \in \mathbb{Z}$ *and* $x \in G$, *then*

$$(kx)\phi = \begin{cases} k(x\phi)+0\phi, & \text{if } k \text{ is even;} \\ k(x\phi), & \text{if } k \text{ is odd.} \end{cases}$$

Lemma 3.2. *Let* $\phi \in \mathcal{S}End\,G$. *If* $k \in \mathbb{Z}$ *and* $x,y \in G$, *then*

$$(kx+y)\phi = \begin{cases} k(x\phi)+y\phi, & \text{if } k \text{ is even;} \\ (k-1)(x\phi)+(x+y)\phi, & \text{if } k \text{ is odd.} \end{cases}$$

Corollary 3.3. *Let* $\phi \in \mathcal{S}End\,G$. *If* $x,y \in G \setminus \{0\}$ *are of orders* m,n, *respectively, then*

(i) $m(x\phi) = 0$ *if* m *is even, and* $m(x\phi) = 0\phi$ *if* m *is odd.*

(ii) $(x+y)\phi = x\phi+y\phi+0\phi$ *if* m *and* n *are not both even.*

Although the semi-endomorphisms ϕ of an abelian group G cannot be determined by their images on the generating elements of G, the generators do help in some way and the following result is clear from a repeated application of (3.1) and (3.2).

Lemma 3.4. *Let* $\phi_1,\phi_2 \in \mathcal{S}End\,G$ *and* $S \subseteq G \setminus \{0\}$ *a generating set. Then* $\phi_1 = \phi_2$ *if and only if* $0\phi_1 = 0\phi_2$ *and* $(\sum_{x \in A} x)\phi_1 = (\sum_{x \in A} x)\phi_2$ *for all nonempty finite subsets* $A \subseteq S$.

The condition that $0\phi_1 = 0\phi_2$ is not superfluous as the following example shows.

Example 3.5. Consider the Klein 4-group $V = \{0,a,b,c\}$. Let $\phi_1 : V \to V$ and $\phi_2 : V \to V$ be given by $x\phi_1 = a$ for all $x \in V$ and $x\phi_2 = a$ if $x \in V \setminus \{0\}$ and $0\phi_2 = 0$. It is easy to see that ϕ_1 and ϕ_2 are semi-endomorphisms of V (cf. (2.3) and (2.8)) which agree on $V \setminus \{0\}$ but they are not equal.

Now, we would like to describe the semi-endomorphisms for any given finitely generated abelian group G using a set of generators.

So, let G be a finitely generated abelian group. Thus, $G = H_1 \oplus \cdots \oplus H_n$ where each H_i is a cyclic group of finite or infinite orders for each $i \in I = \{1,2,\ldots,n\}$. We may assume that $H_i = \langle e_i \rangle$ where each e_i has order $m_i \in \mathbb{N} \cup \{\infty\}$. Let $O = \{j \in I \mid m_j < \infty \text{ and is odd}\}$ and put $E = I \setminus O$. Also, let $\phi \in \mathcal{S}\!End\, G$.

First, we extend a little bit of (2.4).

Lemma 3.6. *If S is a nonempty finite subset of G with $|S| > 1$, then*

$$\left(\sum_{x \in S} x\right)\phi - \sum_{x \in S}(x\phi) \in V_2.$$

Proof. Note that if $|S| \geq 3$ and $x' \in S$, then

$$\left(\sum_{x \in S} x\right)\phi - \sum_{x \in S}(x\phi) = \left(\left(x' + \sum_{x \in S \setminus \{x'\}} x\right)\phi - x'\phi - \left(\sum_{x \in S \setminus \{x'\}} x\right)\phi\right)$$

$$+ \left(\left(\sum_{x \in S \setminus \{x'\}} x\right)\phi - \sum_{x \in S \setminus \{x'\}}(x\phi)\right).$$

The conclusion follows from an induction on $|S|$. $\qquad\qquad\square$

We notice that if J is a nonempty subset of I such that $J \setminus O \neq \varnothing$, then

$$\left(\sum_{j \in J} e_j\right)\phi = \left(\sum_{j \in J \cap O} e_j + \sum_{j \in J \setminus O} e_j\right)\phi$$

$$= \sum_{j \in J \cap O}(e_j\phi) + \left(\sum_{j \in J \setminus O} e_j\right)\phi + |J \cap O|(0\phi)$$

$$= \sum_{j \in J}(e_j\phi) + \left(\left(\sum_{j \in J \setminus O} e_j\right)\phi - \sum_{j \in J \setminus O}(e_j\phi) + |J \cap O|(0\phi)\right);$$

thus, if $J \setminus O \neq \varnothing$, then

(3:1) $\qquad \left(\sum_{j \in J} e_j\right)\phi - \sum_{j \in J}(e_j\phi) = \left(\sum_{j \in J \setminus O} e_j\right)\phi - \sum_{j \in J \setminus O}(e_j\phi) + |J \cap O|(0\phi).$

To continue, we define some subsets of G. For any $m \in \mathbb{N}$ and $y \in G$, let $V_m(y) = \{x \in G \mid mx = y\}$. If $y = 0$, we use V_m for $V_m(0)$. Therefore, $G[2] = \{x \in G \mid 2x = 0\} = V_2$.

The process of obtaining a semi-endomorphism $f : G \to G$ is described in the following:

(1) Choose an element $u \in V_2$ arbitrarily and put $0f = u$.
(2) For each nonempty subset $E' \subseteq E$ with $|E'| \geq 2$, choose an element $a_{E'} \in V_2$. Define $a_\varnothing = u$ and $a_{\{j\}} = 0$ for all $j \in E$.
(3) For each $j \in E$, if m_j is finite, choose an element $v_j \in V_{m_j}$, otherwise, choose $v_j \in G$ arbitrarily. Set $e_j f = v_j$.
(4) For each $j' \in O$, choose an element $v_{j'} \in V_{m_{j'}}(u)$. Set $e_{j'} f = v_{j'}$.
(5) For any nonempty subset $J \subseteq I$, put

$$\left(\sum_{j \in J} e_j\right)f = \left(\sum_{j \in J} v_j\right) + a_{J \cap E} + |J \cap O|u.$$

(6) Finally, for an arbitrary nonempty subset $\{s_i \mid i \in I\} \subseteq \mathbb{N}$, let $A = \{i \in I \mid s_i$ is odd$\}$ and define

$$\left(\sum_{i \in I} s_i e_i\right) f = \left(\sum_{i \in I} s_i v_i\right) + a_{A \cap E} + |A \cap O|u.$$

We now show that such a mapping $f : G \to G$ given by the procedure is indeed a semi-endomorphism of G.

To see that f is well-defined, let $x \in G$ and $x = \sum_{i \in I} s_i e_i = \sum_{i \in I} t_i e_i$. Thus, for $i \in I = \{1, 2, \ldots, n\}$, we have $s_i \equiv t_i \pmod{m_i}$ if $m_i < \infty$, and $s_i = t_i$ if $m_i = \infty$. For those m_i with $m_i < \infty$, let $t_i = s_i + l_i m_i$, where $l_i \in \mathbb{Z}$; hence

$$t_i v_i = (s_i + l_i m_i) v_i = s_i v_i + l_i m_i v_i = \begin{cases} s_i v_i, & \text{if } 2 \mid m_i; \\ s_i v_i + l_i u, & \text{if } 2 \nmid m_i. \end{cases}$$

Therefore,

(3:2)
$$\sum_{i \in I} t_i v_i = \left(\sum_{i \in I} s_i v_i\right) + \left(\sum_{i \in O} l_i u\right).$$

Let $C = \{i \in I \mid l_i$ is odd$\}$. Since $|u| \leq 2$, we have

(3:3)
$$\sum_{i \in O} l_i u = |C \cap O|u.$$

Let $A = \{i \in I \mid s_i$ is odd$\}$ and $B = \{i \in I \mid t_i$ is odd$\}$. For $i \in I$, if $2 \mid m_i$ or $m_i = \infty$, then s_i is even if and only if t_i is even. Therefore,

(3:4)
$$A \cap E = B \cap E.$$

On the other hand, if m_i is odd, then we have that t_i is odd if and only if either (i) s_i is odd and l_i is even, or (ii) s_i is even and l_i is odd; thus,

$$C \cap O = \big((A \cap O) \setminus (B \cap O)\big) \cup \big((B \cap O) \setminus (A \cap O)\big).$$

Therefore, we have

$$|A \cap O| \equiv |B \cap O| + |C \cap O| \pmod{2}$$

and so

(3:5)
$$|A \cap O|u = (|B \cap O| + |C \cap O|)u.$$

Using (3:2), (3:3), (3:4) and (3:5), we obtain

$$\begin{aligned} \left(\sum_{i \in I} t_i e_i\right) f &= \left(\sum_{i \in I} t_i v_i\right) + a_{B \cap E} + |B \cap O|u \\ &= \left(\sum_{i \in I} s_i v_i\right) + |C \cap O|u + a_{A \cap E} + |B \cap O|u \\ &= \left(\sum_{i \in I} s_i v_i\right) + a_{A \cap E} + (|B \cap O| + |C \cap O|)u \\ &= \left(\sum_{i \in I} s_i v_i\right) + a_{A \cap E} + |A \cap O|u = \left(\sum_{i \in I} s_i e_i\right) f. \end{aligned}$$

This shows that f is well-defined.

Finally, let $x = \sum_{i \in I} s_i e_i$ and $y = \sum_{i \in I} t_i e_i \in G$. Set $A = \{i \in I \mid t_i$ is odd$\} = \{i \in I \mid (2s_i + t_i)$ is odd$\}$, and $B = \{i \in I \mid s_i$ is odd$\}$. Since $2a_{B \cap E} = 2(|B \cap O|u) = 0$, we have

$$
\begin{aligned}
(2x+y)f &= \sum_{i\in I}(2s_i+t_i)(e_if)+a_{A\cap E}+|A\cap O|u \\
&= \sum_{i\in I}(2s_i+t_i)(e_if)+a_{A\cap E}+|A\cap O|u+2a_{B\cap E}+2(|B\cap O|u) \\
&= 2(xf)+yf.
\end{aligned}
$$

This shows that f is indeed a semi-endomorphism of G.

We note that any semi-endomorphism of G can be constructed using the above procedure; thus the procedure characterizes semi-endomorphism of G.

REFERENCES

[1] K. I. Beidar, Y. Fong, W.-F. Ke, and W.-R. Wu. On semi-endomorphisms of groups. *Comm. Alg.* **27**, to appear.

[2] J. R. Clay. *Nearrings: Geneses and Applications.* Oxford Univ. Press, Oxford, 1992.

[3] Y. Fong and G. Pilz. Near-rings generated by semi-endomorphisms of groups. *Contributions To General Algebra* **7** (1991), 159–168.

[4] Y. Fong and L. van Wyk. A note on semi-homomorphism of rings. *Bull. Austral. Math. Soc.* **40** (1989), 481–486.

[5] Y. Fong and L. van Wyk. Semi-homomorphisms of near-rings. *Math. Pannonica* **3** (1992), 3–27.

[6] I. N. Herstein. Semi-homomorphisms of groups. *Canad. J. Math.* **20** (1968), 384–388.

[7] S. A. Huq. Semi-homomorphisms of rings. *Bull. Austral. Math. Soc.* **36** (1987), 121–125.

[8] J. D. P. Meldrum. *Near-rings and Their Links with Groups.* No. 134 in Research Note Series. Pitman Publ. Co., 1985.

[9] G. Pilz. *Near-rings.* North-Holland/American Elsevier, second, revised edition, 1983.

DEPARTMENT OF MATHEMATICS, NATIONAL CHENG KUNG UNIVERSITY, TAINAN, TAIWAN

DEPARTMENT OF MATHEMATICS, UNIVERSITY OF SOUTHWESTERN LOUISIANA, LAFAYETTE, LA 70504, U.S.A.

INSTITUTE OF MATHEMATICS, ACADEMIA SINICA, NANKANG, TAIPEI, TAIWAN

A NOTE ON PSEUDO-DISTRIBUTIVITY IN GROUP NEAR-RINGS

R. L. FRAY

ABSTRACT. It is shown that the group near-ring constructed from a pseudo-distributive
near-ring and an arbitrary group is a ring.

1. INTRODUCTION

Let G be any multiplicatively written group and R a near-ring. The group near-ring RG can be defined as the set of all mappings from G into R having finite support with pointwise addition and multiplication via

$$(\alpha \star \beta)(x) = \sum \alpha(y)\beta(y^{-1}x)$$

$x, y \in G, \alpha, \beta \in RG$.

According to Heatherly and Ligh [1] this multiplication will in general not be well-defined, but will depend on the order of the elements in the sum. However, if R is a pseudo-distributive near-ring then RG is also a pseudo-distributive near-ring. By adopting the definition of Le Riche, Meldrum and van der Walt [2], it is shown here that the group near-ring constructed from a pseudo-distributive near-ring turns out to be a ring.

2. PRELIMINARIES AND NOTATION

Let R be a right near-ring with identity 1, G a (multiplicatively written) group with identity e and let R^G denote the cartesian direct sum of $|G|$ copies of $(R, +)$ indexed by the elements of G. The right near-ring of all mappings of the group R^G into itself will be denoted $M(R^G)$. The group near-ring, denoted $R[G]$, of G over R is defined in [2] as a subnear-ring of $M(R^G)$ generated by the set $\{[r, g], r \in R, g \in G\}$, where $[r, g] : R^G \to R^G$ is defined by $([r, g]\mu)(h) = r\mu(hg)$. Since $R[G]$ is a subnear-ring of $M(R^G)$ it follows that $(A + B)\mu = A\mu + B\mu$ and $(AB)\mu = A(B\mu)$ for all $A, B \in R[G]$, $\mu \in R^G$. This makes R^G into an $R[G]$-module. Moreover, R^G is a faithful $R[G]$-module, because $A\mu = 0$ for all $\mu \in R^G$ implies that $A = 0$. In the proofs which follow we shall use induction on the complexity of an element A of $R[G]$, i.e., intuitively speaking, on how far it is removed from being a generator $[r, g]$. We proceed to make this notion precise. A *generating sequence* for an element A of $R[G]$ is a finite sequence $A_1, A_2, \ldots, A_n$ of elements of $R[G]$ such that $A_n = A$, and for all i, $1 \le i \le n$, one of the following three cases applies:

   (i) $A_i = [r, g]$ for some $r \in R$, $g \in G$;
   (ii) $A_i = A_k + A_l$ for some $1 \le k, l < i$;
   (iii) $A_i = A_k A_l$ for some $1 \le k, l < i$.

1991 *Mathematics Subject Classification.* 16Y30 .

Y. Fong et al. (eds.), Near-Rings and Near-Fields, 79–83.
© 2001 *Kluwer Academic Publishers. Printed in the Netherlands.*

The length of a generating sequence of minimal length for A will be called the *complexity* of A and denoted $c(A)$. Clearly $c(A) \geq 1$ for all $A \in R[G]$, and $c(A) = 1$ iff $A = [r, g]$ for some $r \in R$, $g \in G$.

Furthermore, if $c(A) > 1$, then either $A = B + C$ or $A = BC$ with $c(B), c(C) < c(A)$, because if $A_1, A_2, \ldots, A_n$ is a generating sequence for A, then $A_1, A_2, \ldots, A_i$ is a generating sequence for A_i, $i = 1, 2, \ldots, n$, albeit not necessarily one of minimal length. Recall that for a right near-ring R and a positive integer m, R is *m-distributive* if for each $a, b, c, d, r, y_i, z_i \in R$,

(1) $ab + cd = cd + ab$,

(2) $r(\sum_{i=1}^{m} y_i z_i) = \sum_{i=1}^{m} r y_i z_i$.

We say that R is *pseudo-distributive* (hereafter written p. d) if R is m-distributive for all positive integers m.

3. RESULTS

Our main aim in this section is to show that group near-rings over p. d near-rings are in fact rings. A major tool in this is the fact that if R is a p. d near-ring then every element of $R[G]$ is simply the sum of elements of the form $[r, g]$ and $-[r, g]$. To show this we require the following preliminary result.

Lemma 3.1. *If R is p. d, then*

$$[r, g](\sum_{i=1}^{k} [r_i, g_i]) = \sum_{i=1}^{k} [r r_i, g g_i]$$

for all $r, r_1, \ldots, r_k \in R$, $g, g_1, \ldots, g_k \in G$ *and* $k \geq 1$.

Proof. For all $h \in G$, $\mu \in R^G$,

$$([r, g]\sum_{i=1}^{k} [r_i, g_i])(\mu)(h) = ([r, g]((\sum_{i=1}^{k} [r_i, g_i])\mu))(h) = r(((\sum_{i=1}^{k} [r_i, g_i])\mu)(hg))$$

$$= r(\sum_{i=1}^{k} ([r_i, g_i]\mu)(hg)) = r(\sum_{i=1}^{k} r_i \mu(hgg_i)) = \sum_{i=1}^{k} r r_i \mu(hgg_i)$$

$$= \sum_{i=1}^{k} [r r_i, gg_i]\mu(h) = (\sum_{i=1}^{k} [r r_i, gg_i])(\mu).$$

This proves the result. □

Theorem 3.2. *If (h)R is a p. d near-ring then every element of $R[G]$ is simply the sum of elements of the form $[r, g]$ or $-[r, g]$ for some $r \in R$, $g \in G$.*

Proof. Let $A \in R[G]$ and suppose $c(A) = 1$ then $A = [r, g]$ or $-[r, g]$. So the result is trivially true in this case. We therefore only need to consider $A \in R[G]$ with $c(A) \geq 2$. Let $m \geq 2$ and let $c(A) = m$. Suppose that for all $W \in R[G]$ with $c(W) < m$, $W = \sum_{i=1}^{k} \sigma_i \alpha_i$ where $\sigma_i = \pm 1$ and $\alpha_i = [r_i, g_i]$ for some $r_i \in R$, $g_i \in G$ and for $i = 1, 2, \ldots, k$. If $c(A) = m$ then there are two possibilities for A, namely, $A = B + C$ or $A = BC$ with $c(B), c(C) < m$. By the induction hypothesis B and C have the required form and so if $A = B + C$ then A has the required form. If $A = BC$ then by the induction hypothesis, $B = \sum_{i=1}^{l} \sigma_i \alpha_i$ and $C = \sum_{j=1}^{k} \rho_j \beta_j$ where $\alpha_i, \beta_j \in \{[r,g] \mid r \in R, g \in G\}$, $\sigma_i, \rho_j \in \{1, -1\}$, for all $1 \leq i \leq l$, $1 \leq j \leq k$. Therefore,

$$A = \sigma_1 \alpha_1 \left(\sum_{j=1}^{k} \rho_j \beta_j \right) + \sigma_2 \alpha_2 \left(\sum_{j=1}^{k} \rho_j \beta_j \right) + \cdots + \sigma_l \alpha_l \left(\sum_{j=1}^{k} \rho_j \beta_j \right).$$

By using $-[r,g] = [-r,g]$ for all $r \in R$, $g \in G$ and applying lemma 3.1 to each of the terms in the expression for A, it follows that A has the required form in this case as well. This completes our proof. $\qquad \square$

If R is a p. d near-ring we can now, in view of Theorem 3.2, redefine the complexity of $A \in R[G]$, which we also denote by $c(A)$, as the smallest natural number m such that

$$A = \sum_{i=1}^{m} \sigma_i [r_i, g_i]$$

for some $r_i \in R, g_i \in G$ and $\sigma_i = \pm 1$. It is now immediately obvious that if R is a p. d near-ring and $A \in R[G]$ with $c(A) \geq 2$, then $A = A_1 + A_2$ with $c(A_1), c(A_2) < c(A)$ and $A_1, A_2 \in R[G]$.

An immediate consequence of this is that $(R[G], +)$ is abelian because if R is p. d it can easily be shown that $[r,g] + [s,g'] = [s,g'] + [r,g]$ for any $r, s \in R, g, g' \in G$. We therefore have

Theorem 3.3. *If R is a p. d near ring then $(R[G], +)$ is abelian.*

In order to show that $R[G]$ is a ring if R is p.d, it suffices to prove that $R[G]$ is distributive. To this end we require the following preliminary results.

Lemma 3.4. *If R is a p. d near-ring, then $[r,g]\mu_1 + [s,g']\mu_2 = [s,g']\mu_2 + [r,g]\mu_1$ for any $r, s \in R$, $g, g' \in G$, and $\mu_1, \mu_2 \in R^G$,*

Proof. Follows easily from the fact that $ab + cd = cd + ab$ for all $a, b, c, d \in R$. $\qquad \square$

Lemma 3.5. *If R is a p. d near ring then $A\mu_1 + B\mu_2 = B\mu_2 + A\mu_1$ where $A, B \in R[G]$, $\mu_1, \mu_2 \in R^G$.*

Proof. By induction on $c(A)$ and $c(B)$. The case $c(A) = c(B) = 1$ is true by Lemma 3.4. Let $c(A) = m$, $c(B) = n$ with $m, n \geq 2$. Then $A = A_1 + A_2$ and $B = B_1 + B_2$ where $c(A_1) < m$, $c(A_2) < m$, $c(B_1) < n$ and $c(B_2) < n$. Assume that for all A, B with $c(A) < m$, $c(B) < n$, $A\mu_1 + B\mu_2 = B\mu_2 + A\mu_1$. Now we have

$$\begin{aligned}
A\mu_1 + B\mu_2 &= (A_1 + A_2)\mu_1 + (B_1 + B_2)\mu_2 = A_1\mu_1 + A_2\mu_1 + B_1\mu_2 + B_2\mu_2 \\
&= A_1\mu_1 + B_1\mu_2 + A_2\mu_1 + B_2\mu_2 = B_1\mu_2 + A_1\mu_1 + B_2\mu_2 + A_2\mu_1 \\
&= B_1\mu_2 + B_2\mu_2 + A_1\mu_1 + A_2\mu_1 = (B_1 + B_2)\mu_2 + (A_1 + A_2)\mu_1 \\
&= B\mu_2 + A\mu_1,
\end{aligned}$$

by the induction hypothesis.

It remains to consider the case when $c(A) = 1$ and $c(B) > 1$ or $c(A) > 1$ and $c(B) = 1$. If $c(A) = 1$ and $c(B) > 1$ then $A = [s,g]$ for some $s \in R, g \in G$ and $B = B_1 + B_2$ where

$c(B_1), c(B_2) < c(B)$. The required result follows immediately from our induction hypothesis that

$$[s,g]\mu_1 + C\mu_2 = C\mu_2 + [s,g]\mu_1$$

for all C with $c(C) < c(B)$. The case $c(A) > 1$ and $c(B) = 1$ follows similarly. This completes the proof. $\square$

Lemma 3.6. *If R is a p. d near-ring then $[r,g](A\mu_1 + B\mu_2) = [r,g](A\mu_1) + [r,g](B\mu_2)$ where $A, B \in R[G], \mu_1, \mu_2 \in R^G, r \in R, g \in G$.*

Proof. Write $A = \sum_{i=1}^{l} [r_i, g_i]$ and $B = \sum_{j=1}^{k} [r'_j, g'_j]$ then for any $h \in G$,

$$[r,g](A\mu_1 + B\mu_2)(h) = r(\sum_{i=1}^{l} r_i \mu_1(hgg_i) + \sum_{j=1}^{k} r'_j \mu_2(hgg'_j))$$

$$= \sum_{i=1}^{l} rr_i \mu_1(hgg_i) + \sum_{j=1}^{k} rr'_j \mu_2(hgg'_j) = [r,g](A\mu_1)(h) + [r,g](B\mu_2)(h).$$

Hence the result follows. $\square$

Lemma 3.7. *If R is a p. d near-ring, then $A(X\mu_1 + Y\mu_2) = A(X\mu_1) + A(Y\mu_2)$ where $A \in R[G]$, $X, Y \in R[G]$ and $\mu_1, \mu_2 \in R^G$.*

Proof. By induction on $c(A)$. The case $c(A) = 1$ is true by Lemma 3.6. Assume the result is true for any $W \in R[G]$ with $c(W) < c(A)$. Now $A = A_1 + A_2$ with $c(A_1), c(A_2) < c(A)$. Then

$$A(X\mu_1 + Y\mu_2) = (A_1 + A_2)(X\mu_1 + Y\mu_2)$$
$$= A_1(X\mu_1 + Y\mu_2) + A_2(X\mu_1 + Y\mu_2)$$
$$= A_1(X\mu_1) + A_1(Y\mu_2) + A_2(X\mu_1) + A_2(Y\mu_2)$$
$$= A_1(X\mu_1) + A_2(X\mu_1) + A_1(Y\mu_2) + A_2(Y\mu_2)$$
$$= (A_1 + A_2)(X\mu_1) + (A_1 + A_2)(Y\mu_2)$$
$$= A(X\mu_1) + A(Y\mu_2)$$

the third and fourth last steps following from lemma 3.5 and the induction hypothesis, respectively. This completes the proof. $\square$

We are now in a position to show that $R[G]$ is distributive if R is p. d.

Theorem 3.8. *If R is a p. d near-ring, G an arbitrary group, then $R[G]$ is a ring.*

Proof. For all $A, X, Y \in R[G], \mu \in R^G$,

$$(A(X+Y))\mu = A((X+Y))\mu) = A(X\mu + Y\mu)$$
$$= A(X\mu) + A(Y\mu) = (AX)\mu + (AY)\mu = (AX + AY)\mu,$$

where the third last step following from lemma 3.7. Hence $A(X + Y) = AX + AY$ and therefore $R[G]$ is distributive. The result follows by [3, Theorem 9.30]. $\square$

REFERENCES

[1] Heatherly, H. E. and Ligh, S., *Pseudo-distributive near-rings*, Bull.Austral.Math.Soc. **52** (1975), 449–456.

[2] Le Riche, L.R., Meldrum, J.D.P. and van der Walt, A.P.J., *On group near-rings*, Arch.Math. **52** (1989), 132–139.

[3] Meldrum,J.D.P., Near-rings and their links with groups, Res. notes in Math. 134 (1985), Pitman.

DEPARTMENT OF MATHEMATICS, UNIVERSITY OF THE WESTERN CAPE, PRIVATE BAG X17, BELLVILLE 7530, SOUTH AFRICA

THE ALMOST NILPOTENT RADICAL FOR NEAR RINGS

N. J. GROENEWALD

ABSTRACT. In this paper we introduce the concept of an almost nilpotent near-ring. We also define the almost nilpotent radical. The concept of a special radical for near-rings has been treated in several non-equivalent, but related ways in the recent literature. We use the version due to K. Kaarli as basis to define the concept of a weakly special radical in near-rings. We show that the almost nilpotent radical is weakly special on the class $\mathcal{A}$ of all near-rings which satisfy an extended version of the Andrunakievich Lemma. $\mathcal{A}$ includes all d.g. near-rings - and much more. We also give another characterization of the upper nil radical.

1. INTRODUCTION

In [8] Heyman and Van Leeuwen introduced the concept of an almost nilpotent ring. Many authors studied various properties of the class of almost nilpotent rings (see for example, [15], [17] and [14]).

In this note we introduce the concept of an almost nilpotent near-ring. Throughout this paper all near-rings are zero-symmetric right near-rings and R denotes such a near-ring.

The concept of a special radical for near-rings has been treated in several non-equivalent, but related, ways in recent literature. We use the version due to K. Kaarli in [10].

We introduce the concept of a weakly special class to characterize the class of all almost nilpotent near-rings. For associative rings we have the lemma of Andrunakievich, i.e. if K is an ideal of I and I is an ideal of R, then $\langle K \rangle_R^3 \subseteq K$, where $\langle K \rangle_R$ denotes the ideal of R generated by K. For near-rings in general, the above condition is not true. In [2] the concept of an $\mathcal{A}$-ideal (Andrunakievich-ideal) was introduced and a near-ring R is called an $\mathcal{A}$-near-ring if each ideal of R is an $\mathcal{A}$-ideal. The class of $\mathcal{A}$-near-rings is wide and varied, including all d.g. near-rings and all near-rings which are nilpotent or strongly regular. These and many other examples and the basic properties of near-rings in class $\mathcal{A}$ and $\mathcal{A}$-ideals are given in [2]. We establish that the 0-prime radical is weakly special on the class of $\mathcal{A}$-near-rings. We define the almost nilpotent radical as an intersection of almost nilpotent ideals and show that it is a weakly special radical on the class of $\mathcal{A}$-near-rings. In the latter part of this paper we introduce the concept of an almost nil ideal to give another characterization of the upper nil-radical.

2. THE ALMOST NILPOTENT RADICAL

Definition 2.1. A near-ring R is called almost nilpotent if for all $0 \neq I \triangleleft R$ there exists an $n \in \mathbb{N}$ such that $R^n \subset I$. Here $I \triangleleft R$ means I is an ideal of R.

Y. Fong et al. (eds.), Near-Rings and Near-Fields, 84–93.
© 2001 *Kluwer Academic Publishers. Printed in the Netherlands.*

An ideal I of R is almost nilpotent if I is an almost nilpotent near-ring.

Let $C = \{R \in \mathcal{R} : R$ has no nonzero almost nilpotent ideals$\}$. Here $\mathcal{R}$ stands for the class of all near rings.

Let $I \lhd R$. As in [1], I is an $\mathcal{A}$-ideal of R if for each $K \lhd I$ there exists an n such that $\langle K \rangle_R^n \subseteq K$.

Definition 2.2. R is an $\mathcal{A}$-near-ring if every ideal of R is an $\mathcal{A}$-ideal.

For the definitions of different prime ideals in near-rings we refer to [7].

Definition 2.3. Let ρ be a Hoehnke radical on the class $\mathcal{R}_0$ of all zero-symmetric near-rings and let $\mathcal{T}$ be a non-empty subclass of $\mathcal{R}_0$ which is closed under homomorphic images. ρ is called a special (weakly special) $\mathcal{T}$-radical if there exists a non-empty class $\mathcal{K}$ of prime (semiprime) near-rings such that the following hold:

(1) If $R \in \mathcal{K} \cap \mathcal{T}$ and $I \lhd R$ then $I \in \mathcal{K}$.
(2) If $R \in \mathcal{T}$ and $H \lhd I \lhd R$ with $I/H \in \mathcal{K}$ then $H \lhd R$ and $R/(H : I)_R \in \mathcal{K}$. $(H : I)_R = \{r \in R : rI \subseteq H\}$. If it is clear which near-ring we are refering to, we will use $(H : I)$ in stead of $(H : I)_R$.
(3) For each $R \in \mathcal{R}_0$, $\rho(R) = \cap\{I : I \lhd R, R/I \in \mathcal{K}\}$.

The class $\mathcal{K}$ is called a $\mathcal{T}$ -special (weakly special) class.

We now give an alternative definition for a special (weakly special) $\mathcal{T}$-radical. Recall that a class $\mathcal{M}$ is called essentially closed if for any near ring R and any essential ideal I of R (denoted by $I \lhd \cdot R$) with $I \in \mathcal{M}$ it follows that $R \in \mathcal{M}$.

(4) If $R \in \mathcal{T}$ and $H \lhd I \lhd R$ with $I/H \in \mathcal{K}$ then $H \lhd R$ and $\mathcal{K}$ is essentially closed.

Proposition 2.4. Let ρ be a Hoenhke radical on the class $\mathcal{R}_0$ and let $\mathcal{T}$ be a non-empty subclass of $\mathcal{R}_0$ which is closed under homomorphic images. ρ is a special (weakly special) $\mathcal{T}$-radical if there exists a non-empty class $\mathcal{M}$ of 0-prime (0-semiprime) near-rings such that conditions (1), (3) and (4) are satisfied.

Proof. We only have to show that if $\mathcal{M}$ is the class of 0-semiprime near rings, then $\mathcal{M}$ is essentially closed if and only if $R \in \mathcal{T}$ and $K \lhd I \lhd R$ with $I/K \in \mathcal{M}$ implies $R/(K : I) \in \mathcal{M}$. Suppose $\mathcal{M}$ is essentially closed and $K \lhd I \lhd R$ with $I/K \in \mathcal{M}$. We show that $R/(K : I) \in \mathcal{M}$. First we show that $(K : I)$ is maximal relative to having intersection with I equal to K. We have $[(K : I) \cap I]^2 \subseteq (K : I)I \subseteq K$ and since K is a 0−semiprime ideal of I it follows that $(K : I) \cap I \subseteq K$. Equality now follows since $K \subseteq (K : I) \cap I$. If T is any ideal such that $T \cap I = K$, then $TI \subseteq T \cap I = K$ and therefore $T \subseteq (K : I)$. We now show that $[(K : I) + I]/(K : I)$ is an essential ideal of $R/(K : I)$. Suppose $L/(K : I)$ is a nonzero ideal of $R/(K : I)$ i.e. $(K : I) \subset L$. From the above we have $L \cap I \supset K$. Now if $([(K : I) + I]/(K : I)) \cap (L/(K : I)) = 0$, then

$$([(K : I) + I]/(K : I)) \cap (L/(K : I)) = ([(K : I) + I] \cap L)/(K : I))$$
$$= [(I \cap L) + (K : I)]/(K : I) = 0$$

Hence $I \cap L \subseteq (K : I)$ and therefore we have that $I \cap L = I \cap I \cap L \subseteq I \cap (K : I) = K$. Hence $I \cap L = K$ which contradicts our assumption and we have that $[(K : I) + I]/(K : I)$ is an essential ideal of $R/(K : I)$. Furthermore, $([(K : I) + I]/(K : I)) \cong I/[(K : I) \cap I] = I/K \in \mathcal{M}$ and since $\mathcal{M}$ is essentially closed, we have $R/(K : I) \in \mathcal{M}$ which proves one direction.

Now suppose $A \lhd \cdot R$ with $A \in \mathcal{M}$. Since $A \in \mathcal{M}$, we have $A \cong A/0 \in \mathcal{M}$ and from our assumption we have $R/(0:A) \in \mathcal{M}$. Since $[(0:A) \cap A]^2 \subseteq (0:A)A = 0$ and A is 0-semiprime we have $(0:A) \cap A = 0$. But $A \lhd \cdot R$ and consequently $(0:A) = 0$ and it follows that $R \in \mathcal{M}$.   □

Let $\mathcal{P}_0$ denote the 0-prime radical i.e., the intersection of all the 0-prime ideals of a near-ring. In what follows, $\mathcal{A}$ denotes the class of all $\mathcal{A}$-near-rings.

Proposition 2.5. $\mathcal{P}_0$ *is a weakly special $\mathcal{A}$-radical.*

Proof.   (1) Let $\mathcal{M}$ be the class of 0−semiprime near-rings in $\mathcal{R}_0$ and let $I \lhd R \in \mathcal{M} \cap \mathcal{A}$. Consider $X \lhd I$ such that $X^2 = 0$. Since R is an $\mathcal{A}$-near-ring we can find an m such that $\langle X \rangle_R^{2m} \subseteq X^2 = 0$. Hence $\langle X \rangle_R = 0$ and, therefore $X = 0$.
   (3) This follows from [13], Theorem 2.95.
   (4) It follows from [1] Proposition 2.4 that if $R \in \mathcal{M}$ and $H \lhd I \lhd R$ with $I/H \in \mathcal{M}$ then $H \lhd R$. Next, take $R \in \mathcal{A}$ and $I \lhd \cdot R$ with $I \in \mathcal{M}$. We show that R is 0−semiprime. Consider $A \lhd R$ such that $A^2 = 0$. If $A \neq 0$, then since $I \lhd \cdot R$ we have $A \cap I \neq 0$. Since I is 0−semiprime and $(A \cap I)^2 \subseteq A^2 = 0$ we have $A \cap I = 0$. This is a contradiction and consequently $R \in \mathcal{M}$.

                                      □

Definition 2.6. Let $\mathbb{A}_3(R) = \cap \{I : I \lhd R \text{ and } R/I \in \mathcal{C}\}$.

Theorem 2.7. $\mathbb{A}_3$ *is a weakly special $\mathcal{A}$-radical.*

Proof. Let $R \in \mathcal{A} \cap \mathcal{C}$ and suppose $0 \neq I \lhd R$. We show $I^2 \neq 0$. If $I^2 = 0$ then for all $0 \neq J \lhd I$, we have $I^2 \subset J$. Hence I is a nonzero almost nilpotent ideal of R. This is not possible since $R \in \mathcal{C}$. Hence R is 0-semiprime.
   (1) Let $R \in \mathcal{A} \cap \mathcal{C}$ and suppose $0 \neq I \lhd R$. Suppose $I \notin \mathcal{C}$, i.e. it has a nonzero almost nilpotent ideal J. Since $R \in \mathcal{A} \cap \mathcal{C}$ we can find a k such that $\langle J \rangle_R^k \subseteq J$. Furthermore, from above, R is 0-semiprime and it follows that $\langle J \rangle_R^k \neq 0$. Let $0 \neq K \lhd \langle J \rangle_R$. Now $K \cap J \lhd J$ and $K \cap J \lhd K$. $K \cap J \neq 0$, for if $K \cap J = 0$ then $K \cap \langle J \rangle_R^k \subseteq K \cap J = 0$. Hence also $\left(\langle K \cdot \langle J \rangle_R \rangle_{\langle J \rangle_R} \right)^{k+1} \subseteq K \cdot (\langle J \rangle_R)^k \subseteq K \cap \langle J \rangle_R^k = 0$.
      If $Q = \langle K \cdot \langle J \rangle_R \rangle_{\langle J \rangle_R} \neq 0$ then $\langle J \rangle_R$ has a nonzero nilpotent ideal. This is impossible since $\langle J \rangle_R \lhd R$ and because from Proposition 2.5, $\langle J \rangle_R$ is a 0-semiprime near-ring. Hence $K \cdot \langle J \rangle_R = 0$. Now we have $K^2 \subseteq K \cdot \langle J \rangle_R = 0$. Hence K is a nonzero nilpotent ideal of $\langle J \rangle_R$ which is again contradicting the fact that $\langle J \rangle_R$ is 0-semiprime. Hence we may assume $K \cap J \neq 0$. Since J is almost nilpotent, we have $J^n \subset K \cap J \subseteq K$ for some n. Hence $\left(\langle J \rangle_R^k \right)^n \subset (J)^n \subseteq K \cap J \subseteq K$. This shows that $\langle J \rangle_R$ is an almost nilpotent ideal of R. This contradicts the fact that $R \in \mathcal{C}$. Hence $I \in \mathcal{C}$ and we are done.
   (4) Since each element of $\mathcal{A} \cap \mathcal{C}$ is 0-semi-prime, it follows from [1] Proposition 2.4 that if $R \in \mathcal{A} \cap \mathcal{C}$ and $H \lhd I \lhd R$ with $I/H \in \mathcal{A} \cap \mathcal{C}$ then $H \lhd R$. We now show that if $R \in \mathcal{A}$ and $B \lhd \cdot R$ with $B \in \mathcal{C}$, then $R \in \mathcal{C}$. Suppose $R \notin \mathcal{C}$, i.e. there exists $0 \neq J \lhd R$ which is almost nilpotent. $J \cap B \neq 0$ since B is an essential ideal of R.

We show that $J \cap B$ is almost nilpotent. Let I be a proper nonzero ideal of $J \cap B$. Now we have $I \lhd J \cap B \lhd R$ and since $R \in \mathcal{A}$ we can find $k \in \mathbb{N}$ such that $\langle I \rangle_R^k \subseteq I$. But then also $0 \neq \langle I \rangle_J^k \subseteq I$. Since J is almost nilpotent, we can find a t such that $(J)^t \subset \langle (\langle I \rangle_J)^k \rangle_J$. Now $((J \cap B))^{t+k} \subseteq (J)^{t+k} \subset \left(\langle \langle I \rangle_J^k \rangle_J \right)^k \subseteq (\langle I \rangle_J)^k \subseteq I$. Hence $J \cap B$ is an almost nilpotent ideal of B. The rest follows from Definition 2.6. $\qquad\square$

Remark 2.8. If R is an almost nilpotent near-ring, then $\mathcal{P}_0(R) = R$ or $\mathcal{P}_0(R) = 0$.

Proof. First we show that if R is almost nilpotent then it has no nonzero proper $0-$prime ideals. Suppose $0 \neq I \lhd R$ and I is $0-$prime. Since $R^k \subset I$ for some k, and I a $0-$semi prime ideal we have $R = I$. If R is $0-$prime, then 0 is a $0-$prime ideal and $\mathcal{P}_0(R) = 0$. If 0 is not a prime ideal, then we have $\mathcal{P}_0(R) = R$ since R is the only prime ideal. $\qquad\square$

Example 2.9. In Divinsky [4], p103, we have a ring W such that the only nonzero ideals of W are of the form $(2)^n$, $n = 1, 2, \ldots$ and $W = (2)$. It is clear that every nonzero ideal contains a power of W. This containment is strict since $(2)^{n+1} \subset (2)^n$ for $2^n \in (2)^n$ but $2^n \notin (2)^{n+1}$. Therefore every nonzero ideal of W contains a power of W so that W is an almost nilpotent near-ring. W is not a nilpotent near-ring.

Following [9], we define a left almost nilpotent near-ring.

Definition 2.10. A near-ring R is called left almost nilpotent if for any nonzero left R subgroup A of R we can find a $t \in \mathbb{N}$ such that $R^t \subset A$.

Denote the class of all left almost nilpotent near-rings by $\mathbb{A}_2$.

Theorem 2.11. *If $R \in \mathbb{A}_2$ and L is a left R subgroup of R, then $L \in \mathbb{A}_2$.*

Proof. Let L be a nonzero left R subgroup of $R \in \mathbb{A}_2$. If R is nilpotent, then clearly L is also nilpotent and we are done. If $R^n \neq 0$ for all n, then R is 3-prime. Suppose not, and let $0 \neq a, b \in R$ such that $aRb = 0$. Since $R \in \mathbb{A}_2$ we can find $m, n \in \mathbb{N}$ such that $R^m \subseteq Rb$ and $R^n \subseteq Ra$. Hence $R^{m+m} \subseteq RaRb = 0$ a contradiction. Hence R is 3-prime. Let T be any nonzero left L subgroup. We show that $L^t \subset T$ for some t. Since $R \in \mathbb{A}_2$ we can find $k \in \mathbb{N}$ such that $R^k \subset L$. Now $R^k T \neq 0$. If $R^k T = 0$ then since R is 3-prime, we can show that this forces $T = 0$. But T was chosen $\neq 0$ and, therefore, we have $R^k T \neq 0$. Hence we can get $0 \neq t \in T$ such that $Rt \neq 0$. Since $R \in \mathbb{A}_2$ we can get $m \in \mathbb{N}$ such that $R^m \subset Rt$. Hence $L^m \subseteq R^m \subset Rt \subseteq T$ and we are done. $\qquad\square$

Let $C_2 = \{R : R$ has no nonzero $\mathbb{A}_2$-ideals$\}$. We have the following:

(1) If $R \in C_2$ then R is 0-semiprime. Let $0 \neq I \lhd R$ and suppose $I^2 = 0$. For every nonzero left I-subgroup J of I we have $I^2 \subset J$. Hence I is a nonzero ideal of R which is an element of $\mathbb{A}_2$, contradicting the fact that $R \in C_2$.

(2) C_2 is essentially closed:

Let $R \in \mathcal{R}_0$ and $A \lhd \cdot R$ with $A \in C_2$. Let $0 \neq I \lhd R$ and suppose $I \in \mathbb{A}_2$. Since $A \lhd \cdot R$ we have $A \cap I \neq 0$. Since $\mathbb{A}_2$ is hereditary, we have that the nonzero ideal $A \cap I$ of A is an element of $\mathbb{A}_2$. This contradicts the choice of A and hence $R \in C_2$.

We could not show that C_2 is hereditary and therefore it is still an open question if C_2 is weakly special or not.

Definition 2.12. I f $R \in \mathcal{R}$ we define the left almost nilpotent radical $\mathbb{A}_2(R)$ as follows:

$$\mathbb{A}_2(R) = \cap \{I \lhd R : R/I \in C_2\}.$$

Remark 2.13. From Sands [15] we have examples to show that $\mathbb{A}_2 \subsetneqq \mathbb{A}_3$ and also $\mathbb{A}_2(R) \subsetneqq \mathbb{A}_3(R)$. $\mathcal{P}_0 \subsetneqq \mathbb{A}_2$ because for any near-ring R $\mathbb{A}_2(R) = \cap \{I : R/I \in C_2\}$ and $\mathcal{P}_0(R)$ coinsides with the intersection of all the 0-semiprime idaels of R and since each element of C_2 is also a 0-semiprime near-ring. The containment is strict since the near-ring W in Example 2.13 is such that $\mathbb{A}_2(W) = W$ but $\mathcal{P}_0(W) = 0$.

In general the class of 3-prime near-rings is not hereditary. We have the following for left almost nilpotent near-rings.

Proposition 2.14. *Every nonzero left R-subgroup of a 3-prime $\mathbb{A}_2$ near-ring R is also 3-prime.*

Proof. Let $R \in \mathbb{A}_2$ and suppose R is also 3-prime. Let L be any nonzero left R-subgroup of R. Since $R \in \mathbb{A}_2$ there exists a k such that $R^k \subset L$. Suppose L is not 3-prime, i.e. we can find $0 \neq a, b \in L$ such that $aLb = 0$. But now $aR^k b = 0$. Since R is 3-prime, we have $R^n \neq 0$ for all n. Now we can show from the 3-primeness of R and $aR^k b = 0$ that $a = 0$ or $b = 0$, a contradiction. Hence L is 3-prime. $\square$

Proposition 2.15. *If $R \in \mathbb{A}_2$ and it is also 3-prime, then for every nonzero left R-subgroup L of R the right annihilator $r(L)$ of L is zero.*

Proof. Suppose $R \in \mathbb{A}_2$ and it is also 3-prime. Let L be a nonzero R-subgroup of R. Since $R \in \mathbb{A}_2$ we can find a $k \in \mathbb{N}$ such that $R^k \subset L$. Now $Lr(L) = 0$ and, therefore, $R^k r(L) = 0$. Since R is 3-prime we have $r(L) = 0$. $\square$

Theorem 2.16. *If $R \in \mathbb{A}_2$ then:*

    (a) *R is 3-prime if and only if R is completely prime.*
    (b) *R is nilpotent if and only if R has divisors of zero.*

Proof. (a) We only have to show that if R is 3-prime then R is completely prime. Let $x, y \in R$ such that $xy = 0$. Suppose $x \neq 0$ and $y \neq 0$. Now R 3-prime forces $yRx \neq 0$ and hence also $Rx \neq 0$. Since Rx is a nonzero left R-subgroup of R and $RxyR = 0$, it follows from the previous proposition that $yR = 0$. Hence $y = 0$, which contradicts our assumption and we have that R is completely prime

    (b) is clear. $\square$

3. ANOTHER CHARACTERIZATION FOR THE NIL-RADICAL

In [16] APJ van der Walt introduced the concept of an s-prime ideal. We need:

Definition 3.1. A set S of R is called an s-system if and only if S contains a multiplicative system S^*, called the kernel of S, such that for every $s \in S$ we have $\langle s \rangle \cap S^* \neq \phi$. ϕ is defined to be an s-system.

Definition 3.2. A n ideal P in R is an s-prime ideal if and only if $C(P)$ is an s-system.

In [5] the concept of an semi-s-prime ideal was introduced. Denote the class of s-prime near-rings by S (i.e. a near-ring is s-prime if (0) is an s-prime ideal). We also need the following:

Definition 3.3. For any near ring R we have the following:

(a) A set $T \subseteq R$ is called a complete system if $a^n \in T$ for all $a \in T$ and each natural number n.

(b) A set U of R is a u-system if U contains a complete system U_* such that for all $u \in U$ we have $\langle u \rangle \cap U_* \neq \phi$. ϕ is defined to be an u-system.

(c) An ideal Q in R is a semi-s-prime ideal if and only if $C(Q)$ is a u-system.

Denote the class of semi-s-prime near-rings by $\mathcal{U}$, i.e. near-rings for which (0) is a semi-s-prime ideal.

We have:

An s-prime ideal is a semi-s-prime ideal and a semi-s-prime ideal is a semiprime ideal. We also have:

Theorem 3.4. *If $\mathcal{N}(R)$ denotes the nil radical of R, i.e. the sum of all nil ideals of R, then we have*

$$\mathcal{N}(R) = \cap \{I \lhd R : R/I \in S\} \qquad ([16])$$

$$= \cap \{Q \lhd R : R/Q \in \mathcal{U}\} \qquad ([5])$$

Motivated by unpublished work of H le Roux[11] for rings, we have :

Proposition 3.5. *An ideal Q of R is semi-s-prime if and only if for any ideal A in R such that $a^m \in Q$ for all $a \in A$ and some $m \in \mathbb{N}$, it follows that $A \subseteq Q$.*

Proof. Let $Q \lhd R$ such that if for all $A \lhd R$ such that $a^m \in Q$ for all $a \in A$ and some $m \in \mathbb{N}$, it follows that $A \subseteq Q$. We show $C(Q)$ is a u-system. If $Q = R$ we are done, so suppose $Q \neq R$, i.e. $C(Q) \neq \phi$. Let $U_* = \{r \in C(Q) : r^n \in C(Q) \text{ for all } n\}$. $U_* \neq \phi$. If $s^m \notin C(Q)$ for every $s \in C(Q)$ and some $m \in \mathbb{N}$, then we have $s^m \in Q$ for every $s \in R$ and some $m \in \mathbb{N}$. From our assumption it follows that $R \subseteq Q$ which contradicts our assumption, hence $U_* \neq \phi$. Now $U_* \subseteq C(Q)$ and it is clearly a complete system. Let $a \in C(Q)$. We can find $a_1 \in \langle a \rangle$ such that $a_1^n \notin Q$ for all n. (If not, then $a_1^n \in Q$ for all $a_1 \in \langle a \rangle$ and some $n \in \mathbb{N}$.) Hence from our assumption we will have $a \in \langle a \rangle \subseteq Q$, a contradiction. Therefore, $a_1 \in \langle a \rangle \cap U_*$ and it follows that Q is semi-s-prime.

Conversely, suppose Q is a semi-s-prime ideal. From this we have $C(Q)$ is a u-system. Hence there exists a complete system $U_* \subseteq C(Q)$ such that $\langle c \rangle \cap U_* \neq \phi$ for all $c \in C(Q)$. Let A be any ideal of R and suppose $a^m \in Q$ for all $a \in A$ and some $m \in \mathbb{N}$. Let $x \in A$ and suppose $x \notin Q$. Hence $x \in C(Q)$. From the above we have $\langle x \rangle \cap U_* \neq \phi$. Let $d \in \langle x \rangle \cap U_*$. Now $d^m \in U_* \subseteq C(Q)$ for all m. Hence for $d \in \langle x \rangle \subseteq A$ we have $d^m \notin Q$ for all m. This contradicts our assumption. Hence $x \in Q$. Since x was arbitrary, we have $A \subseteq Q$. $\square$

Definition 3.6. Q is called an n-prime ideal of R if for $A, B \lhd R$ and for all $x \in AB$, $x^m \in Q$ for some $m \in \mathbb{N}$ implies $A \subseteq Q$ or $B \subseteq Q$.

Q n-prime $\Rightarrow$ Q semi-s-prime. Suppose for all $A \lhd R$ and for all $a \in A$ there exists an $m \in \mathbb{N}$ such that $a^m \in Q$. Now for $x \in A^2 \subseteq A$ we can find m such that $x^m \in Q$. Hence $A \subseteq Q$ since Q is n-prime.

Definition 3.7. A set T in a near-ring R is called a z-system if T contains a complete system U such that for every $t_1, t_2 \in T$ it follows that $\langle t_1 \rangle \langle t_2 \rangle \cap U \neq \phi$.

Proposition 3.8. *An ideal Q of R is an n-prime ideal if and only if $C(Q)$ is a z-system.*

Proof. Let Q be an n-prime ideal.. Now, as in Proposition 3.5, we have that

$$U = \{r \in C(Q) \mid r^n \in C(Q) \,\forall n \in \mathbb{N}\}$$

is a complete system contained in $C(Q)$. Let $t_1, t_2 \in C(Q)$. We show that $\langle t_1 \rangle \langle t_2 \rangle \cap U \neq \phi$. If $\langle t_1 \rangle \langle t_2 \rangle = 0$ then for all $x \in \langle t_1 \rangle \langle t_2 \rangle$, $x^m = 0 \in Q$. Hence $\langle t_1 \rangle \subseteq Q$ or $\langle t_2 \rangle \subseteq Q$, i.e. $t_1 \in Q$ or $t_2 \in Q$. This is not possible. Hence $\langle t_1 \rangle \langle t_2 \rangle \neq 0$. Suppose $\langle t_1 \rangle \langle t_2 \rangle \cap U = \phi$. Now for every $z \in \langle t_1 \rangle \langle t_2 \rangle$ we have $z \notin U$. Hence there are two possibilities, $z \notin C(Q)$ or $z \in C(Q)$. If $z \notin C(Q)$, then $z \in Q$ and from our assumption, $\langle t_1 \rangle \subseteq Q$ or $\langle t_2 \rangle \subseteq Q$. If $z \in C(Q)$ then, since $z \notin U$, we must have $z^m \in Q$ for some m, and again from our assumption we have $\langle t_1 \rangle \subseteq Q$ or $\langle t_2 \rangle \subseteq Q$. Hence for both cases we have a contradiction. This imples that $\langle t_1 \rangle \langle t_2 \rangle \cap U \neq \phi$ and consequently, $C(Q)$ is a z-system.

Conversely, let $C(Q)$ be a z-system. Suppose for all $I, J \lhd R$ and for all $z \in IJ$ we can find an m such that $z^m \in Q$. Suppose $I \not\subseteq Q$ and $J \not\subseteq Q$, i.e. we can find $a \in I$ and $b \in J$ such that $a, b \in C(Q)$. $C(Q)$ is a z-system, hence $\langle a \rangle \langle b \rangle \cap U \neq \phi$ for a complete system $U \subseteq C(Q)$. Let $0 \neq l \in \langle a \rangle \langle b \rangle \cap U$. From our assumption we can get $n \in \mathbb{N}$ such that $l^n \in Q$. Hence $l^n \in Q \cap C(Q)$ since $l^n \in U \subseteq C(Q)$. This is not possible and, therefore, $I \subseteq Q$ or $J \subseteq Q$. Hence Q is an n-prime ideal. $\qquad \square$

Theorem 3.9. *If R is 0-prime and semi-s-prime then it is n-prime.*

Proof. Clearly if R is n-prime then it is semi-s-prime and 0-prime. Let R be 0-prime and semi-s-prime. Since R is semi-s-prime we have, from [5], Theorem 3.7 that $\mathcal{N}(R) = 0$ and, therefore we can find $0 \neq a \in R$ such that $a^m \neq 0$ for all $m \in \mathbb{N}$. Now $U = \{x \in R : x^n \neq 0, n \in \mathbb{N}\}$ is a non-empty complete system and $U \subseteq C(0)$. Let $x, y \in C(0)$. Since R is 0-prime, we have $\langle x \rangle \langle y \rangle \neq (0)$. Since $\mathcal{N}(R) = 0$, we can find $0 \neq a \in \langle x \rangle \langle y \rangle$ such that $a^n \neq 0$ for all n. This means that $0 \neq a \in \langle x \rangle \langle y \rangle \cap U$. Hence (0) is a n-prime ideal. $\qquad \square$

In [3] Birkenmeier et al defined an ideal I to be nilprime if and only if $\mathcal{N}(R/I) = 0$ and I is 0- prime.

Proposition 3.10. *Let R be a near-ring. $I \lhd R$ is a nilprime ideal if and only if I is an n-prime ideal.*

Proof. Let P be an n-prime ideal. Since it is also semi-s-prime, we have from [5], Theorem 3.7 that $\mathcal{N}(R/I) = 0$. From the previous theorem, I is 0-prime. Suppose now $I \lhd R$ such that $\mathcal{N}(R/I) = 0$ and I 0-prime. Now R/I is a semi-s-prime near-ring from [5], Theorem 3.7. Also, R/I is a 0-prime near-ring. Hence R/I is 0-prime and semi-s-prime. This implies that R/I is an n-prime near-ring. Hence I is an n-prime ideal. $\qquad \square$

Corollary 3.11. *Let $\mathcal{W}$ be the class of all n-prime near-rings. $\mathcal{W}$ is an $\mathcal{A}$-special class.*

Proof. This follows from the previous proposition and [3], Theorem 6. $\qquad \square$

4. THE H-RADICAL AND ALMOST NIL NEAR-RINGS

Definition 4.1. L et J be any ideal of R. $Q \lhd R$ is called J-nil if for all $q \in Q$ we can find an n such that $q^n \in J$.

Lemma 4.2. *If Q_1 and Q_2 are J-nil ideals, then $Q_1 + Q_2$ is also a J-nil ideal.*

Proof. Let $x \in Q_1 + Q_2$. Now $x = q_1 + q_2$. Since Q_1 and Q_2 are J-nil ideals, we can find $m, n \in \mathbb{N}$ such that $q_1^n \in J$ and $q_2^m \in J$. Now

$$
\begin{aligned}
(q_1 + q_2)^2 &= (q_1 + q_2)(q_1 + q_2) \\
&= (q_1 + q_2)(q_2 + q_1') && \text{(where } q_1 + q_2 = q_2 + q_1', q_1' \in Q_1) \\
&= q_1(q_2 + q_1') + q_2(q_2 + q_1') - q_2^2 + q_2^2 \\
&= t + q_2^2 && \text{(for } t \in Q_1)
\end{aligned}
$$

Now:

$$
\begin{aligned}
(q_1 + q_2)^3 &= (q_1 + q_2)(t + q_2^2) = (q_1 + q_2)(q_2^2 + t') && (t' \in Q_1) \\
&= q_1(q_2^2 + t') + q_2(q_2^2 + t') - q_2^3 + q_2^3 = z + q_2^3 && \text{(for } z \in Q_1)
\end{aligned}
$$

Like this we can carry on to get:

$$
(q_1 + q_2)^m = l + q_2^m, \quad \text{where } l \in Q_1.
$$

Since Q_1 is J-nil we can get a $k \in \mathbb{N}$ such that $l^k \in J$. Now, as above, we can show that for $j = q_2^m \in J$ we have

$$
(l + j)^k = y + l^k, \quad y \in J.
$$

Hence $[x^m]^k = y + l^k \in J$ and we are done. □

Corollary 4.3. *The sum of a finite number of J-nil ideals of a near-ring is again J-nil.*

Corollary 4.4. *If for all $\alpha \in A$, Q_α is J-nil, then $\sum_{\alpha \in A} Q_\alpha$ is J-nil.*

Proof. Let $q \in \sum_{\alpha \in A} Q_\alpha$, $q = q_{\alpha_1} + \dots + q_{\alpha_k}$ (say). Then $q \in Q_{\alpha_1} + \dots + Q_{\alpha_k}$ and since $Q_{\alpha_1} + \dots + Q_{\alpha_k}$ is J-nil, we can find an $m \in \mathbb{N}$ such that $q^m \in J$. But q was arbitrary, hence $\sum_{\alpha \in A} Q_\alpha$ is J-nil. □

Definition 4.5. L et R be any near-ring and J any ideal of R. Define a sequence of ideals associated with J as follows:

$$
J = H_0(J) \subseteq H_1(J) \subseteq \dots \subseteq H_i(J) \subseteq H_{i+1}(J) \subseteq \dots,
$$

where, if $H_i(J)$ has been defined, we define $H_{i+1}(J)$ to be the ideal in R generated by all $H_i(J)$-nil ideals, i.e. all ideals Q of R such that $q^k \in H_i(J)$ for all $q \in Q$ and some $k \in \mathbb{N}$.

Define $H(J) = \bigcup_{i=0}^{\infty} H_i(J)$ and call it the *h-radical* of J. Now, from the above, it is clear that $H_{i+1}(J)$ is an $H_i(J)$-ideal.

Lemma 4.6. $H_i(J) = H_1(J) = H(J)$ for all $i \geq 1$.

Proof. From the definition of $H_i(J)$ we have $H_1(J) \subseteq H_i(J)$ for all $i \geq 1$. Let now $x \in H_i(J)$. Hence we can find an m such that $x^m \in H_{i-1}(J)$ and again we can find a k such that $(x^m)^k \in H_{i-2}(J)$. Carrying on like this, we get an $l \in \mathbb{N}$ such that $x^l \in J$. Hence $H_i(J)$ is J-nil. Hence $H_i(J) \subseteq H_1(J)$. Now it follows from the definition of $H(J)$ that $H_i(J) = H_1(J) = H(J)$. □

Theorem 4.7. *If $H(J)$ is the h-radical of an ideal J of a ring R, then*

  (i) $H(J)$ *is a semi-s-prime ideal of R which contains J.*
  (ii) $H(J)$ *is contained in every semi-s-prime ideal of R which contains J.*

Proof.   (i) Clearly $J \subseteq H(J)$. We show $H(J)$ is semi-s-prime. Let Q be any ideal of R such that for any $q \in Q$ there is a $k \in \mathbb{N}$ such that $q^k \in H(J)$. From the fact that $H(J) = H_1(J)$ we have $[q^k]^n \in J$ for some n. Hence Q is J-nil and we have $Q \subseteq H_1(J) \subseteq H(J)$. It follows that $H(J)$ is semi-s-prime.
  (ii) Let S be any semi-s-prime ideal such that $J \subseteq S$ and let $t \in H(J)$. Since $H(J) = H_1(J)$ we can find a $k \in \mathbb{N}$ such that $t^k \in J \subseteq S$. Since S is a semi-s-prime ideal and t an arbitrary element from $H(J)$ we have $H(J) \subseteq S$ and we are done.   □

Let $\mathcal{T} = \{\text{nilpotent near-rings}\}$, $\mathcal{N} = \{\text{nil near-rings}\}$ and $\mathcal{U} = \{\text{almost nilpotent near-rings}\}$. We now define:

Definition 4.8. A n element $x \in R$ is called almost nil if $(x^n) \subset I$ for every nonzero ideal I and some $n \in \mathbb{N}$. If every element of R is almost nil, then R is called an almost nil near-ring.

Examples of almost nil near-rings are nilpotent near-rings, nil near-rings and almost nilpotent near-rings.

Example 4.9. $\{x_\alpha : \alpha$ real and $0 < \alpha < 1\}$. Let F be some field, and let A be the commutative algebra over F with these x_α as a basis. Define multiplication as follows

$$x_\alpha x_\beta = \begin{cases} x_{\alpha+\beta}, & \text{if } \alpha+\beta < 1; \\ 0, & \text{if } \alpha+\beta \geq 1. \end{cases}$$

A is a nil ring. Hence almost nil also. $A = A^2$. From this we have that A is not almost nilpotent because for every $n \in \mathbb{N}$ we have $A^n = A$. Hence for any proper ideal $I \lhd A$, $A^n \not\subseteq I$.

In [6] we introduced the concepts of completely prime (completely semiprime)- ideals for near-rings i.e. if $a, b \in R$ and $P \lhd R$ such that $ab \in P$ ($a^2 \in P$) then $a \in P$ or $b \in P$ ($a \in P$). If $C(I)$ is the completely prime radical of an ideal I i.e. the intersection of all the completely prime ideals of R containing I, then we have the following:

Corollary 4.10.   (i) $H(0) = \mathcal{N}(R)$ *for any ring R.*
  (ii) *If $I \lhd R$ then $H(I) = I$ if and only if I is semi-s-prime.*
  (iii) *Every ideal of a Boolean near-ring is semi-s-prime.*
  (iv) *If R is almost nil, then $H(I) = C(I) = R$ for any nonzero ideal $I \lhd R$.*

Proof.   (i) Follows from the definition of $H(0)$.
  (ii) Suppose I is semi-s-prime. That $I = H(I)$ follows from Theorem 4.7 (i) and (ii). Conversely, if $H(I) = I$, then from Theorem 4.7 (ii) I is semi-s-prime and we are done.

(iii) From [13], Definition 9.30 we know R is Boolean if and only if $x^2 = x$ for all $x \in R$. Let I be any ideal of R and suppose that for every $A \lhd R$ and $a \in A$ there exists $m \in \mathbb{N}$ such that $a^m \in I$. Now since $a^2 = a$ we have $a = a^m \in I$. Hence $A \subseteq I$ and therefore I is semi-s-prime.

(iv) Suppose R is almost nil. Let x be any element from R. Let I be any nonzero ideal of R. Now we can find an n such that $x^n \in \langle x^n \rangle \subset I \subseteq H(I)$. Since $H(I)$ is a semi-s-prime ideal it follows that $R \subseteq H(I)$. Hence $R = H(I)$. Also, $x^n \in I \subseteq C(I)$ and $C(I)$ completely semiprime implies $x \in C(I)$. Hence $R = C(I) = H(I)$. $\square$

Lemma 4.11. *For any almost nil near-ring we always have* $\mathcal{N}(R) = 0$ *or* $\mathcal{N}(R) = R$.

Proof. Suppose $\mathcal{N}(R) = H(0) \neq 0$. For every $x \in R$ we can find an $n \in \mathbb{N}$ such that $x^n \in \langle x^n \rangle \subset H(0)$. Since $H(0)$ is semi-s-prime, we have $R \subseteq H(0)$. Hence $\mathcal{N}(R) = R$ and we are done. $\square$

REFERENCES

[1] G F Birkenmeier, *Andrunakievich's Lemma for near-rings*, Contrib. to Gen. Alg. **9**, Holder-Pichler-Tempsky, Wien (1995), 1–12.

[2] G F Birkenmeier, H E Heatherly and E K Lee, *Andrunakievich's Lemma for near-rings,* Comm. Algebra **23** (1995), 2825–2850.

[3] G F Birkenmeier, H E Heatherly and E K Lee, *Special radicals for near-rings*, Tamkang Journal of Mathematics **27** (1996), 281–288.

[4] N J Divinsky, *Rings and Radicals*, Univ. of Toronto Press (1965).

[5] N J Groenewald, *Strongly semiprime ideals in near-rings*, Chin. Journal of Mathematics **11** (1983), 221–227.

[6] N J Groenewald, *The completely prime radical in near-rings*, Acta Math. Acad.Sci. Hungar. **51** (1988), 301–305.

[7] N j Groenewald, Different prime ideals in near-rings, Comm. in Algebra **19** (1991), 2667–2675.

[8] G A P Heyman and L C A van Leeuwen, *A radical determined by a class of almost nilpotent rings*, Acta Math. Acad.Sci. Hungar. **26** (1975), 259–262.

[9] G A P Heyman, T L Jenkins and H J Le Roux, *Variations on almost nilpotent rings, their radicals and partitions*, Acta Math. Acad. Sci. Hungar. **39** (1982), 11–15.

[10] K Kaarli, *Special radicals of near-rings* (in Russian) Tartu Riikl. Ül. Toimetised **610** (1982), 53–68.

[11] H J Le Roux, *Nil- and almost nil rings*, Manuscript.

[12] N H McCoy, *Completely prime and completely semi -prime ideals*, Rings Modules and Radicals, Coll. Math. Soc. J Bolyai 6, North Holland (1973), 147–152.

[13] G Pilz, *Near-Rings*, rev. ed., (North Holland, Amsterdam, 1983).

[14] E R Puczylowski, *A note on almost nilpotent rings*, Acta Math. Hungar. **48** (1986), 289–291.

[15] A D Sands, *Almost nilpotent rings*, Acta Math. Acad. Sci. Hungar. **45** (1985), 41–43.

[16] A P J van der Walt, *Prime ideals and nil radicals in near-rings*, Arch. Math. (Basel) **15** (1964), 408–414.

[17] R Wiegandt, *Characterizations of the Baer radical class by almost nilpotent rings*, Pub. Math. Deb. **23** (1976), 15–17.

DEPARTMENT OF MATHEMATICS, UINVERSITY OF PORT ELIZABETH, P.O. BOX 1600, PORT ELIZABETH, SOUTH AFRICA

POLYNOMIAL NEAR-RINGS IN SEVERAL VARIABLES

JAIME GUTIERREZ* AND CARLOS RUIZ DE VELASCO

ABSTRACT. In this paper we define and study some properties of the multivariate polynomial composition ring $R[\bar{x}]$. In particular, we explicitly describe the semigroup of left identities of the near-ring $R[\bar{x}]$. We also investigate the ideal structure and we give a complete description of all maximal ideals, for some special rings R.

1. INTRODUCTION

J. R. Clay on his recent book (see [2]) poses some "Exploratory problems" on the composition of two polynomials in several variables. These problems can be stated as follows: Let R be a commutative ring with identity. It is well known that the set $R[x]$ of all polynomials over R in the variable x is a near-ring under usual addition and substitution of polynomials "$\circ$" (i.e., $f(x) \circ g(x) = f(g(x)) = f \circ g$). Moreover, the structure $(R[x], +, ., \circ)$ has the following properties :

 1) $(R[x], +, .)$ is a ring.
 2) $(R[x], +, \circ)$ is a near-ring with identity x, and
 3) $(f.g) \circ h = (f \circ h).(g \circ h)$.

In other words, $(R[x], +, ., \circ)$ is a tri-operation algebra, or a composition ring, (see [10]).

Several natural questions arise when we consider the set $R[\bar{x}] = R[x_1, \ldots, x_n]$ of the polynomials in the variables $x_1, \ldots, x_n$ over R:

What does one do with the set $R[\bar{x}]$? Certainly, $(R[\bar{x}], +, .)$ is a ring with identity. For $f(\bar{x}), g(\bar{x}) \in R[\bar{x}]$ what should $f(\bar{x}) \circ g(\bar{x})$ be ? What are some significant problems concerning $(R[\bar{x}], +, \circ)$?

We start with the following binary operation:

Definition 1.1. The *composition* of two polynomials $f(\bar{x}), g(\bar{x}) \in R[\bar{x}]$ is the polynomial

$$f(\bar{x}) \circ g(\bar{x}) = f(g(\bar{x}), \ldots, g(\bar{x})).$$

Thus $R[\bar{x}] = (R[\bar{x}], +, \circ)$ becomes a near-ring which we will call the polynomial near-ring in the indeterminates $x_1, \ldots, x_n$ with coefficients from R or the multivariate polynomial near-ring $R[\bar{x}]$. In fact, $(R[\bar{x}], +, ., \circ)$ is a composition algebra. The polynomials with zero constant term constitute the zero-symmetric part of the near-ring $R[\bar{x}]$ and the constant polynomials are the constant part of the near-ring $R[\bar{x}]$.

1991 *Mathematics Subject Classification.* 16 A 20, 16 A 30, 16 A 96, 12.

Key words and phrases. Near-ring, composition ring, radical, polynomials.

*Partially supported by Caja Cantabria and the Spanish grant PB97-0346.

94

Y. Fong et al. (eds.), Near-Rings and Near-Fields, 94–102.
© 2001 *Kluwer Academic Publishers. Printed in the Netherlands.*

Remark 1.2. (1) For $n = 1$, we get the well known polynomial near ring in one variable.

(2) For $n > 1$ and r, $1 \leq r \leq n$, we can consider the following alternative definition:

$$f(\bar{x}) \circ_r g(\bar{x}) = f(g(\bar{x}),\ldots,g(\bar{x}),x_{r+1},\ldots,x_n).$$

Thus $(R[\bar{x}],+,\circ_r)$ is a near-ring. We also note that $(R[\bar{x}],+,\circ_r)$ is isomorphic to the near-ring:

$$(R[x_{r+1},\ldots,x_n][x_1,\ldots,x_r],+,\circ),$$

that is, the polynomial near-ring in the indeterminates $x_1,\ldots,x_r$ over the polynomial ring $R[x_{r+1},\ldots,x_n]$.

In order to motivate the above definition by relating it to functions from R^n to R, we consider $M = M(R^n,R)$, the set of all mappings from R^n to the ring R. If $f,g \in M(R^n,R)$, we define the binary operation $\star$ by

$$f \star g = f \circ \delta \circ g$$

where δ is the diagonal injection of R into R^n : $\delta(a) = (a,\ldots,a)$. Thus $(M,+,\star)$ becomes a near-ring: the near-ring of all mappings from the set R^n to the ring R. Again, $(M,+,.,\star)$ is a composition algebra.

Every polynomial $f(\bar{x}) \in R[\bar{x}]$ defines a function $\tilde{f}$ from R^n to R, thus, we obtain the well known evaluation homomorphism:

$$E : R[\bar{x}] \to M(R^n,R); f(x) \mapsto \tilde{f},$$

where $\tilde{f}(\bar{a}) = f(\bar{a}),(\bar{a} \in R^n)$.

The image $E(R[\bar{x}])$ is a subnear-ring (subalgebra) of $M(R^n,R)$, which is called the *near-ring of polynomial maps of R^n to R*, and is denoted by $P(R^n,R)$.

The ring $R[\bar{x}]$ is embedded in $R[\bar{x}]^n = R[\bar{x}] \times \ldots \times R[\bar{x}]$ through the diagonal inclusion δ such that

$$\delta(f(\bar{x})) = (f(\bar{x}),\ldots,f(\bar{x})).$$

On the other hand, the composition algebra $M(R^n,R)$ is embedded into the composition algebra $M(R^n,R^n)$ through the corresponding diagonal:

$$\delta : M(R^n,R) \to M(R^n,R^n)$$

such that $\delta(f) = (f,\ldots,f)$. Here we are considering $M(R^n,R^n)$ as the near-ring (respectively composition algebra) of all mappings from R^n to the abelian group (respectively ring) R^n, with the usual binary operations.

Now, we consider the following extension of E:

$$E_n : R[\bar{x}] \times \ldots \times R[\bar{x}] \to M(R^n,R^n)$$

such that

$$E_n\big(f_1(\bar{x}),\ldots,f_n(\bar{x})\big) = \big(E(f_1(\bar{x})),\ldots,E(f_n(\bar{x}))\big),$$

then, we have that E_n is a ring homomorphism. So, we have the following commutative diagram:

$$
\begin{array}{ccccc}
R[\bar{x}] & \xrightarrow{\ E\ } & P(R^n,R) & \xrightarrow{\ i\ } & M(R^n,R) \\
\delta\downarrow & & \delta\downarrow & & \downarrow\delta \\
R[\bar{x}] \times \ldots \times R[\bar{x}] & \xrightarrow{\ E_n\ } & P(R^n,R^n) & \xrightarrow{\ i_n\ } & M(R^n,R^n)
\end{array}
\qquad [1]
$$

Finally, we introduce a new composition in $R[\bar{x}] \times \ldots \times R[\bar{x}]$ in a way such that the above diagram becomes a commutative diagram as composition algebras. In order to do that, we are obliged to define

$$
(1:1) \quad \big(f_1(\bar{x}),\ldots,f_n(\bar{x})\big) \circ \big(g_1(\bar{x}),\ldots,g_n(\bar{x})\big)
$$
$$
= \big(f_1(g_1(\bar{x}),\ldots,g_n(\bar{x})),\ldots,f_n(g_1(\bar{x}),\ldots,g_n(\bar{x}))\big),
$$

for all $f_i(\bar{x}), g_i(\bar{x}) \in R[\bar{x}]$, $(i = 1,\ldots,n)$. Thus, $(R[\bar{x}] \times \ldots \times R[\bar{x}], +, \circ)$ is a near-ring with identity $(x_1,\ldots,x_n)$. We also note that this near-ring is not the direct product of n copies of the near-ring $R[\bar{x}]$; in fact, the near-ring $R[\bar{x}]$ has no identity for $n > 1$.

The paper is divided into five sections. In section 2 we study the annihilator ideal and as a consequence of this, we get the semigroup of the left identities. We also include in this section results on left identities giving partial answers to questions in Clay's book on identities in polynomial near-rings in several variables. Section 3 is a treatment of the ideal structure. Then (Section 4) we study the maximal ideals. We give an explicit description of the Jacobson radical for some special rings R. Finally, we conclude with a discussion of some open questions.

2. THE ANNIHILATOR IDEAL

In this section we study the left identities of the near-ring $R[\bar{x}]$. We start by giving a description of the annihilator ideal. Let

$$
\mathcal{J} = (0 : R[\bar{x}]) = Ann(R[\bar{x}]) = \{f(\bar{x}) \in R[\bar{x}] \mid f(\bar{x}) \circ g(\bar{x}) = 0, \forall g(\bar{x}) \in R[\bar{x}]\}
$$

be the annihilator ideal.

Proposition 2.1. *The set $\mathcal{J}$ is a full ideal of the composition algebra $R[\bar{x}]$ and it is generated as an ideal in the polynomial ring by $\{x_1 - x_n,\ldots,x_{n-1} - x_n\}$; that is,*

$$
\mathcal{J} = (x_1 - x_n, x_2 - x_n,\ldots,x_{n-1} - x_n).
$$

Proof. Obviously, $(x_1 - x_n, x_2 - x_n,\ldots,x_{n-1} - x_n) \subseteq \mathcal{J}$. Conversely, let $f(\bar{x})$ be an element of $\mathcal{J}$. By the uniqueness of the classical euclidean division

$$
f(\bar{x}) = q_1(\bar{x}).(x_1 - x_n) + r_1(x_2,\ldots,x_n)
$$

and

$$
r_1(x_2,\ldots,x_n) = q_2(\bar{x}).(x_2 - x_n) + r_2(x_3,\ldots,x_n).
$$

Now, by recurrence, we have:

$$
f(\bar{x}) = q_1(\bar{x}).(x_1 - x_n) + q_2(\bar{x}).(x_2 - x_n) + \ldots + q_{n-1}(\bar{x}).(x_{n-1} - x_n) + r_n(x_n).
$$

Because $0 = f(\bar{x}) \circ x_n$, that's implies $r_n(x_n) = 0$. □

The above characterization of the annihilator ideal permits us to describe the set of left identities. Let $U_{R[\bar{x}]}$ be the semigroup of left identities of the near-ring $R[\bar{x}]$:

$$U_{R[\bar{x}]} = \{u(\bar{x}) \in R[\bar{x}] \mid u(\bar{x}) \circ f(\bar{x}) = f(\bar{x}), \forall f(\bar{x}) \in R[\bar{x}]\}.$$

For each $u(\bar{x}) \in U_{R[\bar{x}]}$, we consider the set:

$$A_{u(\bar{x})} = \{v(\bar{x}) \in R[\bar{x}] \mid v(\bar{x}) \circ u(\bar{x}) = v(\bar{x})\}.$$

Obviously $(A_{u(\bar{x})}, \circ)$ is a semigroup with identity $u(\bar{x})$. Now, we are interested in finding the group of units of this semigroup.

Corollary 2.2. *With the above notation we have:*

1. $U_{R[\bar{x}]} = \{x_1 + f(\bar{x}) \mid f(\bar{x}) \in \mathcal{I}\} = x_1 + \mathcal{I}$.
2. *For every* $u(\bar{x}) \in U_{R[\bar{x}]}$, *we have* $A_{u(\bar{x})} = R[u(\bar{x})]$, *the composition subalgebra generated by* $u[\bar{x}]$ *over R, and the mapping*

$$\varphi : R[u(\bar{x})] \to R[t]$$

where $\varphi(u(\bar{x})) = t$, *is a composition algebra isomorphism.*

Remark 2.3. (1) As a consequence of the above corollary, the units of the near-ring $R[u(\bar{x})]$ are corresponding in a natural way with the units of the near-ring $R[t]$. In particular, if R is an integral domain, then the units of the near-ring $R[u(\bar{x})]$ are the polynomials of the form $\alpha u(\bar{x}) + \beta$, where $\alpha, \beta \in R$ and such that α is an unit of the ring R.

(2) To determine the units of the near-ring $R[\bar{x}] \times \ldots \times R[\bar{x}]$ for the case $n > 1$ seems more complicated. For instance, if R is a field of characteristic zero, the charaterization problem of the units is strongly related to the famous Jacobian conjecture. That is, an element $F = (f_1(x_1, \ldots, x_n), \ldots, f_n(x_1, \ldots, x_n))$ in the near-ring $R[\bar{x}] \times \ldots \times R[\bar{x}]$ is a unit if and only if the Jacobian matrix has a nonzero constant determinant if and ony if the polynomial map F is an automorphism.

(3) Assume that the ring R is an arbitrary field. Now, from a computational point of view, we want to know when a multivariate polynomial $f(\bar{x}) \in R[\bar{x}]$ has a non-trivial decomposition, that is, when do there exist polynomials $g(\bar{x}), h(\bar{x}) \in R[\bar{x}]$ where $g(\bar{x})$ is not a left identity and such that $f(\bar{x}) = g(\bar{x}) \circ h(\bar{x})$ in the polynomial near-ring. Also, in the affirmative case, how do we compute g and h. Using the above ideas, it is very easy to see that this computational problem is equivalent to the well known multivariate polynomial decomposition problem, (see [4]):

Given a polynomial $f(x_1, \ldots, x_n) \in R[\bar{x}]$ of (total) degree $m = rs$, to determine when there exist $g(y) \in R[y]$ and $h(x_1, \ldots, x_n) \in R[\bar{x}]$ of degrees r, s respectively, such that

$$f(x_1, \ldots, x_n) = g(h(x_1, \ldots, x_n))$$

and in the affirmative case, to compute them.

Assume the field has more than $n + 1$ elements. An algorithm which can be perfomed with $O(mn(m+1)^n Log(m))$ operations is presented in [3]. See [5] for several multivariate decomposition problems.

3. THE IDEAL STRUCTURE OF $R[\bar{x}]$.

Let $u(\bar{x})$ be an element of $U_{R[\bar{x}]}$. We consider the composition algebra epimorphism

$$\varphi_{u(\bar{x})} : R[\bar{x}] \to R[u(\bar{x})]$$

such that $\varphi_{u(\bar{x})}(f(\bar{x})) = f(\bar{x}) \circ u(\bar{x})$. The kernel of $\varphi_{u(\bar{x})}$ is precisely the annihilator ideal $\mathcal{J}$. We can state the following version for an arbitrary composition ring:

Lemma 3.1. *Let N be a composition algebra with a left identity u. Then*
1 $N \circ u = \{n \circ u \mid n \in N\}$ is a composition subalgebra with identity u.
2 The map $\varphi_u : N \to N \circ u$, such that $\varphi_u(n) = n \circ u$ is a composition algebra epimorphism, and the kernel is the annihilator $(0 : N)$.

Using the above Lemma and the Correspondence Theorem, there exists a bijection between the ideals L of $R[x_n]$ and the ideals K of $R[\bar{x}]$ which contain $\mathcal{J}$:

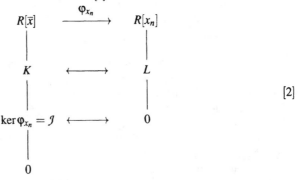

[2]

It is well known that if R is an infinite field, then the polynomial near-ring in one variable $R[x_n]$ is simple, (see [10]). On this issue polynomial near-rings in several variables differ from the one variable case. Now, the natural question is: How many ideals will $R[\bar{x}]$ have ? First, we have the following result:

Proposition 3.2. *Every ideal I of the ring $R[\bar{x}]$ is a left ideal of the near-ring $R[\bar{x}]$. Moreover, if I is contained in $\mathcal{J}$, then I is a full ideal.*

Proof. We consider a monomial $ax_1^{\alpha_1} \ldots x_n^{\alpha_n} \in R[\bar{x}]$, $p(\bar{x}) \in R[\bar{x}]$ and $i(\bar{x}) \in I$. We have

$$ax_1^{\alpha_1} \ldots x_n^{\alpha_n} \circ (p(\bar{x}) + i(\bar{x})) - ax_1^{\alpha_1} \ldots x_n^{\alpha_n} \circ p(\bar{x})$$

$$= a(p(\bar{x}) + i(\bar{x}))^{\alpha_1 + \ldots + \alpha_n} - ax_1^{\alpha_1} \ldots x_n^{\alpha_n} \circ p(\bar{x})$$

$$= a(p(\bar{x}) + i(\bar{x}))^{\alpha_1 + \ldots + \alpha_n} - ap(\bar{x})^{\alpha_1 + \ldots + \alpha_n},$$

which is in the form $i(\bar{x})q(\bar{x})$ for a $q(\bar{x}) \in R[\bar{x}]$. Thus I is a left ideal.
If $f(\bar{x}) \in I \subset \mathcal{J}$ and $g(\bar{x}) \in R[\bar{x}]$, then $f(\bar{x}) \circ g(\bar{x}) = 0 \in I$, because $f(\bar{x}) \in \mathcal{J}$. $\square$

In order to characterize the ideals of the near-ring $R[\bar{x}]$ when R is an infinite field, we need the following result which is a consequence of the proof of the main theorem in [7].

Lemma 3.3. *Let R be a commutative ring R with identity which contains a unit u, such that $u - 1$ (or $u + 1$) is also a unit. If I is an ideal of the near-ring $R[\bar{x}]$ which contains R, then $I = R[\bar{x}]$.*

The following result characterizes the ideals of the near-ring $R[\bar{x}]$ when R is an infinite field.

Theorem 3.4. *Let R be an infinite field, then every proper ideal I of the near-ring $R[\bar{x}]$ is a full ideal contained in $\mathcal{J}$. Conversely, any ideal I of the ring $R[\bar{x}]$ contained in $\mathcal{J}$ is a full ideal.*

Proof. We know from the above diagram and remarks that there are no ideals between the annihilator ideal and $R[\bar{x}]$. We suppose $I \neq 0$ is an ideal of the near-ring $R[\bar{x}]$ which is not contained in $\mathcal{J}$ and we will prove that $I = R[\bar{x}]$. First we prove that $R \subseteq I$.

For $0 \neq f \in I$ such that $f \notin \mathcal{J}$ and for each $\beta \in R \subseteq R[\bar{x}]$, we have $f \circ \beta = f(\beta, \ldots, \beta) = \alpha \in I$, for some $\alpha \in R$. We claim that there is a nonzero $\alpha = f \circ \beta$, for some $\beta \in R$. Otherwise, if $f \circ \beta = 0$ for all $\beta \in R$, then f belongs to the ideal $I(\{(\beta, \ldots, \beta), \forall \beta \in R\})$ of the polynomial ring $R[\bar{x}]$, where

$$I(\{(\beta, \ldots, \beta), \forall \beta \in R\}) = \{h \in R[\bar{x}] \mid h(\beta, \ldots, \beta) = 0, \forall \beta \in R\}$$

Since R is an infinite field, it is very easy to prove that the above ideal is exactly $\mathcal{J}$, so $f \in \mathcal{J}$, which is a contradiction. Now, let $\alpha \in R$ be a nonzero element of I and let β be an arbitrary element of R, since $x_1 \beta \in R[\bar{x}]$, we have $x_1 \beta \circ (0 + \alpha) - x_1 \circ 0 = \beta \alpha \in I$, so we have $R \subseteq I$. Now, we apply the above Lemma, then $I = R[\bar{x}]$. The rest of the proof is an immediate corollary of the above Proposition. □

We can not relax the condition that R is infinite, as the following example illustrates.

Example 3.5. Let $R = F_q$ be the finite field with q elements. Let $Kernel(E)$ be the kernel of the evaluation homomorphism E, (see Section 1). Then $Kernel(E)$ is a full ideal, moreover $Kernel(E)$ is generated as an ideal in the ring $R[\bar{x}]$ by:

$$Kernel(E) = (x_1^q - x_1, \ldots, x_n^q - x_n).$$

It is easy to see that $\mathcal{J} \not\subset Kernel(E)$ and $Kernel(E) \not\subset \mathcal{J}$.

Over a finite field F_q, it is well known (see [10]) that every ideal of the univariate polynomial near-ring $F_q[x]$ is a full ideal if and only if the characteristic of F_q is bigger that 2. In the following proposition we extend this result.

Proposition 3.6. *Let $R = F_q$ be the finite field with q elements. Then every ideal I of the polynomial near-ring $R[\bar{x}]$ is a full ideal if and only if the characteristic of R is bigger than 2.*

Proof. Assume that the characteristic of the field R is 2. The set

$$(x_n^q + x_n)^2 . F_q[x_n^2] + (x_n^q + x_n)^4 . F_q[x_n]$$

is an ideal of $R[x_n]$ which is not a full ideal (see [11]). Now, by Lemma 3.1 and the Correspondence Theorem we have that

$$(x_n^q + x_n)^2 . F_q[x_n^2] . R[\bar{x}] + (x_n^q + x_n)^4 . F_q[x_n] . R[\bar{x}] + \mathcal{J}.$$

is an ideal of $R[\bar{x}]$ which is not a full ideal.

For the converse, let I be an ideal of the near-ring $R[\bar{x}]$. Then for all $f \in I$ and all $p \in R[\bar{x}]$, we have that $f.p = 1/2(x_1^2 \circ (f+p) - x_1^2 \circ p - (x_1^2 \circ (f+0) - x_1^2 \circ 0)) \in I$. □

As in the univariate polynomial near-ring, the proof of the converse implication in the above Proposition -which is also valid for any ring such that 2 is a unit-, together with Proposition 3.2 shows that if R is a finite field with characteristic bigger than 2, then the left ideals of the polynomial near-ring $R[\bar{x}]$ are exactly the ideals of polynomial ring $R[\bar{x}]$.

4. MAXIMAL IDEALS OF $R[\bar{x}]$

The maximal ideals of the univariate polynomial near-ring $K[x]$ over a field K are as follows:

  (a) If K is infinite, then $K[x]$ is simple, so $Kernel(E) = \{0\}$ is the only maximal ideal of $K[x]$ (see [10]).
  (b) If K is finite, say, $K = F_q$, but $q \neq 2$, then $Kernel(E)$ is the unique maximal ideal of $F_q[x]$, (see [2], [10]).
  (c) $F_2[x]$ has exactly two maximal ideals: $V_1 = \{f \in F_2[x] \mid, f(0) = f(1)\}$ and $T_1 = \{f \in F_2[x] \mid f(c)^2 + f(c) = 0\}$, where $1 + c + c^2 = 0$, (see [1], [2]).

In [7] the following result is obtained:

> If R is a ring as in Lemma 3.3, then all maximal ideals of the near-ring in one variable $R[x]$ are given by the ideals $M<x>$,
>
> $$M<x> = \{f(x) \in R[x] \mid f(a) \in M, \forall a \in R\},$$
>
> where M is a maximal ideal of the ring R.

This result determines all maximal ideals of $R[x]$ for a large class of commutative rings. In particular, all fields with more than two elements are examples of such rings. All commutative rings with characteristic an odd positive number are also examples of such rings, but not Z. The maximal ideal of the univariate polynomial near-ring $Z[x]$ are determined in [6].

Next, we obtain results concerning maximal ideals of the near-ring $R[\bar{x}]$ when R is an arbitrary field. First, we present a result that is valid for arbitrary near-rings.

Proposition 4.1. *Let N be a near-ring with a left identity. Then every maximal ideal of N contains the annihilator ideal.*

Proof. Let i be a left identity of N, J the annihilator ideal and let M be a maximal ideal. Suppose $J \not\subseteq M$. Then, we have $N = M + J$. Moreover, there exists $m \in M, j \in J$ such that $i = m + j$. But then for $x \in J$ such that $x \notin M$, we have $x = mx + jx = mx \in M$. □

As a consequence of the the above Proposition and the Correspondence Theorem we have:

Theorem 4.2. *Let R be a ring as in Lemma 3.3. The maximal ideals of the near-ring $R[\bar{x}]$ are given by the ideals*

$$M<\bar{x}> = \{f(\bar{x}) \in R[\bar{x}] \mid f(a, \dots, a) \in M, \forall a \in R\}$$

where M is a maximal ideal of R.

These last results permit us to characterize the Jacobson radical of $R[\bar{x}]$: the intersection of all maximal ideals, for some particular rings R.

Corollary 4.3. *Let R be a commutative ring R identity, which contains a unit u, such that $u - 1$ (or $u + 1$) is again a unit. Then, the Jacobson radical of the near-ring $R[\bar{x}]$ is*

$$Rad(R[\bar{x}]) = Rad(R)\langle\bar{x}\rangle,$$

We also have:

(1) *If R is an infinite field, then*

$$Rad(R[\bar{x}]) = J,$$

the unique maximal ideal of $R[\bar{x}]$.

(2) *If $R = F_q$ is the finite field with $q > 2$ elements, then $Rad(R[\bar{x}])$ is the unique maximal ideal which is generated as ring ideal by:*

$$Rad(R[\bar{x}]) = (x_1 - x_n, \ldots, x_{n-1} - x_n, x_n^q - x_n).$$

(3) *If $R = F_2$, then $R[\bar{x}]$ has exactly two maximal ideals V_n, the ideal generated by $x_n^3 + x_n + 1$ plus J, and T_n, the ideal generated by x_n^3 plus J. So $Rad(R[\bar{x}])$ is the ideal generated by 1 plus J.*

Proof. It is straightforward to prove that $Rad(R[\bar{x}]) = Rad(R)\langle\bar{x}\rangle$. Now, the items (1) and (2) are immediate. Finally, we have that $T_1 \cap V_1$ is the near-ring ideal generated by 1, (see [6]).   □

Finally, it is interesting to remark that in the univariate polynomial near-ring $F_q[x]$ where $q > 2$, the maximal ideal is the full ideal given by the kernel of the evaluation homomorphism $Kernel(E)$, that is, the set of all zero polynomial mappings. However, the maximal ideal of the multivariate polynomial ring $R[\bar{x}]$ (when $R = F_2, q > 2$) is the full ideal $Rad(R[\bar{x}])$. Now, it is easy to see in the multivariate case that $Kernel(E)$ (see Example 3.5) is strictly contained in the maximal ideal $Rad(R[\bar{x}])$, that is, we have the following:

$$(x_1^q - x_1, \ldots, x_n^q - x_n) \subset (x_1 - x_n, \ldots, x_{n-1} - x_n, x_n^q - x_n).$$

5. Conclusions 5

We have presented the multivariate polynomial near-ring for motivating several important issues related to polynomials and polynomial functions. For many purposes it would be valuable to have a better knowledge of the ideal of all polynomials wich induce the zero function, that is, $Kernel(E)$ or more generally $Kernel(E_n)$. This kernel decides if one can identify polynomials and polynomial functions. As we have seen in this paper, it also has several connections with the radicals of polynomial near-rings. In [9] these questions are investigated for the univariate case and in [8] for the multivariate one, but they are far from being solved. Looking at diagram 2 in Section 1, we observe many interesting open questions.

REFERENCES

[1]   J. L. Brenner, *Maximal ideals in the nearring of polynomials modulo 2*. Pacific J. Math. **52** (1974), 595–600.

[2]   J. Clay, *Nearrings: Geneses and Applications*. Oxford Science Publications, 1992.

[3]   J. Gathen, *Functional decomposition of polynomials: the tame case*. J. of Symbolic Computation 9, (1990), 281–299.

[4]   J. Gutierrez, *A polynomial decomposition algorithm over factorial domains*. Comptes Rendues Acad. Science, Vol. XIII-2 (1991), 437–452.

[5]   von zur J. Gathen, J. Gutierrez and R. Rubio San Miguel, *On multivariate polynomial Decomposition*. Proc. CASC'99. Lect. Notes in Computer Science. Springer-Verlag.

[6]   J. Gutierrez and C. Ruiz de Velasco, *Ideals in the near-ring of polynomials $Z[x]$*. Proc. Conf. Near-rings and Near-fields, Oberwolfach 1989, pp. 91–95 (Edit. G. Betsch, G. Pilz, and H. Wefelscheid).

[7]   H. Kautschitsch, *Maximal ideals in the near-rings of polynomials*. Proc. 1st. Conf. Radical Theory, Eger(1982). Colloqu. Math. Soc. Janos Bolyai **38** (1985), 183–194.

[8]   H. Lausch and W. Nöbauer *Algebra of Polynomials*. North Holland. American Elsevier, Amsterdam, 1973.

[9]   J. Meldrum and G. Pilz, *Polynomial algebras and polynomial maps*. Proc. Conf. Uni. Algebra, Klagenfurt, 1982, pp. 263–272. Teubner ed. (1983).

[10] G. Pilz, *Near-rings*. 2nd ed., North-Holland, Amsterdam, 1983.

[11] E. Straus, *Remark on the preceding paper, ideals in near rings of polynomials over a field*. Pacific J. Math. **52** (1974), 601–603.

DPTO. MATEMÁTICAS, ESTADÍSTICA Y COMPUTACIÓN, UNIVERSIDAD DE CANTABRIA, SANTANDER 39071, SPAIN

s-PRIMITIVE IDEALS IN MATRIX NEAR-RINGS

J. F. T. HARTNEY AND S. MAVHUNGU

ABSTRACT. Let R be a near-ring and $\mathbb{M}_n(R)$ its associated Meldrum-van der Walt matrix near-ring. If A is an ideal of R such that A^* is s-primitive in $\mathbb{M}_n(R)$, then A is an intersection of s-primitive ideals of R. If R satisfies the descending chain condition for left ideals, then A is s-primitive. Examples of finite near-rings R exist such that s-primitive ideals of $M_n(R)$ are not necessarily of the form A^* for A s-primitive in R.

1. INTRODUCTION

In what follows R is a zero symmetric, right distributive near-ring and possesses a multiplicative identity. Our R-groups are assumed to be unitary and the kernels of R-homomorphisms from Ω will be called R-submodules of Ω. Subgroups of Ω which are also R-groups will be called R-subgroups of Ω. R^n denotes the direct sum of n copies of $(R, +)$ and its elements will be written as $\langle r_1, r_2, \ldots, r_n \rangle$, $r_i \in R$. We write $\bar{r}$ for the n-tuple $\langle r_1, r_2, \ldots, r_n \rangle$. $M(\Omega)$ will denote the zero symmetric near-ring of all zero fixing maps of Ω into itself. The mappings in $M(\Omega)$ are added pointwise and the multiplication in $M(\Omega)$ is mapping composition.

The near-ring of $n \times n$ matrices over R is defined to be the sub-near-ring of $M(R^n)$ generated by the set of functions $\left\{ f_{ij}^r : R^n \to R^n \mid r \in R, \ 1 \leq i \leq n, \ 1 \leq j \leq n \right\}$, where $f_{ij}^r(\langle r_1, \ldots, r_n \rangle) = \langle s_1, \ldots, s_n \rangle$ with $s_i = r r_j$ and $s_k = 0$ if $k \neq i$. The near-ring of $n \times n$ matrices over R is denoted by $\mathbb{M}_n(R)$. $\mathbb{M}_n(R)$ is a right distributive, zero symmetric near-ring with identity $I = f_{11}^1 + f_{22}^1 + \cdots + f_{nn}^1$. The elements of $\mathbb{M}_n(R)$ are called matrices over R. The reader is referred to [3], [6], [5] for further details.

Definition 1.1 ([1]). An R-group Ω of type-0 is said to be of *type-s* if for all $\omega \in \Omega$, $\omega \neq 0$. $R\omega = \bigoplus_{i=1}^{k} \Omega_i$, where each Ω_i is of type-0 and an R-submodule of $R\omega$.

An ideal is *s-primitive* if it is the annihilating ideal of an R-group of type-s. The *s-radical*, $J_s(R)$ is the intersection of all s-primitive ideals of R. R itself is said to be *s-primitive* if the zero ideal is an s-primitive ideal. R is *s-semisimple* if $J_s(R) = (0)$.

We note the following facts concerning s-primitivity and the s-radical. [1]

Fact (i) An s-primitive ideal is an 0-primitive ideal which is an intersection of maximal left ideals. If R satisfies the descending chain condition for left ideals (DCCL), then any 0-primitive ideal which is an intersection of maximal left ideals is s-primitive.

1991 *Mathematics Subject Classification.* 16Y30 .

Y. Fong et al. (eds.), *Near-Rings and Near-Fields*, 103–107.
© 2001 *Kluwer Academic Publishers. Printed in the Netherlands.*

Fact (ii)   If R is s-semisimple and satisfies the DCCL, then

$$R = \bigoplus_{i=1}^{k} L_i$$

where each left ideal L_i is an R-group of type s and $L_i = Re_i$ with $\{e_i\}$ $i = 1,\ldots,k$ an orthogonal idempotent set.

Fact (iii)   If R is as in (ii) and $A \neq (0)$ is any ideal of R, then

$$A = \bigoplus_{i=1}^{\ell} Re'_i \text{ with } \ell \leq k, \ e'_i \in \{e_1,\ldots,e_k\}, \ i = 1,\ldots,\ell.$$

Fact (iv)   If R is as in (ii) and A and B are ideals such that $AB = (0)$, then $A \cap B = (0)$.

We list the following well known results which will be used in the sequel.

Theorem 1.2 ([6]).   *If $G = Rg$ is a monogenic R-group, then G^n is a monogenic $\mathbb{M}_n(R)$-group under the action*

$$A\langle g_1,\ldots,g_n\rangle := \left(A\langle r_1,\ldots,r_n\rangle\right)g$$

for $A \in \mathbb{M}_n(R)$, $g_i \in G$, $r_i \in R$ and $r_i g = g_i$ for $i = 1,\ldots,n$.

Theorem 1.3 ([6]).   *Let Γ be a monogenic $\mathbb{M}_n(R)$-group and let $G = \left\{f_{11}^l \gamma : \gamma \in \Gamma\right\}$. Then G is an R-group under the action $rf_{11}^l \gamma := f_{11}^l f_{11}^r \gamma$ and the mapping $\phi\Gamma \to G^n$ given by $\phi(\gamma) = \langle f_{11}\gamma,\ldots,f_{1n}\gamma\rangle$ is a group isomorphism.*

Theorem 1.4 ([6]).   *If G is a monogenic R-group, then any $\mathbb{M}_n(R)$-subgroup (resp. $\mathbb{M}_n(R)$-submodule) of G^n is of the form H^n, where H is an R-subgroup (resp. R-submodule) of G.*

Theorem 1.5 ([6]).   *If G is a monogenic R-group, then G has no nontrivial R-subgroups (respectively has no nontrivial R-submodules) iff G^n has no nontrivial $\mathbb{M}_n(R)$-subgroups (respectively has no nontrivial $\mathbb{M}_n(R)$-submodules).*

2. A^* IDEALS IN $\mathbb{M}_n(R)$

In this section we seek to relate s-primitive ideals in $\mathbb{M}_n(R)$ to ideals in R. For this purpose we need the following definition.

Definition 2.1.   Let A be an ideal of R we define

$$A^* = \left\{ U \in \mathbb{M}_n(R) \, U(\bar{r}) \in A^n \text{ for all } \bar{r} \in R^n \right\}.$$

The following facts about A^* are known. [3], [6], [2].

  (i) A^* is a two-sided ideal of $\mathbb{M}_n(R)$.
 (ii) If A^* is the annihilating ideal of the $\mathbb{M}_n(R)$-group Γ, then A is the annihilating ideal of $G = \{f_{11}^l \gamma\gamma \in \Gamma\}$.
(iii) If G is a monogenic R-group and A is the annihilating ideal of G then A^* is the annihilating ideal of the $\mathbb{M}_n(R)$-group G^n.
 (iv) If A is v-primitive, $v = 0, s, 2$, then A^* is v-primitive.
  (v) Any 2-primitive ideal of $\mathbb{M}_n(R)$ is of the form A^*, where A is a 2-primitive ideal of R.

For s-primitive ideals we have the following:

Theorem 2.2. *Let A be an ideal of R such that A^* is an s-primitive ideal of $\mathbb{M}_n(R)$. Then A is an intersection of s-primitive ideals of R.*

Proof. Let Γ be an $\mathbb{M}_n(R)$-group of type-s of which A^* is the annihilator. Then A is the annihilating ideal of $G = \{f^1_{11}\gamma\gamma \in \Gamma\}$. We need only show that every monogenic R-subgroup of G is a direct sum of submodules each of which is of type-0. From this we deduce that every type-0 R-subgroup of G is of type-s and it readily follows that A is an intersection of s-primitive ideals. Since Γ is monogenic, Theorem 1.3 gives us a group isomorphism $\phi\Gamma \to G^n$. For any $g \in G$, let Ω be the subgroup of Γ which corresponds to $(Rg)^n$ under the isomorphism ϕ. An easy calculation shows that for any $f^r_{ij} \in \mathbb{M}_n(R)$ and any $\gamma \in \Omega$, $\phi f^r_{ij}\gamma = f^r_{ij}\phi(\gamma)$. Using induction on the *weight* $\omega(U)$ of the matrix $U \in \mathbb{M}_n(R)$ we see that $\phi(U\gamma) = U\phi(\gamma)$ and hence ϕ restricted to Ω is an $\mathbb{M}_n(R)$-isomorphism from Ω onto $(Rg)^n$. Now $(Rg)^n$ is a monogenic $\mathbb{M}_n(R)$-group hence Ω is a monogenic $\mathbb{M}_n(R)$-subgroup of Γ and consequently $(Rg)^n$ is a direct sum of $\mathbb{M}_n(R)$-submodules each of which is an $\mathbb{M}_n(R)$-group of type-0. By Theorem 1.4 $(Rg)^n$ has the form

$$(Rg)^n = H^n_1 \oplus \cdots \oplus H^n_k$$

where H_i is an R-submodule of Rg.

It is not hard to see that

$$Rg = H_1 \oplus \cdots \oplus H_k.$$

Now g has a unique expression of the form

$$g = g_1 + g_2 + \cdots + g_k, \quad g_i \in H_i, \quad i = 1,\ldots k.$$

Hence using left distribution over a direct sum of submodules we have $Rg_i = H_i$, $i = 1,\ldots,k$. Since each $(Rg_i)^n = H^n_i$ is an $\mathbb{M}_n(R)$-group of type-0, Theorem 1.4 tells us that each $Rg_i = H_i$ is an R-group of type-0. This completes the proof. $\square$

In order to prove the next theorem we need the following:

Lemma 2.3. *Let R be a near-ring satisfying the DCCL and A be an ideal of R such that $A \supseteq J_s(R)$. If the ideal A^* of $\mathbb{M}_n(R)$ is prime, then A is prime.*

Proof. We put $J_s(R) = \bar{0}$ and $R/J_s(R) = \bar{R}$. Then by Fact (ii) $\bar{R} = \bigoplus^k_{i=1} \bar{L}_i$, where $\bar{R}\bar{e}_i = \bar{L}_i$, $\bar{L}_i$ of type-s and $\{\bar{e}_i\}$ an orthogonal idempotent set $i = 1,\ldots,k$.

Let C and D be ideals of R such that $CD \subseteq A$. Then $\bar{C}\bar{D} \subseteq \bar{A}$, where

$$\bar{C} = \frac{C + J_s(R)}{J_s(R)}, \quad \bar{D} = \frac{D + J_s(R)}{J_s(R)} \quad \text{and} \quad \bar{A} = \frac{A}{J_s(R)}.$$

If $\bar{C} \cap \bar{D} \neq \bar{0}$, then by Fact (iii) it is a direct sum of some of the $\bar{R}\bar{e}_i = \bar{L}_i$ appearing in the direct sum decomposition of $\bar{R}$. But for these idempotents $\bar{e}_i$ we have $\bar{e}_i = \bar{e}_i\bar{e}_i \in \bar{C}\bar{D}$ so that $\bar{C}\bar{D} \subseteq \bar{A}$ implies that $\bar{C} \cap \bar{D} \subseteq \bar{A}$. That is to say $CD \subseteq A$ implies that $C \cap D \subseteq A$. Consequently we have

$$C^*D^* \subseteq C^* \cap D^* = (C \cap D)^* \subseteq A^*.$$

But A^* is prime so that either $C^* \subseteq A^*$ or $D^* \subseteq A^*$. It follows that $C \subseteq A$ or $D \subseteq A$. $\square$

Theorem 2.4. *Let R be a near-ring which satisfies the DCCL. If A is an ideal of R such that A^* is an s-primitive ideal in $\mathbb{M}_n(R)$, then A is an s-primitive ideal of R.*

Proof. By Theorem 2.2 A is an intersection of s-primitive ideals of R and since R satisfies the DCCL we may write $A = \bigcap_{i=1}^{k} A_i$, where A_i is s-primitive, $i = 1, \ldots, \ell$. By Lemma 2.3 A is prime because A^* is prime. Thus $A_1 A_2 \ldots A_k \subseteq \bigcap_{i=1}^{k} A_i = A$ implies $A_i \subseteq A$ for some i. That is $A = A_i$ is s-primitive. $\square$

Corollary 2.5. *If* $\mathbb{M}_n(R)$ *is* s-*primitive, then* R *is* s-*semisimple. If* R *satisfies the DCCL, then* $\mathbb{M}_n(R)$ s-*primitive implies that* R *is* s-*primitive.*

The following is Example 2.3 of [4] in which it is shown that $J_0(R) = J_2(R)$ yet $J_0(M_2(R)) \neq J_2(M_2(R))$. We show that in this example $J_0(M_2(R)) = J_s(M_2(R))$ and that an s-primitive ideal of $M_2(R)$ is not necessarily of the form A^* for A s-primitive in R.

Example 2.6. Let $G := Z_2 \oplus Z_2 \oplus Z_2$. Put

$$H_1 := \big\{(0,0,0), (0,1,0)\big\}, \; H_2 := \big\{(0,0,0), (1,0,0)\big\}, \; H_3 := \big\{(0,0,0), (1,1,0)\big\}$$

and let $H = H_1 \oplus H_2$. H is isomorphic to the four-group with subgroups H_i, $i = 1, 2, 3$. Define R as follows

$$R := \Big\{ f \in M(G) : f(H_i) \subseteq H_i, i = 1, 2, 3 \text{ and } f(g_1) - f(g_2) \in H \text{ if } g_1 - g_2 \in H \Big\}.$$

R is a finite, zero-symmetric near-ring with identity and G a faithful, monogenic R-group with R-generator $(1,1,1)$. H is an R-submodule of G and using Theorem 2.1 of [2] one sees that $J_0(R) = J_s(R) = J_2(R) \neq (0)$.

Define the R-subgroup K of $(R, +)$ as follows:

$$K := \Big\{ f \in R : f(0,0,1) \in H \text{ and } f(g) = (0,0,0) \text{ if } g \neq (0,0,1) \Big\}.$$

Then K and H are isomorphic as groups and $K = \{k_0, k_1, k_2, k_3\}$, where k_0 is the zero map, $k_1(0,0,1) = (1,0,0)$, $k_2(0,0,1) = (0,1,0)$, $k_3(0,0,1) = (1,1,0)$ and $k_i(g) = (0,0,0)$ if $g \neq (0,0,1)$, $i = 1, 2, 3$.

Clearly, K is *not* a monogenic R-group but each of its non-zero elements generates an R-subgroup which is of type-2 and (group) isomorphic to Z_2. The only proper non-zero R-subgroups of K are $K_1 = \{k_0, k_1\}$, $K_2 = \{k_0, k_2\}$, $K_3 = \{k_0, k_3\}$. Let $h, h_1 \in R$ be such that $h_1(0,0,1) \rightarrow (0,1,1)$, $h(0,1,1) \rightarrow (1,0,0)$, $h(0,0,1) \rightarrow (0,0,0)$.

Then $\big[h(h_1 + k_2) - hh_1\big](0,0,1) = (1,0,0)$ so that $h(h_1 + k_2) - hh_1 \notin K_2$ and K_2 is not an R-submodule of K. Similarly one shows that K_1 and K_3 are not R-submodules of K so that K is irreducible i.e. it has no proper, non-zero R-submodules. However K is not of type-0 since it is not monogenic.

We now consider K^2 as an $M_2(R)$-group under the *Action 2* defined in [4] as follows: $K = Rk_1 \oplus Rk_2$ as additive groups and $k_3 = k_1 + k_2$. If $\langle k, k' \rangle \in K^2$, then $k = r_1 k_1 + r_2 k_2$, $k' = r_1' k_1 + r_2' k_2$, $r_i, r_i' \in R$, $i = 1, 2$ and we have an isomorphism

$$\psi\langle k, k' \rangle = \big\langle \langle r_1 k_1, r_1' k_1 \rangle, \langle r_2 k_2, r_2' k_2 \rangle \big\rangle \in (Rk_1)^2 \oplus (Rk_2)^2.$$

$U \in \mathbb{M}_n(R)$ acts under action 2 as follows:

$$U\langle k, k' \rangle = \psi^{-1} \Big\langle \big(U\langle r_1, r_1 \rangle\big) k_1, \big(U\langle r_2, r_2' \rangle\big) k_2 \Big\rangle.$$

Taking $U = f_{11}^{r_1} + f_{21}^{r_2}$ we obtain as in [4]

$$U\langle k_3, 0 \rangle = \langle r_1 k_1 + r_1 k_2, r_2 k_1 + r_2 k_2 \rangle.$$

Thus taking $r_1, r_2 \in \{0, k_1, k_2, 1\}$ we obtain all 16 elements of K^2 so that $\langle k_3, 0 \rangle$ is an $M_2(R)$-generator of K^2 via action 2. Similarly taking $U = f_{11}^1 + f_{12}^1$ we have via (action 2)

$$
\begin{aligned}
U\langle k_1, k_2 \rangle &= \psi^{-1} U \psi \langle k_1, k_2 \rangle = \psi^{-1} U \langle \langle k_1, 0 \rangle, \langle 0, k_2 \rangle \rangle \\
&= \psi^{-1} \langle \langle k_1, 0 \rangle, \langle k_2, 0 \rangle \rangle = \langle k_1 + k_2, 0 \rangle \\
&= \langle k_3, 0 \rangle.
\end{aligned}
$$

Thus $\langle k_1, k_2 \rangle$ is also an $M_2(R)$-generator of K^2 and we have in fact the following 9 generators for K^2 under action 2. (k_i, k_3), $i = 0, 1, 2, 3$, (k_3, k_i), $i = 0, 1, 2$, (k_1, k_2), (k_2, k_1).

The only proper $M_2(R)$-subgroups of K^2 are K_1^2 and K_2^2 each of which is of type-2. It follows that K^2 is of type-s and as pointed out in [4] it is not $M_2(R)$-isomorphic to an $M_2(R)$-group L^2 where L is an R-group of type-s.

Now the annihilating ideal of K^2 is s-primitive but not 2-primitive in $M_2(R)$ and hence does not contain $J_2(M_2(R))$. Thus $J_s(M_2(R)) \neq J_2(M_2(R))$ and we have $(J_s(R))^* = (J_2(R))^* = J_2(M_2(R)) \supsetneq J_s(M_2(R))$.

ACKNOWLEDGEMENT

This paper forms part of the second author's M.Sc. research report to be submitted to the University of the Witwatersrand. He gratefully acknowledges the support from the German Academic Exchange Service.

REFERENCES

[1] J. F. T. Hartney, A radical for near-rings, *Proc̄Royal SocĖdinburgh* **93A** (1982), 105–110.

[2] J. F. T. Hartney, s-Primitivity in matrix near-rings, *QuaestionesMath.* **18** (1995), 487–500.

[3] J. D. P. Meldrum and A. P. J. van der Walt, Matrix near-rings, *ArchMath.* **47** (1986), 312–319.

[4] J. D. P. Meldrum and J. H. Meyer, Modules over matrix near-rings and the J_0-radical, *Monash Math.* **112** (1991) No2, 125–139.

[5] J. H. Meyer, Left ideals and 0-primitivity in matrix near-rings, *ProcĒdin. MathŚoc.* **35** (1992), 173–187.

[6] A. P. J. van der Walt, Primitivity in matrix near-rings, *Quaestiones Math.* **9** (1986), 459–469.

DEPARTMENT OF MATHEMATICS, UNIVERSITY OF THE WITWATERSRAND, P.O. WITS 2050, JOHANNESBURG, SOUTH AFRICA

ESSENTIAL IDEALS AND R-SUBGROUPS IN NEAR-RINGS

AHMED A. M. KAMAL

1. INTRODUCTION

Throughout this paper, R will denote a right near-ring. Undefined notions can be found in Pilz [5]. By $[x,y]$ and (x,y) we will denote the commutator $xy - yx$ and the additive-group commutator $x + y - x - y$, while the symbol (x,y,z) will denote the distributor $x(y + z) - xz - xy$, for all $x,y,z \in R$. Finally, if M be an R-module and X,Y are non-empty subsets of M, we will define $(X : Y) = \{r \in R : ry \in X \text{ for all } y \in Y\}$.

Our aim in this paper is to extend some results due to G.F. Birkenmeier [1] and J.W. Brewer [2] concerning the property of essentiality of ideals in rings to the essentiality of R-subgroups and ideals in near-rings. First we give a simple example to show that these results are not true in near-rings. We also state and prove analogous results in the case of right near-ring under some suitable conditions. These results in fact either extend or generalize the ring case results.

2. PRELIMINARIES

In this section we collect some well-known facts concerning R-subgroups and ideals in near-rings which can be found in Pilz [5]. We also provide a few auxiliary results that are needed throughout this paper.

(2.1) *Let R be a near-ring, M an R-module, N an R-ideal of M. Then N is an R_0-module.*

As an immediate consequence of (2.1) we have the following result

(2.2) *Let R be a zero-symmetric near-ring, M an R-module. Then every R-ideal of M is an R-submodule. In particular every left ideal of R is a left R-subgroup.*

(2.3) *Let M be an R-module and let X and Y be subsets of M*

  (i) *If $(X, +)$ is a subgroup of $(M, +)$, then $((X : Y), +)$ is a subgroup of $(R, +)$.*
  (ii) *If $(X, +)$ is a normal subgroup of $(M, +)$, then $((X : Y), +)$ is a normal subgroup of $(R, +)$.*
  (iii) *If X is an R-submodule, then $(X : Y)$ is a left R-subgroup of R.*
  (iv) *If X is an R-ideal of M, then $(X : Y)$ is a left ideal of R.*

(2.4) *Let R be a near-ring. If I is a right ideal of R and J is a right R-subgroup (right ideal) of R, then $I + J$ is a right R-subgroup (right ideal) of R.*

We now begin our work with the following discussion concerning the conditions under which $xR\,(Rx), x \in R$ becomes a right (left) ideal (R-subgroup) of R. This discussion will be useful in the sequel.

(2.5) *Let R be a near-ring. Then Rx is a left R-subgroup of R for all $x \in R$.*

Y. Fong et al. (eds.), Near-Rings and Near-Fields, 108–117.
© 2001 Kluwer Academic Publishers. Printed in the Netherlands.

Proof. Follows directly from the facts that the right distributive law is satisfied, $0x = 0$ and $(-y)x = -yx$ for all $x \in R$. □

We give here an example to show that (2.5) fails for right ideals (right R-subgroups) in a right near-ring R.

(2.6) *Let R be the zero-symmetric regular simple near-ring with identity $M_0(Z_3)$ and let f be the element of R which maps $0, 1$ to 0 and 2 to 1. Then fR is not a right R-subgroup of R and so fR is not a right ideal of R. Notice also that, f is not a homomorphism of Z_3 and so f is not a distributive element of R.*

(2.7) *Let R be a near-ring. If x is a distributive element of R, then xR is a right R-subgroup of R.*

Proof. Routine. □

We notice that example (2.6) shows that the condition x is distributive in (2.7) is not redundant, however, it is not necessary as the following example shows.

(2.8) *Let R be the simple regular near-ring with identity $M(Z_3)$ and let f be the element of R which maps 0 to 1, 1 to 0 and 2 to 2. Then $fR = R$, although f is not a distributive element of R.*

We give an example of an abelian near-ring R which contains an element x such that xR is not a right ideal of R.

(2.9) Consider the abstract affine near-ring $R = \{0,a,b,c\}$, [3, Near-ring of lower order (E20)] defined by the tables

+	0	a	b	c
0	0	a	b	c
a	a	0	c	b
b	b	c	0	a
c	c	b	a	0

·	0	a	b	c
0	0	0	0	0
a	a	a	a	a
b	0	a	b	c
c	a	0	c	b

If we take $x = a$, then $xR = \{a\}$ is not a right ideal of R. Notice that a is not a distributive element of R.

(2.10) *Let R be a near-ring and x be a distributive element of R such that $(a,b) = 0$ for all $a \in xR$ and $b \in R$. Then xR is a right ideal of R.*

Proof. Follows directly by using (2.7) and our hypothesis. □

Notice that example (2.9) shows that the condition x is a distributive element in (2.10) is not redundant. Notice also that our hypotheses in (2.10) are sufficient for xR to be a right ideal of R, however it is not necessary as the following example shows.

(2.11) Consider $R = \{0,a,b,c,d,e\}$ with addition and multiplication operations on R defined as

+	0	a	b	c	d	e
0	0	a	b	c	d	e
a	a	0	e	d	c	b
b	b	d	0	e	a	c
c	c	e	d	0	b	a
d	d	b	c	a	e	0
e	e	c	a	b	0	d

·	0	a	b	c	d	e
0	0	0	0	0	0	0
a	0	0	0	0	a	a
b	0	0	0	0	c	b
c	0	0	0	0	b	c
d	0	0	0	0	e	d
e	0	0	0	0	d	e

R is a zero-symmetric simple near-ring, $eR = \{0,d,e\}$ is a right ideal of R, but e is not a distributive element of R and $(e,a) \neq \emptyset$.

We now give an example to illustrate that Rx is not necessary a left ideal of the near-ring R, where x is an element of R

(2.12) Take the near-ring $R = \{0,a,b,c,x,y,z,w\}$, [3, Near-ring of lower order (K 132)] defined by the tables

+	0	a	b	c	x	y	z	w
0	0	a	b	c	x	y	z	w
a	a	b	c	0	y	z	w	x
b	b	c	0	a	z	w	x	y
c	c	0	a	b	w	x	y	z
x	x	w	z	y	0	c	b	a
y	y	x	w	z	a	0	c	b
z	z	y	x	w	b	a	0	c
w	w	z	y	x	c	b	a	0

·	0	a	b	c	x	y	z	w
0	0	0	0	0	0	0	0	0
a	x	z	x	z	x	a	x	a
b	0	0	0	0	0	b	0	b
c	x	z	x	z	x	c	x	c
x	x	x	x	x	x	x	x	x
y	0	b	0	b	0	y	0	0
z	x	x	x	x	x	z	x	z
w	0	b	0	b	0	w	0	w

$Rb = \{0,x\}$ is a subgroup of $(R,+)$ which is not normal, moreover $c(W+ab) = cw = y$ is not in Rb. Therefore Rb is not a left ideal of R, although b is a distributive element of R.

3. ESSENTIAL IDEALS AND R-SUBGROUPS

In this section we are concerned with essential ideals and R-subgroups in near-rings. The main purpose of this section is to show that some results concerning the essentiality of right ideals in rings due to G.F. Birkenmeier [1] are not true in the near-ring case. We also give the conditions under which we can extend these results in the case of near-rings which is our main aim in this section.

Before doing so, we pause to refresh the reader on the definition of an essential submodule.

Definition 3.1. Let R be a near-ring and M a left (right) R-module. A submodule N of M is called essential if and only if, for each nonzero submodule K of M, $K \cap N \neq 0$.

In particular in the ring case, notice that, a right (left) ideal I of the ring R is essential if, for each nonzero element r of R, there is an element t of R such that rt (tr) is a nonzero element of I. We give here a counter-example to show that this is not true for right ideals in near-rings.

Example 3.2. Consider the near-ring in example (2.11) and let $I = \{0, e, d\}$. Then I is an essential right ideal of R. If we consider the nonzero element a of R, then there is no $t \in R$ such that $0 \neq at \in I$, since $aR = \{0, a\}$.

Lemma 3.3. *Let R be a near-ring and I be an essential left R-subgroup of R. Then for all x in R for which $Rx \neq 0$, there exists an element $r \in R$ such that $0 \neq rx \in I$.*

Proof. Follows directly from definition 3.1 and (2.5). □

Lemma 3.4. *Let R be a near-ring and I be an essential right R-subgroup of R. Then for each distributive element $x \in R$ such that $xR \neq 0$, there exists an element $r \in R$ such that $0 \neq xr \in I$.*

Proof. Follows directly from definition 3.1 and (2.7). □

We now turn our attention to the central idempotents in near-rings and give the following lemma which will be used in the sequel.

Lemma 3.5. *Let b and c be central idempotents in a near-ring R with identity such that $(1, b) = 0$. Then $bR + cR = bR \oplus c(1 - b)R$ and $e = b + c(1 - b)$ is an idempotent in R. Moreover, if $(X, a, -b) = 0, (x, b, c(1 - b)) = 0$ and $x(-b) = -xb$ for all $x \in R$, then e is central and $bR + cR = eR$.*

Proof. Let y be an element of $bR + cR$. Then $y = br + cs$ for some $r, s \in R$. By using the identity $(1, b) = 0$, we can write $y = br + c[b + (1 - b)]s$. Hence $y = b(r + cs) + c(1 - b)s \in bR + c(1 - b)R$, by the distributivity of b and c. Let $y \in bR \cap c(1 - b)R$. Then $y = by = (1 - b)y$, since $(1 - b)$ is an idempotent and b, c are central idempotents. Thus $y = by = b(1 - b)y = 0$. Therefore $bR + cR = bR \oplus c(1 - b)R$. Now let $e = b + c(1 - b)$. Then $e^2 = be + c(1 - b)e = b + (1 - bc[b + (1 - b)] = e$. If we assume the identities $(x, 1 - b) = 0$ and $x(-b) = -xb$ for all $x \in R$, then $(1 - b)$ is a central idempotent, so the identity $(x, b, c(1 - b)) = 0$ assures that e is central. It is clear that eR is a subset of $bR + cR$. Notice that $eb = b$ and $ec(1 - b) = c(1 - b)$, so bR and $c(1 - b)R$ are contained in eR. By using the distributivity of e and (2.7), eR is a right R-subgroup of R, thus by $bR \oplus c(1 - b)R = eR$. □

Corollary 3.6. *Let b and c be central idempotents in the abelian near-ring R with identity. Then $bR + cR = bR \oplus c(1 - b)R = eR$, where $e = b + c(1 - b)$ is a distributive idempotent element in R.*

Proof. Follows directly from the fact that the set of all distributive elements of R form a subnear-ring of R (see [4, proposition 1]). □

Example 3.7. This example illustrates that the condition in lemma 3.5 and corollary 3.6 that both of the idempotents b and c should be central is not redundant. In example (2.9), b is the identity element and a is an idempotent which is not central. Notice that $e = b + a(b - b) = b + a0 = b + a = c$ is not a distributive element, moreover e is neither idempotent nor central element.

Proposition 3.8. *Let b and c be central idempotents in a near-ring with identity which satisfies the following identities $(x, 1, -b) = 0, (x, b, c(1 - b)) = 0, x(-b) = -xb$ for each $x \in R$ and $(1, b) = 0$. If $f : bR \to X$ and $g : cR \to X$ are monomorphisms where X is a right*

R-subgroup of R such that $(\alpha, \beta) = 0$ *for each* $\alpha \in$ *Im f and* $\beta \in$ *Im g, then there exists a monomorphism* $h : bR + cR \to X$.

Proof. From lemma 3.5, $bR + cR = bR \oplus c(1-b)R = eR$, where $e = b + c(1-b)$ is a central idempotent in R. Define $h : eR \to X$ by $h(et) = f(bt) + g(c(1-b)t$. So h is well defined, since the sum $bR + c(1-b)R$ is direct and f, g are well defined maps. The condition $(\alpha, \beta) = 0$ for each $\alpha \in$ Im f and $\beta \in$ Im g implies that h is a homomorphism of groups. Now assume that $et \in$ Ker$h$ so $f(bt) = -g(c(1-b)t)$. Thus $f(bt) = 0$. Therefore $et = 0$ from the injectivity of f and g. Hence h is a monomorphism. $\square$

Corollary 3.9. *Let b and c be central idempotents in abelian near-ring with identity. If* $f : bR \to X$ *and* $g : cR \to X$ *are monomorphisms, where X is a right R-subgroup of R, then* $bR + cR = eR$ *and there exists a monomorphism* $h : eR \to X$ *where* $e = b + c(1-b)$ *is a distributive idempotent in R.*

Proof. Follows directly from Corollary 3.6 and the proof of Proposition 3.8. $\square$

We are now ready to discuss some results of G.F. Birkenmeier [1] concerning the essentiality of right ideal in rings. We begin by giving an example to illustrate that the following result is not true in near-rings.

Proposition 3.10 ([1, Proposition 1(ii)]). *Let R be a ring and x be any element of R. If xR is a reduced prinicipal right ideal which is essential in a right ideal Y, then Y is isomorphic to xY and* $(xR)Y$ *is essential in Y.*

Example 3.11. Consider the zero symmetric simple right near-ring R of example (2.11). The element e is a right identity of R and $eR = \{0, d, e\}$ is a reduced right ideal of R which is essential in the right ideal $Y = R$. But $eY = eR$ is not isomorphic to $R = Y$.

We need the following two lemmas to give an analogous result of proposition 3.10 in near-rings.

Lemma 3.12. *Let R be a zero symmetric near-ring with identity and I be an essential left R-subgroup of R. Then* $(I : k) = \{r \in R : rk \in I\}$ *is an essential left R-subgroup of R for all nonzero elements k in R.*

Proof. From [(2.3), (iii)], $(I : k)$ is a left R-subgroup of R. Let s be a nonzero element of R. Then either $sk = 0$ or $sk \neq 0$. In the first case $a(sk) = 0$ for all $a \in R$, since R is zero symmetric. Choose $t \in R -$ Ann$_1(s) \neq 0$, so $t(sk) = 0 \in I$ and $0 \neq ts \in (I : k)$. Now we deal with the second case, by the essentiality of I, there exists $t \in R$ such that $0 \neq t(sk) \in I$. Thus $0 \neq ts \in (I : k)$. Therefore $(I : k)$ is an essential left R-subgroup of R. $\square$

Lemma 3.13. *Let I be a reduced left R-subgroup of the zero symmetric near-ring with identity R. If J is a left R-subgroup of R which is an essential extension of I, then J is reduced.*

Proof. Let y be a nonzero element of J such that $y^2 = 0$. From (2.5) Ry is a nonzero left R-subgroup of R. So by using Lemma 3.3, there exists $a \in R$ such that $0 \neq ay \in I$. $y^2 = 0$ implies that $(yay)^2 = 0$. Thus $yay = 0$, since I is reduced. Consequently $(ay)^2 = 0$, which contradicts $ay \neq 0$ and I is reduced. Therefore J is reduced. $\square$

Proposition 3.14. *Let R be a zero symmetric near-ring with identity and Rx be a reduced left R-subgroup of R. If Rx is essential in a left R-subgroup I of R, then I is isomorphic to Ix and $I(Rx)$ is essential in I.*

Proof. Let $i : Rx \to I$ be the inclusion homomorphism and define $f : I \to Rx$ by $f(a) = ax$ for each $a \in I$. Let $bx \in \text{Ker}(fi)$. Then $bx^2 = 0$, so $(xbx)^2 = 0$. Thus $xbx = 0$, since Rx is reduced. Consequently $bx = 0$ and $\text{Ker}(fi) = 0$. For each $c \in Rx \cap \text{Ker} f, f(i(c)) = f(c) = 0$, thus $c = 0$ and so $\text{Ker} f = 0$ by the essentiality of Rx. Therefore I is isomorphic to Ix.

Now let a be a nonzero element of I. Then $(Rx : a) = \{t \in I : ta \in Rx\}$ is essential in I by lemma 3.12. Hence there exists $b \in R$ such that $0 \neq ba \in (Rx : a)$. By using lemma 3.13 I is reduced. Now if $ba^2 = 0$, then $(aba)^2 = 0$. Thus $aba = 0$, which imply that $ba = 0$. Therefore ba^2 is a nonzero element of Rx. Consequently by the same argument $0 \neq aba^2 \in I(Rx)$. Hence $I(Rx)$ is essential in I. □

The following example illustrates that the condition R is zero symmetric in proposition 3.14 is not redundant.

Example 3.15. Consider the near-ring in example (2.9). $Ra = \{0, a\}$ is a reduced ideal of R which is essential in R as a left R-subgroup, since $0, Ra$ and R are the only left R-subgroups of R. But clearly R is not isomorphic to Ra.

Remark 3.16.   (1) By a distributive left (right) R-subgroup of a near-ring R, we mean a left (right) R-subgroup of R in which every element is distributive in R. It is not difficult to show that lemma 3.13 is true for distributive right R-subgroups of R. That is "If I is a reduced right R-subgroup of a zero symmetric near-ring R with identity such that I is essential in a distributive right R-subgroup J of R, then J is reduced".

(2) Proposition 3.14 is true for left ideals which are essential as left R-subgroups. (Since the near-ring R is zero-symmetric) that is "Let R be a zero-symmetric near-ring with identity and Rx is a reduced left ideal of R. If Rx is essential as a left R-subgroup in the left ideal I of R, then I is isomorphic to Ix and $I(Rx)$ is essential in I."

(3) The first part of proposition 3.14 is true for right R-subgroups (right ideals) of R under some condition on the element x, that is "Let R be a zero-symmetric near-ring with identity and xR be a reduced right R-subgroup (right ideal) of R where x is distributive in R. If xR is essential in a right R-subgroup (right ideal) I of R, then I is isomorphic to xI.

Now we show that the following result due to G.F. Birkenmeier is not true for near-rings.

Proposition 3.17 ([1, Proposition 1, (iii)]). *Let R be a ring such that for each finitely generated reduced right ideal X, there exists an idempotent d such that X is essential in dR. Then there exists $x \in X$ such that xR is essential in X and xR is isomorphic to dR.*

Example 3.18. Let $R = \{0, a, b, c\}$ be the abstract affine near-ring of (2.9) and $X = \{0, a\}$. Then X is a reduced ideal of R which is essential as a right ideal in $R = bR$, where b is the identity element of R. But there is no $x \in X$ such that xR is isomorphic to $bR = R$. Therefore proposition 3.17 is not true for near-rings.

It is well known that a Hamiltonian group is a non-abelian group in which every sub-group is normal. Moreover, a finite group G is Hamiltonian if and only if $G = A \times B \times C$, where A is a quaternion group, B is an elementary abelian 2-group, and C is an abelian group of odd order. Now we attempt to give an analogous result of proposition 3.17 for right near-rings. In order to do this, we need the following lemma and definitions.

Definition 3.19. A near-ring R will be called Hamiltonian near-ring if $(R, +)$ is a Hamiltonian group.

Lemma 3.20. *Let R be a Hamiltonian near-ring and x_i, $i = 1, \ldots, n$ distributive elements of R. Then $\sum_{i=1}^{n} x_i R$ is a right ideal of R.*

Proof. Follows directly from (2.7) and (2.4).   □

Definition 3.21. Let R be a Hamiltonian near-ring. A right ideal of R of the form $\sum_{i=1}^{n} a_i R$, where a_i, $i = 1, \ldots, n$ are distributive elements of R and $n \in Z^+$ will be called right ideal of type fd.

Now we are ready to conclude this section by the following result in which we add a sufficient condition to extend proposition 3.17 for right near-rings.

Theorem 3.22. *Let R be a zero-symmetric Hamiltonian near-ring with identity such that $(x, b, c(1 - b)) = 0$ for all central idempotents $b, c \in R$ and all $x \in R$. If for each reduced right ideal of type fd, X of R such that $(X, +)$ is abelian, there exists a distributive right ideal of the form dR, where d is a distributive idempotent in R satisfying the following identities, $(X, 1, -d) = 0, x(-d) = -xd$ for all $x \in R$ and $(1, d) = 0$ such that X is essential in dR. Then there exists $a \in X$ such that aR is essential in X and aR is isomorphic to dR.*

Proof. From lemma 3.13 dR is reduced near-ring, so dR is a near-ring with identity d, since dR is zero symmetric and d is a distributive idempotent. If $X = xR$ for some distributive element $x \in R$, then by using the third part of Remarks 3.16 with $I = dR$, we obtain that dR is isomorphic to xdR as a right R-submodules of R and xdR is an R-submodule of xR. Moreover, since xR is contained in dR, we get that $xR = xdR$. Now assume that $X = sR + tR$ where s and t are distributive elements of R, so there exist distributive ideals bR and cR such that sR and tR are essential in bR and cR respectively. Since b and c are distributive idempotents in the zero-symmetric reduced near-ring dR, then b and c are central idempotents in dR. Again by using the third part of remarks 3.16, we obtain that sbR is isomorphic to bR and tcR is isomorphic to cR, which imply the existence of monomorphisms $f : bR \to X$ and $g : cR \to X$. Proposition 3.8 implies the existence of a monomorphism $h : eR \to X$, where $e = b + c(1 - b)$ is a central idempotent in R. If $eR \neq dR$, then $dR = eR \oplus Y$. But $X \subseteq eR$, so $X \cap Y = 0$ contradicts the fact that X is essential in dR therefore h is a monomorphism from dR to X and $dR = bR + cR$. Let $h(d) = a$. Hence aR is isomorphic to dR. If aR is not essential in X, then aR is not essential in dR. So there are nonzero distributive idempotents u and v which are central in dR such that $dR = uR \oplus vR$ and aR is essential in uR. Hence $h(v) = h(dv) = av$. Since v is central in $dR, av \in aR \cap vR = 0$ which contradicts that h is a monomorphism. Therefore,

aR is essential in X. The proof for $X = \sum_{i=1}^{n} x_i R$ is a routine generalization of the above case $X = sR + tR$. $\qquad\qquad\square$

4. ENDOMORPHISMS FIXING ESSENTIAL IDEALS

In this section we are concerned with the following question "Let R be a near-ring and M be a left R-module. If N is an essential R-submodule of M and f is an R-endomorphism of M such that f is the identity map on N, must f be an automorphism?"

We shall attempt to answer the question by providing a counter example and we follow it by two results to show that the question has an affirmative answer in the ideal theoretical case, under some conditions.

Example 4.1. Let R be the near-ring considered in example (2.9), R is reduced Noetherian abstract affine right near-ring with identity b. The nontrivial subgroups of $(R, +)$ are $I = \{0, a\}, J = \{0, b\}$ and $K = \{0, c\}$, moreover

1) I is an ideal of R and also is a left R-submodule of R.
2) J is a left ideal of R but is neither a right ideal of R nor a left R-submodule of R.
3) K is a left ideal of R and a quasi-ideal of R but it is neither a right ideal of R nor a left R-submodule of R.

Therefore I is an essential left R-submodule of R. Let $f : R \rightarrow R$ be defined by $f(x) = xa$ for each $x \in R$. Then f is an R-endomorphism of R which is the identity on I but f is not an automorphism.

We now discuss the conditions under which our question has an affirmative answer. Our first positive result is the following proposition:

Proposition 4.2. *Let R be a zero-symmetric reduced near-ring with identity and I be an essential left ideal of R. If f is an R-endomorphism of the R-module $_R R$ for which $f(x) = x$ amd $[x, f(1)] = 0$ for all $x \in I$, then f is an automorphism.*

Proof. Let x be an element in $I \cap \operatorname{Ker} f$. Then $x = f(x) = 0$. From the essentiality of I, $\operatorname{Ker} f = 0$ and f is injektief. For all $r \in R, f(r) = f(r1) = rf(1)$, so f is a multiplication on the right by $f(1)$. If $f(1)$ has a left inverse $s \in R$, then $r = r1 = rsf(1) = f(rs)$ for all $r \in R$, thus f is an automorphism. Now assume that f is not an automorphism, so $f(1)$ has no left inverse and $b = 1 - f(1)$ is a nonzero element of R. $bx = (1 - f(1))x = x - f(1)x = x - xf(1) = x - f(x) = 0$ for all $x \in I$. Thus b is in the left annihilator J of I. By (2.3), (iv) J is a left ideal of R, so $J \cap I$ contains nonzero element a by the essentiality of I. Therefore $a^2 = 0$ which contradicting the fact that R is reduced. $\qquad\square$

Remark 4.3. We notice that example 4.1 shows that the conditions $f(1)$ commutes with elements of I and R is zero-symmetric of proposition 4.2 are not redundant, since $0 = 0a \neq a0 = a$.

Remark 4.4. By using the facts that near-rings with identity which satisfying I.F.P. are zero-symmetric and regular near-ring with identity which satisfying I.F.P. are right strongly regular and therefore reduced. We can conclude that proposition 4.2 holds for regular right near-rings with identity which satisfying I.F.P.

Corollary 4.5. *Let R be a reduced ring with identity and I be an essential left ideal of R. If f is an R-endomorphism of $_RR$ which is the identity on I, then f is an automorphism.*

It is well known that, if R is a right near-ring, then $(-x)y = -xy$ for all $x,y \in R$. We shall show that the identity $x(-y) = -xy$ may be not satisfied in right near-rings. But if R is a zero-symmetric and x is a distributive element of R, then $x(-y) = -xy$ for all $y \in R$. Also we give here two examples to show that the conditions, R is zero-symmetric and x is a distributive element of R are not redundant.

Example 4.6. Let $R = \{0,a,b,c,x,y,z,w\}$ be the near-ring [3, Near-rings of low order (K1)] defined by the tables

+	0	a	b	c	x	y	z	w
0	0	a	b	c	x	y	z	w
a	a	b	c	0	y	z	w	x
b	b	c	0	a	z	w	x	y
c	c	0	a	b	w	x	y	z
x	x	w	z	y	0	c	b	a
y	y	x	w	z	a	0	c	b
z	z	y	x	w	b	a	0	c
w	w	z	y	x	c	b	a	0

·	0	a	b	c	x	y	z	w
0	0	0	0	0	0	0	0	0
a	0	a	a	a	0	a	a	0
b	0	b	b	b	0	b	b	0
c	0	c	c	c	0	c	c	0
x	x	x	x	x	x	x	x	x
y	x	y	y	y	x	y	y	x
z	x	z	z	z	x	z	z	x
w	x	w	w	w	x	w	w	x

The distributive elements of R are $0,a,b$ and c. Take $r = s = a$, then $a = ac = a(-a) = r(-s) \neq -rs = -a^2 = -a = c$. Therefore $r(-s) \neq -rs$, since R is not zero-symmetric.

Example 4.7. Consider the zero-symmetric simple near-ring with identity $R = M_0(Z_3)$ and

$$\begin{array}{ll} 0 \mapsto 0 & 0 \mapsto 0 \\ \text{let } f: 1 \mapsto 0 \text{ and } g: 1 \mapsto 1 \\ 2 \mapsto 2 & 2 \mapsto 1 \end{array}$$

be two elements of R. Then f is not a homomorphism of the additive group Z_3 and so not a distributive element in R. Simple calculations show that, $0 = -fg \neq f(-g) = -g \neq 0$.

Lemma 4.8. *Let R be a near-ring with identity and $a,b \in R$. If $b = 1 - a$ and $(1,a) = 0$, then $a = 1 - b$.*

Proof. The proof is routine.   □

Lemma 4.9. *Let R be a near-ring with identity and b be a nonzero nilpotent element of R. If one of the following two conditions, 1) $[b, 1-b] = 0$ or 2) R is a zero-symmetric and b is distributive holds, then $1 - b$ has a left inverse.*

Proof. Assume that $[b, 1-b] = 0$, clearly $[b^n, 1-b] = 0$ for each positive integer n. If b is of index k, then $(1 + b + \ldots + b^{k-1})(1 - b) = (1-b) + b(1-b) + \ldots + b^{k-1}(1-b) = (1-b) + (1-b)b + \ldots + (1-b)b^{k-1} = 1 - b + b + (-b)b + \ldots + b^{-1} + (-b)b^{k-1} = 1 - b^2 + b^2 + \ldots + b^{k-1} - b^k = 1 - b^k = 1$. Thus $1 + b + \ldots + b^{k-1}$ is a left inverse of $1 - b$. If condition (2) holds, then the proof is similarly by using the notice after corollary 4.5.   □

Example 4.10. This example illustrates that conditions (1) and (2) in lemma 4.9 are not redundant. Consider the near-ring $R = M_0(Z_3)$ of example 4.7 and let $f : \begin{matrix} 0 & \mapsto & 0 \\ 1 & \mapsto & 0 \\ 2 & \mapsto & 1 \end{matrix}$ be an element of R. f is a nilpotent element of index 2. But neither f is a distributive element nor f commute with $1 - f$. Moreover it easily verified that $1 - f$ has neither left inverse nor right inverse.

We are now ready to prove our second positive result.

Theorem 4.11. *Let R be a left Noetherian zero-symmetric near-ring with identity and I be an ideal of R which is essential as a left R-subgroup of R (i.e. as an R-submodule of R). If f is an R-endomorphism of $_R R$ which is the identity on I and the following identities $(1, f(1)) = 0, [f(1), 1 - f(1)] = 0, (x, 1, -f(1)) = 0$ and $x(-f(1)) = -xf(1)$ for all $x \in I$ are satisfied, then f is an automorphism.*

Proof. As in the proof of proposition 4.2 f is injective and it is an automorphism if and only if $f(1)$ has a left inverse. Now assume that f is not automorphism, so $b = 1 - f(1)$ is a nonzero element of R. By using lemma 4.8 $f(1) = 1 - b$ has no inverse, so lemma 4.9 implies that b is not a nilpotent element of R. Therefore for each positive integer n, Rb^n is a nonzero left R-subgroup of R by using (2.5). Thus the essentially of I shows that, for each positive integer n, there exists an element r_n of R for which $0 \neq r_n b^n \in I$; that is $r_n \in (I : b^n)$ for each positive integer n, since I is a right ideal of R. We have the ascending chain of left ideals $(I : b) \subseteq (I : b^2) \subseteq \cdots \subseteq (I : b^{n-1}) \subseteq \cdots$. By using the two identities $(x, 1, -f(1)) = 0$ and $x(-f(1)) = -xf(1), xb = x(1 - f(1)) = x + x(-f(1)) = x - xf(1) = x - f(x) = 0$ for each $x \in I$. Now if $r_n b^{n-1}$ were in I, then the fact that $Ib = 0$ would force $r_n b^n = 0$, a contradiction to our chaise of r_n. So $r_n \notin (I : b^{n-1})$ and our chain form a properly ascending chain of left ideals of R, contradicting the fact that R is a left Noetherian.   □

Remark 4.12. We notice that example 4.1 shows that the identities $[f(1), 1 - f(1)] = 0$ and $(x, 1, -f(1)) = 0$ for each $x \in I$ in theorem 4.11 are not redundant, since $[f(1), 1 - f(1)] = ac - ca = a - 0 = a \neq 0$ and if we take $x = a \in I$, then $(x, 1, -f(1)) = (a, b, a) = a(b + a) - a^2 - ab = ac - a - a = a - a - a = -a = a \neq 0$.

Corollary 4.13. *Let R be a left Noetherian ring with identity and I an ideal of R which is essential as a left ideal. If f is an R-endomorphism of $_R R$ which is the identity on I, then f is an automorphism.*

REFERENCES

[1] G. F. Birkenmeieer, *Reduced right ideals which are strongly essential indirect summands*, Arch. Math. **52** (1989), 223–225.

[2] J. W. Brewer, D. C. Lantz and L.R. le Riche, *R-Endomorphisms fixing essential ideals need not be automorphisms*, Communications in Algebra. **10** (1982), 1907–1910.

[3] J. R. Clay, *Near-rings on groups of low order*, Math. Z. **104** (1968), 364–371.

[4] H. Gonshor, *On abstract affine near-rings*. Pacific J. Math. **14** (1964), 1237–1240.

[5] G. Pilz, Near-rings, North-Holland, Amsterdam, 1983.

DEPARTMENT OF MATHEMATICS, CAIRO UNIVERSITY, GIZA, EGYPT

CONDITIONS THAT $M_{\mathcal{A}}(G)$ IS A RING

FU-AN LI

ABSTRACT. This note presents some necessary and sufficient conditions, without finiteness assumption, for the centralizer near-ring $M_{\mathcal{A}}(G)$ to be a ring. The result illustrates that whether $M_{\mathcal{A}}(G)$ is a ring is a "local" behaviour.

Let $(G,+)$ be a group with identity 0, written additively, but not necessarily abelian, and let S be a subsemigroup of $\text{End}\,G$. The set of functions from G to G which fix 0 and commute with each $\alpha \in S$

$$M_S(G) = \{f : G \longrightarrow G \mid f(0) = 0 \ \text{and} \ f\alpha = \alpha f \ \forall \alpha \in S\}$$

is a zero-symmetric (right) near-ring with identity under pointwise addition and composition of functions, called the *centralizer near-ring* determined by the pair (S, G). Centralizer near-rings are very general, indeed, any zero-symmetric near-ring with identity is isomorphic to some $M_S(G)$ (cf. [5]). When $S = \mathcal{A}$ is a subgroup of $\text{Aut}\,G$, we obtain

$$M_{\mathcal{A}}(G) = \{f : G \longrightarrow G \mid f(0) = 0 \ \text{and} \ f\alpha = \alpha f \ \forall \alpha \in \mathcal{A}\}.$$

The study of centralizer near-rings is important and interesting. Many authors are devoted to find necessary and/or sufficient conditions such that $M_S(G)$ or $M_{\mathcal{A}}(G)$ is a distributively generated near-ring, a near-field, or a ring etc. (see [2] for the literature). Among those, C. J. Maxson [3] derived a necessary and sufficient condition in terms of the pair $(\mathcal{A}, G)$ for $M_{\mathcal{A}}(G)$ to be a ring under the assumption that $(\mathcal{A}, G)$ satisfies a finiteness condition. The purpose of this note is to discuss this problem without finiteness assumption.

We denote $M = M_{\mathcal{A}}(G)$ for short. For $a \in G$, let

$$\text{st}(a) = \{\alpha \in \mathcal{A} \mid \alpha a = a\}$$

be the stabilizer of a in $\mathcal{A}$, which is a subgroup of $\mathcal{A}$. Clearly, if $\beta \in \mathcal{A}$, then $\text{st}(\beta a) = \beta \text{st}(a)\beta^{-1}$ is a congugate of $\text{st}(a)$. A simple and well-known result due to Betsch says that, for any $a, b \in G$, $\text{st}(a) \subseteq \text{st}(b)$ if and only if there exists some $f \in M$ such that $f(a) = b$ (see [1] and also [6, Proposition 9.199]). In fact, if $\text{st}(a) \subseteq \text{st}(b)$, one can define

$$f(c) = \begin{cases} \beta b & \forall c = \beta a \in \mathcal{A}a; \\ 0 & \forall c \notin \mathcal{A}a, \end{cases}$$

1991 *Mathematics Subject Classification.* 16Y30 entralizer near-ring .
Supported by the National Natural Science Foundation of China.

Y. Fong et al. (eds.), *Near-Rings and Near-Fields*, 118–121.
© 2001 *Kluwer Academic Publishers. Printed in the Netherlands.*

which is well-defined since $\beta_1 a = \beta_2 a \Longrightarrow \beta_2^{-1}\beta_1 a = a \Longrightarrow \beta_2^{-1}\beta_1 \in \mathrm{st}(a) \subseteq \mathrm{st}(b) \Longrightarrow$
$\beta_2^{-1}\beta_1 b = b \Longrightarrow \beta_1 b = \beta_2 b$. Obviously, $f(0) = 0$, $f(a) = b$, and for any $\alpha \in \mathcal{A}$, since
$c \in \mathcal{A}a \Longleftrightarrow \alpha c \in \mathcal{A}a$,

$$f\alpha(c) = \begin{cases} \alpha\beta b & \forall c = \beta a \in \mathcal{A}a; \\ 0 & \forall c \notin \mathcal{A}a, \end{cases}$$

which is equal to $\alpha f(c)$, so $f\alpha = \alpha f$, and hence, $f \in M$. The converse is evidently true.

For $0 \neq a \in G$, let

$$H_a = \{b \in G \mid \mathrm{st}(a) \subseteq \mathrm{st}(b)\} = \{f(a) \mid f \in M\}$$

be the set of images of a under M.

Lemma 1. (1) H_a is a subgroup of G.
(2) $b \in H_a \Longleftrightarrow \mathrm{st}(a) \subseteq \mathrm{st}(b) \Longleftrightarrow H_a \supseteq H_b$.
(3) $\alpha H_a = H_{\alpha a} \, \forall \alpha \in \mathcal{A}$.
(4) $g(H_a) \subseteq H_a \, \forall g \in M$ (and so $g|_{H_a}$ makes sense).

Proof. (1) It is clear since $f(a) - g(a) = (f + (-g))(a) \in H_a$.
(2) It is also clear by definition.
(3) $\alpha H_a = \{\alpha b \in G \mid \mathrm{st}(a) \subseteq \mathrm{st}(b)\} = \{\alpha b \in G \mid \alpha \mathrm{st}(a)\alpha^{-1} \subseteq \alpha \mathrm{st}(b)\alpha^{-1}\}$
$= \{\alpha b \in G \mid \mathrm{st}(\alpha a) \subseteq \mathrm{st}(\alpha b)\} = H_{\alpha a}$.
(4) $g(H_a) = \{gf(a) \mid f \in M\} \subseteq H_a$. □

Now denote $\mathcal{A}_a = \{\alpha \in \mathcal{A} \mid \alpha H_a \subseteq H_a\} = \{\alpha \in \mathcal{A} \mid H_{\alpha a} \subseteq H_a\}$, which is a subsemigroup of $\mathcal{A}$. Thus, we have a near-ring $M_{\mathcal{A}_a}(H_a)$ determined by the pair $(\mathcal{A}_a, H_a)$.

Lemma 2. $\mathcal{A}_a a = H_a \cap \mathcal{A}a$.

Proof. By definition, $\mathcal{A}_a a = \{\alpha a \mid \alpha \in \mathcal{A}, H_{\alpha a} \subseteq H_a\}$. So $\mathcal{A}_a a \subseteq H_a$ and $\mathcal{A}_a a \subseteq H_a \cap \mathcal{A}a$. On the other hand, let $b \in H_a \cap \mathcal{A}a$. Then $b = f(a) = \alpha a$ for some $f \in M$ and $\alpha \in \mathcal{A}$. Hence, $H_{\alpha a} = H_{f(a)} \subseteq H_a$. Thus, $b = \alpha a \in \mathcal{A}_a a$. □

Proposition 3. If M is a ring and $0 \neq a, b \in G$, then

(1) $\mathrm{st}(a) \subseteq \mathrm{st}(b) \Longrightarrow \mathcal{A}a = \mathcal{A}b$,
(2) H_a is an abelian group, and
(3) $H_a = \mathcal{A}_a a \cup \{0\}$.

Proof. (1) This is a result of Maxson [3] and his proof does not need the finiteness condition. In fact, by Betsch's Lemma, if $\mathrm{st}(a) \subseteq \mathrm{st}(b)$, then $b = f(a)$ for some $f \in M$. Let e_x denote the orbit idempotent on $\mathcal{A}x$, i.e., $e_x(c) = c$ if $c \in \mathcal{A}x$, and $e_x(c) = 0$ if $c \notin \mathcal{A}x$. It is straightforward to check that $e_x \in M$. If $a + b = 0$, then $0 = e_a(a + b) = e_a(e_a + f)(a) = (e_a^2 + e_a f)(a) = a + e_a(b)$ since M is a ring. Thus, $e_a(b) = -a = b$, which implies that $b \in \mathcal{A}a$. If $a + b \neq 0$, then $a + b = e_{a+b}(e_a + f)(a) = e_{a+b}e_a(a) + e_{a+b}f(a) = e_{a+b}(a) + e_{a+b}(b)$, which implies that both a and b belong to $\mathcal{A}(a + b)$ and so $\mathcal{A}a = \mathcal{A}(a + b) = \mathcal{A}b$.
(2) For any $b, c \in H_a$, write $b = f(a)$ and $c = g(a)$ where $f, g \in M$. Since M is a ring, $f + g = g + f$, whence $b + c = c + b$.
(3) By Lemma 2, $H_a \supseteq \mathcal{A}_a a \cup \{0\}$. On the other hand, for any $0 \neq b \in H_a$, we have $\mathrm{st}(a) \subseteq \mathrm{st}(b)$. From (1), $b \in \mathcal{A}a$. It follows that $b \in H_a \cap \mathcal{A}a = \mathcal{A}_a a$ by Lemma 2 again. □

Now denote

$$M_a = \{f \in M \mid f(b) = 0 \; \forall b \notin \mathcal{A}a\}.$$

It is easy to see that M_a is a subnear-ring, even a left ideal of M since, for any $f \in M_a$, $g, h \in M$ and $b \notin \mathcal{A}a$, $(g(h+f) - gh)(b) = g(h+f)(b) - gh(b) = g(h(b)+0) - gh(b) = 0$, i.e., $g(h+f) - gh \in M_a$. Notice that each $f \in M_a$ is uniquely determined by $f(a)$.

As a left M-module, M has a direct product decomposition

$$M = \prod_{i \in I} M_{a_i}, \tag{$*$}$$

where $\{a_i\}_{i \in I}$ is an arbitrary fixed set of representatives of the non-zero orbits in G under the action of $\mathcal{A}$.

By Lemma 1(4), the restriction induces a homomorphism of near-rings

$$M_a \longrightarrow M_{\mathcal{A}_a}(H_a), \quad f \longmapsto f|_{H_a}.$$

This homomorphism is in fact an embedding, for $f, g \in M_a$, $f|_{H_a} = g|_{H_a} \Longrightarrow f(a) = g(a) \Longleftrightarrow f = g$.

Theorem 4. *If M is a ring, then*

(1) M_a *it is an ideal of M (hence a subring of M),*

(2) $M_a \hookrightarrow M_{\mathcal{A}_a}(H_a)$ *is a ring isomorphism, and*

(3) *the above $(*)$ is a ring decomposition.*

Proof. (1) Since M_a is a left ideal of M, it suffices to show that M_a is also a right ideal of M. Let $f \in M_a$, $g \in M$, and $b \notin \mathcal{A}a$. We have to show that $fg(b) = 0$. Without loss of generality, we may assume $g(b) \neq 0$. Since $\mathrm{st}(b) \subseteq \mathrm{st}(g(b))$, we have $\mathcal{A}b = \mathcal{A}g(b)$ by Proposition 3(1). Thus $g(b) \notin \mathcal{A}a$, and so $fg(b) = f(g(b)) = 0$. Hence $fg \in M$.

(2) By Proposition 3(3), $H_a = \mathcal{A}_a a \cup \{0\}$. For any $g \in M_{\mathcal{A}_a}(H_a)$, define $f \in M_a$ as follows:

$$f(c) = \begin{cases} \alpha g(a) & \forall c = \alpha a \in \mathcal{A}a, \\ 0 & \forall c \notin \mathcal{A}a. \end{cases}$$

Then $f|_{H_a} = g$. So the embedding is surjective, hence, is an isomorphism of rings.

(3) It follows from (1) and $(*)$.   □

Theorem 5. *The following statements are equivalent:*

(1) $M_{\mathcal{A}_a}(H_a)$ *is a ring.*

(2) M_a *is a ring.*

(3) $H_a = \mathcal{A}_a a \cup \{0\}$ *is an abelian group and $f|_{H_a} \in \mathrm{End}\, H_a$ for each $f \in M$. Under the above equivalent conditions, $M_a = M_{\mathcal{A}_a}(H_a)$ is a subring of $\mathrm{End}\, H_a$.*

Proof. (1)$\Longrightarrow$(2): It is obvious since M_a is embedded into $M_{\mathcal{A}_a}(H_a)$.

(2)$\Longrightarrow$(3): By an argument similar to the proof of Proposition 3(2)(3), we can establish that $H_a = \mathcal{A}_a a \cup \{0\}$ and it is abelian.

For $f \in M$, we define $f' \in M$ as follows:

$$f'(c) = \begin{cases} \beta f(a) & \forall c = \beta a \in \mathcal{A}a, \\ 0 & \forall c \notin \mathcal{A}a. \end{cases}$$

Since $H_a = \mathcal{A}_a a \cup \{0\}$, f' and f coincide on H_a, and f' belongs to M_a. So without loss of generality, we may assume $f \in M_a$. For any $b, c \in H_a$, there exist $g, h \in M$ such that $b = g(a)$ and $c = h(a)$. As above, we may also assume $g, h \in M_a$. ¿

$$f(b \pm c) = f(g \pm h)(a) = (fg \pm fh)(a) = fg(a) \pm fh(a) = f(b) \pm f(c).$$

Hence, $f|_{H_a} \in \operatorname{End} H_a$.

(3)$\Longrightarrow$(1): It follows that $(M_{\mathcal{A}_a}(H_a), +)$ is an abelian group from the fact that H_a is abelian. Since $H_a = \mathcal{A}_a a \cup \{0\}$, as in the proof of Theorem 4(2), each $f \in M_{\mathcal{A}_a}(H_a)$ can be extended to an element of M (even an element of M_a) and so $f \in \operatorname{End} H_a$ by the assumption of (3). Thus, for any $g, h \in M_{\mathcal{A}_a}(H_a)$ and $b \in H_a$,

$$f(g + h)(b) = f(g(b) + h(b)) = fg(b) + fh(b) = (fg + fh)(b),$$

which implies the left distributivity of $M_{\mathcal{A}_a}(H_a)$. Therefore, $M_{\mathcal{A}_a}(H_a)$ is a ring.

Finally, under the above equivalent conditions, we have seen that each $f \in M_{\mathcal{A}_a}(H_a)$ can be extended to an element of M_a. So the embedding $M_a \hookrightarrow M_{\mathcal{A}_a}(H_a)$ is surjective, hence, $M_a = M_{\mathcal{A}_a}(H_a)$ is a subring of $\operatorname{End} H_a$.   □

Combining the above results, we easily derive:

Theorem 6. *The following statements are equivalent:*

(1) *M is a ring.*
(2) *$M_{\mathcal{A}_a}(H_a)$ is a ring for each $a \in G$.*
(3) *M_a is a ring for each $a \in G$.*
(4) *For any $a \in G$ and $f \in M$, $H_a = \mathcal{A}_a a \cup \{0\}$ is an abelian group and $f|_{H_a} \in \operatorname{End} H_a$.*
(5) *$M_{\mathcal{A}_{a_i}}(H_{a_i})$ is a ring for each $i \in I$.*
(6) *M_{a_i} is a ring for each $i \in I$.*
(7) *For any $i \in I$ and $f \in M$, $H_{a_i} = \mathcal{A}_{a_i} a_i \cup \{0\}$ is an abelian group and $f|_{H_{a_i}} \in \operatorname{End} H_{a_i}$.*
Under the above equivalent conditions, $M = \prod_{i \in I} M_{a_i}$ is a ring decomposition.

Remark 7. From Theorem 6 we see that whether $M_{\mathcal{A}}(G)$ is a ring is a "local" behaviour.

Remark 8. If the finiteness condition for $(\mathcal{A}, G)$ is satisfied (i.e., for any $a \in G$ and $\alpha \in \mathcal{A}$, $\operatorname{st}(a) \subseteq \operatorname{st}(\alpha a) \Longrightarrow \operatorname{st}(a) = \operatorname{st}(\alpha a)$), then by [4, Theorem II.1 and Corollary II.2], the results of [3] are consequences of our Theorems 4–6.

REFERENCES

[1] G. Betsch, *Near-rings of group mappings*, in: Oberwolfach Conference, 1976.
[2] Y. Fong, A. Oswald, G. Pilz and K. C. Smith (eds.), *Near-Ring Newsletter*, National Cheng Kung Univ., Tainan, Taiwan, 1995.
[3] C. J. Maxson, *When is $M_{\mathcal{A}}(G)$ a ring?* in: Near-Rings and Near-Fields, Proc. Conf. on Near-Rings and Near-Fields, 1993, edited by Y. Fong et al., Kluwer Acad. Publ., Dordrecht-Boston-London, 1995, pp. 199–202.
[4] C. J. Maxson and J. D. P. Meldrum, *Centralizer representations of near-fields*, J. Algebra **89** (1984), 406–415.
[5] C. J. Maxson and K. C. Smith, *Near-ring centralizers*, in: Proc. Ninth USL Math. Conf., Univ. Southwestern Louisiana, Lafayette, 1979.
[6] G. Pilz, *Near-Rings*, 2nd revised ed., North-Holland/American Elsevier, Amsterdam, 1983.

INSTITUTE OF MATHEMATICS, CHINESE ACADEMY OF SCIENCES, BEIJING 100080, CHINA. E-MAIL: fal@math08.math.ac.cn

ON DEPENDENCE AND INDEPENDENCE IN NEAR-RINGS

HEATHER MCGILVRAY AND C. J. MAXSON

Dedicated to Professor Dov Tamari

ABSTRACT. We generalize Tamari's 1948 classification of rings to near-rings. As suggested by Tamari, we find the situation is indeed different for "ring-like domains".

1. INTRODUCTION

In [2], O. Ore defines a ring R to be regular if R has no divisors of zero and every pair of elements in R has a common right multiple (CRM), i.e. for each $x, y \in R$, there exist $a, b \in R$, not both zero such that $xa = yb$. If this is not the case there are two elements $x_1, y_1 \in R$ such that $x_1 a_1 + y_1 b_1 = 0$ if and only if $a_1 = b_1 = 0$. Ore then goes on to say that it would be natural to characterize rings by the maximum number of elements $x_1, x_2, \ldots, x_n$ in a ring such that if $x_1 a_1 + \cdots + x_n a_n = 0$ then $a_1 = \cdots = a_n = 0$ where n could be finite or infinite and n would be called the order of right irregularity of the ring.

In 1948 (see [4]) Dov Tamari, following Ore's suggestion showed that the collection of rings is partitioned into 9 classes. Tamari states that the situation is different for ring-like domains but does not pursue this comment further. In 1964, Tamari mentioned this problem to the second author who considered it briefly in his dissertation written under the supervision of Tamari. In 1991 the present authors considered this classification problem for ring-like domains called near-rings as part of an undergraduate research project. This paper is a report of the 1991 collaboration. It is perhaps worth mentioning that the results discussed here intersect the work of three generations of relatives of a mathematical family.

Let $(R, +, \cdot)$ be a ring and let $a_i, b_i \in R$, $i = 1, \ldots, n$, where we allow $a_i = a_j$ if $i \neq j$. If $a_1 b_1 + \cdots + a_n b_n = 0$ with at least one of the b_i's not zero, we say that the a_i's are *linearly right dependent*. If at least one of the a_i's is non-zero, then we say the b_i's are *linearly left dependent*. If, for a given set $\{a_1, \ldots, a_n\} \subseteq R$, $a_1 x_1 + \cdots + a_n x_n = 0$ implies that $x_i = 0$ for all i, we say that $\{a_1, \ldots, a_n\}$ is *linearly right independent*. Similarly, if $x_1 a_1 + \cdots + x_n a_n = 0$ implies that $x_i = 0$ for all i, we say that $\{a_1, \ldots, a_n\}$ is *linearly left independent*. Two elements, $x, y \in R$, are said to have a *CLM (common left multiple)* if there are a and b, not both zero, in R so that $ax = by$. Similarly, they have a *CRM (common right multiple)* if there are a and b, not both zero, in R so that $xa = yb$. We say that R is of *left order n_ℓ* if there is a linearly left independent subset of R having n_ℓ elements and if any subset of R having $n_\ell + 1$ elements is linearly left dependent. R is *of right order n_r* if there

1991 *Mathematics Subject Classification.* 16Y30 .

Y. Fong et al. (eds.), *Near-Rings and Near-Fields*, 122–129.
© 2001 *Kluwer Academic Publishers. Printed in the Netherlands.*

is a linearly right independent subset of R having n_r elements and if any subset of R with $n_r + 1$ elements is linearly right dependent. In this case, we say R is of *type* (n_ℓ, n_r).

The ring-like domains we consider in this paper are zero-symmetric abelian near-rings. Recall that a right near-ring $N = (N, +, \cdot)$ is a set N with binary operations $+, \cdot$, such that $(N, +)$ is a group with zero, 0, $(N, \cdot)$ is a semigroup, and for $a, b, c \in N$, $(a + b) \cdot c = a \cdot c + b \cdot c$. The near-ring N is zero-symmetric if $a \cdot 0 = 0$ for each $a \in N$ and is abelian if $(N, +)$ is an abelian group. If $(G, +)$ is an abelian group, then $M_0(G) = \{f \colon G \to G \mid f(0) = 0\}$ is a zero-symmetric abelian near-ring under the operations of function addition and function composition. In the sequel we consider only zero-symmetric abelian near-rings and we just refer to them as "near-rings". A non-ring is a near-ring that is not a ring.

In the next section we briefly sketch some of Tamari's work and in Section III we discuss the situation for near-rings.

2. TAMARI'S RESULTS

We first present Tamari's Theorem. Since the proof is so elegant we repeat it here.

Theorem 2.1 (Tamari, [4]). *There are no rings of finite (left or right) order $n > 1$.*

Proof. Suppose R is a ring having finite left order $n_\ell > 1$. Then, we can find at least one set having two elements which is linearly left independent. Let $S = \{x, y\}$ denote this set. Thus, if $a, b \in R$ and $ax + by = 0$, then we have that $a = b = 0$. Now consider the set $\{xx, xy, yx, yy\}$. Suppose we have $a, b, c, d \in R$ so that $axx + bxy + cyx + dyy = 0$, then, $(ax + cy)x + (bx + dy)y = 0$, which means that $ax + cy = 0$ and $bx + dy = 0$. Again we have that $a = c = b = d = 0$ since S is linearly left independent. We continue in this manner to get a linearly left independent subset having 2^n elements for any positive integer n. Thus, $n_\ell(R) = \infty$. ∎

Corollary 2.2. *There exist at most the following nine types of rings:* (∞, ∞), $(0, 0)$, $(1, 1)$, $(\infty, 1)$, $(1, \infty)$, $(\infty, 0)$, $(0, \infty)$, $(0, 1)$, *and* $(1, 0)$.

Corollary 2.3.   (i) *A ring R has right (left) order 0 if and only if all elements are left (right) divisors of zero.*
   (ii) *A ring R has right (left) order 1 if and only if every pair of elements has a non-trivial CRM (CLM).*
   (iii) *If R is a ring with $n_r \neq 0$ $(n_\ell \neq 0)$ then $n_r = \infty$ $(n_\ell = \infty)$ if and only if not every pair of elements has a non-trivial CRM (CLM).*

Corollary 2.4. *A commutative ring will be either type $(0, 0)$ or type $(1, 1)$.*

Proof. If R is a commutative ring, then, $n_r(R) = n_\ell(R)$. Thus, R is of type $(0, 0)$, $(1, 1)$ or (∞, ∞). For $x, y \in R$, $xy + y(-x) = xy + (-x)y = xy + -(xy) = 0$. So, $\{x, y\}$ is linearly right dependent for all $x, y \in R$, hence $n_r(R) \leq 1$. Therefore, R is of type $(1, 1)$ or $(0, 0)$. ∎

Tamari [4] presents examples of all 9 types of rings. We sketch only 3 of these examples which we use in the next section.

Example 2.5 (Type $(0, \infty)$).

Define a semigroup, F, to be all of the possible "words" from the alphabet $\{x, y\}$ where the operation is defined by connecting one word to another. Define a semigroup, S, to be all the words, f, in F as well as the symbols ε and $f\varepsilon$, for all $f \in F$. Composition is defined by connecting words where an ε not at the end of a word is suppressed. That is, for $xy\varepsilon, yxxy\varepsilon \in S$, $(xy\varepsilon) \cdot (yxxy\varepsilon) = xyyxxy\varepsilon$. Consider the semigroup ring $Z_2(S) = \left\{ \sum_i a_i f_i + \left(\sum_i b_i g_i \right) \cdot \varepsilon \text{ where } a_i, b_i \in Z_2 \text{ for all } i \text{ and } f_i, g_i \in F \text{ for all } i \right\}$. One shows that all elements in $Z_2(S)$ are right divisors of zero, so, $n_\ell(Z_2(S)) = 0$. Now, if we consider the set $\{x, y\} \subseteq Z_2(S)$, we can see that there are no non-zero $P = P_1 + P_2\varepsilon$, $Q = Q_1 + Q_2\varepsilon \in Z_2(S)$ so that $xP = yQ$, since xP is a word beginning with x and yQ is a word beginning with y. Thus, not all pairs of elements in $Z_2(S)$ have a non-trivial CRM, which implies $n_r(Z_2(S)) = \infty$.

Example 2.6 (Type $(\infty, 0)$).

Let F be as in the above example and let S' be the semigroup defined as in the above example except that now we have all words in F together with the symbols ε and εf, where $f \in F$. Composition is essentially the same, but now with ε's not at the start of a word suppressed. Consider the semigroup ring $Z_2(S') = \{(\Sigma a_i f_i) + \varepsilon \cdot (\Sigma b_i g_i)$ where $a_i, b_i \in Z_2$ for all i and $f_i, g_i \in F$ for all $i\}$. We have that every element in $Z_2(S')$ is a left divisor of zero so, $n_r(Z_2(S')) = 0$, and, again using $\{x, y\} \subseteq Z_2(S')$, we get $n_\ell(Z_2(S')) = \infty$.

Example 2.7 (Type $(\infty, 1)$).

Let G be the semigroup consisting of all symbols $x^\alpha y^\beta$ where $\alpha, \beta \in \{0, 1, \ldots\}$, with composition defined by $x^{\alpha_1} y^{\beta_1} \cdot x^{\alpha_2} y^{\beta_2} = x^{\alpha_1 + \alpha_2} y^\gamma$ where $\gamma = \beta_1 + 2^{\alpha_1} \beta_2$. Consider the semigroup ring $Z_2(G) = \left\{ \sum_i a_i g_i \text{ where } a_i \in Z_2 \text{ and } g_i \in G \text{ for all } i \right\}$. Consider $\{xy, xy^2\} \subseteq Z_2(G)$ and suppose there are $P, Q \in Z_2(G)$ such that $P(xy) = Q(xy^2)$. That is, for $P = \Sigma \alpha_i x^{a_i} y^{b_i}$, $Q = \Sigma \beta_i x^{a_i} y^{b_i}$, $\Sigma \alpha_i x^{a_i} y^{b_i} xy = \Sigma \beta_i x^{a_i} y^{b_i} xy^2$, so $\Sigma \alpha_i x^{a_i+1} y^{b_i+2^{a_i}} = \Sigma \beta_i x^{a_i+1} y^{b_i+2^{a_i+1}}$. Hence, we must have $\alpha_i = \beta_i = 0$ for all i and so xy and xy^2 have no CLM which implies $n_\ell(Z_2(G)) = \infty$.

Further, $\{xy\}$ is linearly right independent since for $xyR = 0$, where $R = \Sigma \alpha_i x^{a_i} y^{b_i} \in Z_2(G)$, we have $\Sigma \alpha_i xy x^{a_i} y^{b_i} = 0$, or $\Sigma \alpha_i x^{a_i+1} y^{2b_i+1} = 0$. Thus, $\alpha_i = 0$ for all i, consequently $n_r(Z_2(G)) \geq 1$. One shows that every pair of elements in $Z_2(G)$ has a CRM, hence $n_r(Z_2(G)) \leq 1$, i.e., $n_r(Z_2(G)) = 1$.

In the next section we turn to the left and right order of near-rings.

3. NEAR-RING RESULTS

In this section we consider for near-rings the left and right orders defined by Tamari. Since the near-rings considered satisfy the right distributive law we see from the proof of Theorem 2.1, that for any near-ring N, $n_\ell(N) \in \{0, 1, \infty\}$. We will see that the situation is different for $n_r(N)$. We note that the first two parts of Corollary 2.3 hold, thus if N is a near-ring, $n_r(N) = 0$ $(n_\ell(N) = 0)$ if and only if every element of N is a left (right) divisor of zero and $n_r(N) = 1$ $(n_\ell(N) = 1)$ if and only if every pair of elements in N has a non-trivial CRM(CLM).

If a near-ring has a multiplicative identity, say 1, then the set $\{1\}$ is linearly left and right independent.

Theorem 3.1. *Let N be a right near-ring with unity. Then,* $n_\ell(N) \geq 1$ *and* $n_r(N) \geq 1$.

In the next theorem we show that the orders of the Cartesian product of near-rings are completely determined by the orders of the component near-rings.

Theorem 3.2. *Let N and N' be right near-rings of type* (n_ℓ, n_r) *and* (n'_ℓ, n'_r), *respectively. If* $N'' = N \times N'$, *then* N'' *is of type* (η_L, η_R) *where* $\eta_L = \min\{n_\ell, n'_\ell\}$ *and* $\eta_R = \min\{n_r, n'_r\}$.

Proof. We show that $\eta_L = \min\{n_\ell, n'_\ell\}$ as η_R follows similarly. Suppose that one of n_ℓ or n'_ℓ is finite (without loss of generality, suppose $n_\ell < n'_\ell$ and that n_ℓ is finite). Since N has left order n_ℓ, we can find a linearly left independent set $S \subseteq N$ having n_ℓ elements, but any set $T \subseteq N$ having $n_\ell + 1$ elements will be linearly left dependent. Also, since N' has left order n'_ℓ, we can find a linearly left independent set $S' \subseteq N'$ having n_ℓ elements. Consider the set $\bar{S} \subseteq N''$ having n_ℓ elements defined as $\bar{S} = \{(s_1, s'_1), \ldots, (s_{n_\ell}, s'_{n_\ell})\}$ where $s_i \in S$ and $s'_i \in S'$ for all $i \in \{1, \ldots, n_\ell\}$. Let $(\alpha_1, \alpha'_1), \ldots, (\alpha_{n_\ell}, \alpha'_{n_\ell}) \in N''$. Suppose $(\alpha_1, \alpha'_1)(s_1, s'_1) + \cdots + (\alpha_{n_\ell}, \alpha'_{n_\ell})(s_{n_\ell}, s'_{n_\ell}) = 0$. Then, $(\alpha_1 s_1, \alpha'_1 s'_1) + \cdots + (\alpha_{n_\ell} s_{n_\ell}, \alpha'_{n_\ell} s'_{n_\ell}) = 0$, and so $(\alpha_1 s_1 + \cdots + \alpha_{n_\ell} s_{n_\ell}, \alpha'_1 s'_1 + \cdots + \alpha'_{n_\ell} s'_{n_\ell}) = 0$. Hence, $\alpha_1 s_1 + \cdots + \alpha_{n_\ell} s_{n_\ell} = 0$ and $\alpha'_1 s'_1 + \cdots + \alpha'_{n_\ell} s'_{n_\ell} = 0$ which implies $\alpha_i = 0$ for all $i \in \{1, \ldots, n_\ell\}$ and $\alpha'_i = 0$ for all $i \in \{1, \ldots, n_\ell\}$. Thus, $\bar{S}$ is linearly left independent. Now, let $\bar{T}$ be any subset of N'' having $n_\ell + 1$ elements, say

$$\bar{T} = \{(t_1, t'_1), \ldots, (t_{n_\ell+1}, t'_{n_\ell+1}) \mid t_i \in N, t'_i \in N' \text{ for all } i \in \{1, \ldots, n_\ell + 1\}\}.$$

We can choose α_i's in N (not all zero) such that $\alpha_1 t_1 + \cdots + \alpha_{n_\ell+1} t_{n_\ell+1} = 0$. Then, $(\alpha_1, 0)(t_1, t'_1) + \cdots + (\alpha_{n_\ell+1}, 0)(t_{n_\ell+1}, t'_{n_\ell+1}) = (\alpha_1 t_1, 0) + \cdots + (\alpha_{n_\ell+1} t_{n_\ell+1}, 0) = (\alpha_1 t_1 + \cdots + \alpha_{n_\ell+1} t_{n_\ell+1}, 0) = (0, 0)$. Since not all of the α_i's are zero, not all of the $(\alpha_i, 0)$'s are zero. So, $\bar{T}$ is linear left dependent. This implies $\eta_L = n_\ell = \min\{n_\ell, n'_\ell\}$. Now, suppose $n_\ell = \infty = n'_\ell$, but $\eta_L \neq \infty$, say $\eta_L = i$. Then, for any $\bar{S} \subseteq N''$ having $i + 1$ elements, $\bar{S}$ is linearly left dependent. So, if $\bar{S} = \{(s_1, s'_1), \ldots, (s_{i+1}, s'_{i+1}) \mid s_j \in N, s'_j \in N' \text{ for } j \in \{1, \ldots, i+1\}\}$, then there are $(\alpha_j, \alpha'_j) \in N''$, which are not all zero, such that $(\alpha_1, \alpha'_1)(s_1, s'_1) + \cdots + (\alpha_{i+1}, \alpha'_{i+1})(s_{i+1}, s'_{i+1}) = 0$. That is, $(\alpha_1 s_1, \alpha'_1 s'_1) + \cdots + (\alpha_{i+1} s_{i+1}, \alpha'_{i+1} s'_{i+1}) = 0$ or, $(\alpha_1 s_1 + \cdots + \alpha_{i+1} s_{i+1}, \alpha'_1 s'_1 + \cdots + \alpha'_{i+1} s'_{i+1}) = 0$. Thus, we have $\alpha_1 s_1 + \cdots + \alpha_{i+1} s_{i+1} = 0$ and $\alpha'_1 s'_1 + \cdots + \alpha'_{i+1} s'_{i+1} = 0$ where either not all of the α_j's are zero or not all of the α'_j's are zero. So, either $S = \{s_1, \ldots, s_{i+1}\} \subseteq N$ or $S' = \{s'_1, \ldots, s'_{i+1}\} \subseteq N'$ is linearly left dependent. Hence, either $n_\ell < i + 1$ or $n'_\ell < i + 1$, a contradiction to $n_\ell = \infty = n'_\ell$. Thus, $\eta_L = \infty = \min\{n_\ell, n'_\ell\}$. $\qquad\square$

The next two theorems completely characterize the case in which the near-ring order is finite. Note that in this situation. $n_\ell = 0$ or $n_\ell = 1$.

We first give a well known number theoretic lemma to be used in our main result.

Lemma 3.3. *Let* $\{a_1, \ldots, a_n\} \subseteq Z_n$ *with the* a_i's *not necessarily distinct. For some* $k, \ldots, k + \ell$, *we have* $a_k + \cdots + a_{k+\ell} \equiv 0 (\bmod n)$.

Proof. Let $\{a_1,\ldots,a_n\} \subseteq Z_n$. Consider the partial sums

$$
\begin{aligned}
S_1 &= a_1, \\
S_2 &= (a_2+a_1)(\bmod n), \\
&\vdots \\
S_n &= (a_n+\cdots+a_1)(\bmod n).
\end{aligned}
$$

If all are distinct, then zero must be one of these sums and so we are finished. If not, then for some $i,j \in \{1,\ldots,n\}$, $S_i = S_j$, where we take $i < j$, say $j = i+\ell$. Then, $a_1+\cdots+a_i \equiv (a_1+\cdots+a_j)(\bmod n)$ which in turn gives $a_{i+1}+\cdots+a_j \equiv 0(\bmod n)$, as desired. $\quad\square$

For $M_0(G)$, G an abelian group, the next lemma shows that we always have $n_\ell(M_0(G)) = 1$:

Lemma 3.4. *Let G be an abelian group. Then, $n_\ell(M_0(G)) = 1$.*

Proof. Since $M_0(G)$ has an identity, by Theorem 3.1, $n_\ell(M_0(G)) \geq 1$. Now, let $f,g \in M_0(G)$. Define the functions $y,z \in M_0(G)$ by $y(a_\alpha) = z(b_\alpha)$ (not all zero) where $f(\alpha) = a_\alpha$ and $g(\alpha) = b_\alpha$ for every $\alpha \in G$. Then, for $\alpha \in G$,

$$
(y \circ f)(\alpha) = y(f(\alpha)) = y(a_\alpha) = z(b_\alpha) = z(g(\alpha)) = (z \circ g)(\alpha).
$$

Thus, for every $f,g \in M_0(G)$, f and g have CLM, so $n_\ell(M_0(G)) \leq 1$. $\quad\square$

Theorem 3.5. *There exists a near-ring, not a ring, of type $(1,k)$ for all positive integers k.*

Proof. Consider the near-ring $M_0(Z_{k+1})$ where k is a positive integer. Then, by Lemma 3.4, $n_\ell(M_0(Z_{k+1})) = 1$. Now, let $g \in M_0(Z_{k+1})$ denoted by $g = \begin{pmatrix} 0 & 1 & \cdots & k \\ 0 & g_1 & \cdots & g_k \end{pmatrix}$. Let $f = \begin{pmatrix} 0 & 1 & \cdots & k \\ 0 & 1 & \cdots & 1 \end{pmatrix}$, consider f listed k times and let $\alpha_1,\ldots,\alpha_k \in M_0(Z_{k+1})$. Suppose $f\alpha_1 + \cdots + f\alpha_k = 0$. If $\gamma_i = f\alpha_i$ for all $i = 1,\ldots,k$, then $\gamma_i = \begin{pmatrix} 0 & 1 & \cdots & k \\ 0 & \beta_1^i & \cdots & \beta_k^i \end{pmatrix}$ where $\beta_j^i \in \{0,1\}$. Thus for all i, $\sum_{j=1}^{k} \beta_j^i = 0$, we must have $\beta_j^i = 0$, for all $i,j \in \{1,\ldots,k\}$, since $\sum_j \beta_j^i \leq k$ for all i,j, i.e., $\alpha_i = 0$ for all i. Hence, S is linearly right independent, and so $n_r(M_0(Z_{k+1})) \geq k$. Now we show that any collection $S' \subseteq M_0(Z_{k+1})$ having $k+1$ (not necessarily distinct) elements is linearly right dependent. Let $S' = \{f_1,\ldots,f_{k+1}\} \subseteq M_0(Z_{k+1})$ and $\alpha_1,\ldots,\alpha_{k+1} \in M_0(Z_{k+1})$ where $f_i = \begin{pmatrix} 0 & 1 & \cdots & k \\ 0 & a_1^i & \cdots & a_k^i \end{pmatrix}$ and $\alpha_i = \begin{pmatrix} 0 & 1 & \cdots & k \\ 0 & \alpha_1^i & \cdots & \alpha_k^i \end{pmatrix}$ for $i = 1,\ldots,k+1$. Then,

$$
\begin{aligned}
f_1\alpha_1 + \cdots + f_{k+1}\alpha_{k+1} = {} & \begin{pmatrix} 0 & 1 & \cdots & k \\ 0 & a_1^1 & \cdots & a_k^1 \end{pmatrix} \begin{pmatrix} 0 & 1 & \cdots & k \\ 0 & \alpha_1^1 & \cdots & \alpha_k^1 \end{pmatrix} \\
& + \cdots + \begin{pmatrix} 0 & 1 & \cdots & k \\ 0 & a_1^{k+1} & \cdots & a_k^{k+1} \end{pmatrix} \begin{pmatrix} 0 & 1 & \cdots & k \\ 0 & \alpha_1^{k+1} & \cdots & \alpha_k^{k+1} \end{pmatrix}.
\end{aligned}
$$

Consider the set $\{a_1^1, \ldots, a_1^{k+1}\} \subseteq Z_{k+1}$. Then, by Lemma 3.3, for some $n, \ldots, n+m \in \{1, \ldots, k+1\}$, $a_1^n + \cdots + a_1^{n+m} \equiv 0 (\bmod (k+1))$. So, if we let $\alpha_j^i = 0$ for all i, j except for $i \in \{n, \ldots, n+m\}$ and define $\alpha_1^{n+j} = 1$ for all $j \in \{0, \ldots, m\}$, then

$$f_1 \alpha_1 + \cdots + f_{k+1} \alpha_{k+1}$$

$$= \begin{pmatrix} 0 & 1 & \cdots & k \\ 0 & a_1^n & \cdots & a_k^n \end{pmatrix} \begin{pmatrix} 0 & 1 & 2 & \cdots & k \\ 0 & 1 & 0 & \cdots & 0 \end{pmatrix}$$

$$+ \cdots + \begin{pmatrix} 0 & 1 & \cdots & k \\ 0 & a_1^{n+m} & \cdots & a_k^{n+m} \end{pmatrix} \begin{pmatrix} 0 & 1 & 2 & \cdots & k \\ 0 & 1 & 0 & \cdots & 0 \end{pmatrix}$$

$$= \begin{pmatrix} 0 & 1 & 2 & \cdots & k \\ 0 & a_1^n & 0 & \cdots & 0 \end{pmatrix} + \cdots + \begin{pmatrix} 0 & 1 & 2 & \cdots & k \\ 0 & a_1^{n+m} & 0 & \cdots & 0 \end{pmatrix}$$

$$= \begin{pmatrix} 0 & 1 & 2 & \cdots & k \\ 0 & \sum_{i=0}^{m} a_1^{n+i} & 0 & \cdots & 0 \end{pmatrix} = 0.$$

Consequently, $\{f_1, \ldots, f_{k+1}\}$ is linearly right dependent. Thus $n_r(M_0(Z_{k+1})) \leq k$, hence $n_r(M_0(Z_{k+1})) = k$, and so $M_0(Z_{k+1})$ is of type $(1, k)$. $\square$

Corollary 3.6. *There exists a near-ring, not a ring, of type $(0, k)$ for all positive integers k.*

Proof. From Example 2.5, we have a near-ring $Z_2(S)$ type $(0, \infty)$. From Theorem 3.2, $M_0(Z_{k+1}) \times Z_2(S)$ is of type (η_L, η_R) where $\eta_L = \min\{1, 0\}$ and $\eta_R = \min\{k, \infty\}$, i.e., $M_0(Z_{k+1}) \times Z_2(S)$ is of type $(0, k)$. $\square$

We next construct examples of non-rings of types (∞, ∞), $(\infty, 1)$ and $(\infty, 0)$. Let $A = \{id, -id\}$, a group of fixed point free automorphisms of the additive group $(\mathbb{Z}, +)$ of integers. We know $M_A(\mathbb{Z}) = \{f \in M_0(\mathbb{Z}) \mid f\sigma = \sigma f, \forall \sigma \in A\}$ is a zero-symmetric near-ring, with identity, which is not a ring. The near-ring $M_A(\mathbb{Z})$ is called the *centralizer near-ring* determined by the pair $(A, \mathbb{Z})$. To define a function f in $M_A(\mathbb{Z})$ it suffices to define f on $\mathbb{Z}^+ = \{x \in \mathbb{Z} \mid x > 0\}$ since $f \in M_A(\mathbb{Z})$ implies $f(-x) = -f(x)$, $x \in \mathbb{Z}$.

Theorem 3.7. *The centralizer near-ring $M_A(\mathbb{Z})$ is of type (∞, ∞).*

Proof. Suppose first that $M_A(\mathbb{Z})$ has finite left order. Then from Theorem 3.1, $n_\ell(M_A(\mathbb{Z})) = 1$. Define functions $f, g \in M_A(\mathbb{Z})$ as follows: for $x \in Z^+$

$$f(x) = \begin{cases} \frac{x+1}{2}, & \text{if } x \text{ is odd;} \\ \frac{x}{2}, & \text{if } x \text{ is even,} \end{cases} \quad \text{and} \quad g(x) = \begin{cases} \frac{x+1}{2}, & \text{if } x \text{ is odd;} \\ -\frac{x}{2}, & \text{if } x \text{ is even.} \end{cases}$$

Since f and g have a CLM, there are $y, z \in M_A(Z)$ such that $y \circ f = z \circ g$ where not both y and z are the zero function. Then, $y(f(x)) = z(g(x))$ for all $x \in Z$. For $x = 1$, we obtain $y(1) = z(1)$. For $x = 2$ we get $y(1) = z(-1) = -z(1)$. Thus, $y(1) = 0 = z(1)$. Continuing in this manner we find for each $x \in Z$, $y(x) = 0 = z(x)$, a contradiction. Hence, $M_A(Z)$ has infinite left order. Also, $n_r(M_A(Z)) \geq 1$ since $M_A(Z)$ has a multiplicative identity. Consider the functions $f, g \in M_A(Z)$ defined as follows: for $x \in Z^+$, $f(x) = 10$ and $g(x) = 10^2$. Suppose there are $y,z \in M_A(Z)$ such that $f \circ y + g \circ z = 0$. Then, for all $x \in Z$, $(f \circ y + g \circ z)(x) = 0$ or, $f(y(x)) + g(z(x)) = 0$. Since for every $x \neq 0$ in Z, $f(x) = \pm 10$ and $g(x) = \pm 10^2$ we get $y(x) = z(x) = 0$ for all $x \in Z$. Thus, $\{f, g\}$ is linearly right independent, hence $n_r(M_A(Z)) \geq 2$. Similarly, for any $n \in Z^+$ we define functions $f_1, \ldots, f_n \in M_A(Z)$ as follows: for $x \in Z^+$, $f_1(x) = 10, \ldots, f_n(x) = 10^n$. So, if $f_1 y_1 + \cdots + f_n y_n = 0$ where the y_i's are in $M_A(Z)$, then $f_1(y_1(x)) + \cdots + f_n(y_n(x)) = 0$ for all $x \in Z$. Hence, $y_1 = \cdots = y_n = 0$, so, $\{f_1, \ldots, f_n\}$ is linearly right independent, which implies that $n_r(M_A(Z)) = \infty$. □

From Example 2.5, $Z_2(S)$ is a near-ring of type $(0, \infty)$, from Example 2.6, $Z_2(S')$ is a near-ring of type $(\infty, 0)$ and from Example 2.7, $Z_2(G)$ is a near-ring of type $(\infty, 1)$. Applying Theorem 3.2 we obtain the following corollary.

Corollary 3.8. (i) *The non-ring $M_A(\mathbb{Z}) \times Z_2(S)$ is of type $(0, \infty)$.*
(ii) *The non-ring $M_A(\mathbb{Z}) \times Z_2(S')$ is of type $(\infty, 0)$.*
(iii) *The non-ring $M_A(\mathbb{Z}) \times Z_2(G)$ is of type $(\infty, 1)$.*

We conclude this section and the paper investigation by determining the left order of trivial near-rings. We recall from Pilz [3], that a near-ring $(N, +, *)$ is a *trivial near-ring* if $(N, +)$ is a group (for our purposes, abelian) and multiplication is defined as follows: for

$$F \subseteq N(0 \notin F), \; n_1 * n_2 = \begin{cases} n_1, & \text{if } n_2 \in F; \\ 0, & \text{if } n_2 \notin F. \end{cases}$$

Theorem 3.9. *If $(N, +, *)$ is a trivial near-ring, then $n_\ell(N) \leq 1$.*

Proof. Case 1: $F \neq \phi$. Let $a \in F \subseteq N$. Suppose there is $x \in N$ such that $x * a = 0$. Then, $x = 0$ by the definition of multiplication, so $\{a\}$ is linearly left independent. Thus, $n_\ell(N) \geq 1$. Let $a, b \in N$.

Case 1') $a, b \in F$. Then $(-a) * a + a * b = (-a) + a = 0$ so $\{a, b\}$ is linearly left dependent.
Case 1'') $a, b \notin F$. Choose x, y, not both zero, in N. Then, $x * a + y * b = 0 + 0 = 0$ so $\{a, b\}$ is linearly left dependent.
Case 1''') $a \in F, b \notin F$. Let $x = 0, 0 \neq y \in N$. Then, $x * a + y * b = x + 0 = x = 0$. So again $\{a, b\}$ is linearly left dependent.

Hence, we get $n_\ell(N) \leq 1$ and therefore $n_\ell(N) = 1$.

Case 2: $F = \phi$. Let $x \in N$. Choose $0 \neq a \in N$. We have that $a * x = 0$ so x is a right divisor of zero for all $x \in N$. Thus, $n_\ell(N) = 0$. □

As an application we consider the group of integers $\mathbb{Z}$, and let $F = Z - \{0\}$. Thus, the multiplication is $n_1 * n_2 = \begin{cases} n_1, & \text{if } n_2 \neq 0; \\ 0, & \text{if } n_2 = 0. \end{cases}$ For $x \in F$, $y * x = 0$ implies that $y = 0$. Thus, x is not a right divisor of zero so $n_\ell(Z, +, *) \neq 0$. Hence, $n_\ell(Z, +, *)$ must be 1. Also, $\forall n \in \mathbb{Z}^+$, $\{1, 2, \ldots, n\}$ is right independent so we have a non-ring of type $(1, \infty)$.

If we now let $F = \phi$, then as shown in the previous proof, $n_\ell(Z, +, *) = 0$. We also get $n_r(Z, +, *) = 0$ since every non-zero element is a left divisor of zero. Thus, we have constructed a non-ring of type $(0, 0)$.

In conclusion, we have constructed non-rings of type $(0,0)$, (∞,∞), $(\infty,1)$, $(\infty,0)$, $(0,\infty)$ and $(1,\infty)$ as well as $(1,k)$ and $(0,k)$ for all positive integers k. Our work has completed the finite case. It remains to construct (if possible) non-rings of type (∞,k).

REFERENCES

[1] Dickson, L.E., *Definitions of a group and a field by independent postulates*, Transactions Amer. Math. Soc., **6** (1905), 198–204.

[2] Ore, O., *Theory of non-commutative polynomials*, Ann of Math., **34** (1933), 480–508.

[3] Pilz, Günter, Near-rings, North Holland: Amsterdam, 1983.

[4] Tamari, Dov, *On a certain classification of rings and semigroups*, Bulletin of the Amer. Math. Soc., **54** (1948), 153-158.

2007 63rd ST., SE EVERETT, WA 98203, U.S.A. E-MAIL: **doberman@gte.net**

DEPARTMENT OF MATHEMATICS, TEXAS A & M UNIVERSITY, COLLEGE STATION,TX 77843, U.S.A.
E-MAIL: **cjmaxson@math.tamu.edu**

ON MODULES OF HOMOGENEOUS MAPPINGS

DOROTA NIEWIECZERZAŁ

ABSTRACT. Let R be a ring, G an R-module, $M(G)$ the left near-ring of all homogeneous mappings from G to G under usual addition and composition of mappings ([1]) and let $E(G)$ be the ring of all R-endomorphisms of G. Of course $E(G)$ is contained in $M(G)$ and, as many examples show, ([2], [3], [4], [5]), in general $M(G)$ is larger then $E(G)$.

In this paper we try "to measure a distance" between $M(G)$ and $E(G)$ under some additional assumptions on R and G.

In the paper a ring R be a commutative domain not beeing a field and G be a torsion-free module over R. Let next $rk(G)$ be a rank of an R-module G and for an element a from G let $(a)_*$ be the pure submodule of G generated by a. We can consider a near-ring $M(G)$ as a left module over $E(G)$. Assume that $M(G)$ is not equal to $E(G)$.

Theorem 1. *If G is of finite rank R-module and $M(G)$ is not equal $E(G)$, then $M(G)$ is not finitely generated as $E(G)$-module.*

Proof. Let $rk(G) = n$ and let K be a field of quotients of a domain R. Let next $x_1, \ldots, x_n$ be elements of G independent over R. So the map $(r_1, \ldots, r_n) \longrightarrow r_1 x_1 + \ldots + r_n x_n$ is an embedding of the R-module R^n into the R-module G. If a is an arbitrary element from G, then elements $a, x_1, \ldots, x_n$ are not independent and so there exist elements $r, r_1, \ldots, r_n$ in R such that $0 \neq ra = r_1 x_1 + \ldots + r_n x_n$. Now we can embed G into K^n (as an R-module) defining the image of an element a as $(r_1/r, \ldots, r_n/r)$. So up to isomorphism we have the following inclusions: $R^n \subseteq G \subseteq K^n$.

Moreover for any element $v \in K^n$ there exists a nonzero element $r \in R$ such that $rv \in G$.

Let $h \in E(G)$. Then for any element v from K^n there exists an element r from R such that $h(rv)$ belongs to G. Now we can define a mapping h' on K^n by $h'(v) = r^{-1}h(rv)$. It is easy to check that h' is a K- endomorphism of K^n which extends h. This way we receive an embedding of $E(G)$ into the matrix ring $M_n(K)$. Of course a multiplication of G by an element from R is an endomorphism of G and so R is contained in $E(G)$. By the assumption of this paper R has to be infinite. The cardinalities of R and K are the same. So a cardinalities of R (and so of R^n too) and $M_n(K)$ are also the same. But the ring $E(G)$ is contained between R and $M_n(K)$ and finally the cardinalities of $E(G)$ and R are the same.

As the last step of the proof we will show that the cardinality of a near-ring $M(G)$ is larger then the cardinality of R. Of course $n > 1$ because $M(G)$ is not equal $E(G)$ (comp.[1],[2]). So we can choose two independent elements, namely x and y. We consider the set $S = \{x + ry : r \in R\}$. It is easy to check that elements of the set S are pairwise

---

1991 *Mathematics Subject Classification.* 16Y30 .

Y. Fong et al. (eds.), *Near-Rings and Near-Fields*, 130–132.
© 2001 *Kluwer Academic Publishers. Printed in the Netherlands.*

independent. From this we have that S and R have the same cardinality. Next we take pure submodules $G_r = (x+ry)_*$ generated by elements of S. Since $G_r = \{g \in G \mid tg \in R(x+ry)$ for some $0 \neq t \in R\}$ and the elements of S are pairwise independent it follows that the family $\{G_r\}_{r \in R}$ and R have the same cardinalities.

Homogeneous mapping of G are defined independently on each submodule G_r. We can take a subset M_1 of $M(G)$ considering homogeneous mapping of G beeing on G_r a multiplication by some element a_r of R and beeing 0 on the rest of module G ([1],[2]). Then the cardinality of M_1 is the same as of the set of all function from R to R and so is larger then the cardinality of R. It follows of course that the cardinality of $M(G)$ is larger then the cardinality of $E(G)$ and so $M(G)$ can not be finitely generated as a module over $E(G)$. $\square$

Recall that an R-module G is completely anisotropic if it is torsion-free of rank at least two and for any $x, y \in G$ we have $Hom((x)_*, (y)_*) = 0$ whenever $(x)_* \neq (y)_*$. From [2,3,5] we know that there exist completely anisotropic modules over some domains of arbitrary rank n, when $2 \leq n \leq R$. So we can formulate

Theorem 2. *Let G be a completely anisotropic module over R. Then $M(G)$ is not a finitely generated module over $E(G)$. Moreover $E(G) \subseteq K$ and $M(G)$ is a torsion-free $E(G)$-module not beeing of finite rank.*

Proof. The idea of the proof is the same as of the previous one.

Firstly we will show that the cardinality of $E(G)$ is the same as the cardinality of R. If elements a and b from G have noncomparable types it follows that there is no nonzero endomorphism f of G such that $f(a) \notin (b)_*$. So a restriction of f to a pure submodule $(a)_*$ is an endomorphism of this submodule. The submodule $(a)_*$ has rank 1, so the restriction of f to this submodule is a multiplication by some element from R. Let the restriction of f to $(a)_*$ be a multiplication by r, the restriction of f to $(b)_*$ be a multiplication by s and the restriction of f to $(a+b)_*$ be a multiplication by w. We have: $w(a+b) = f(a+b) = f(a) + f(b) = ra + sb$ and by independence of a and b we receive that $w = r = s$. Therefore endomorphisms of G are multiplications by elements from K. In this way we have a natural embedding of $E(G)$ into K, hence the cardinalities of $E(G)$, K and R are the same. Using arguments as in the proof of the Theorem 1 we receive that the cardinality of $M(G)$ is larger then $E(G)$. So, of course, $M(G)$ can not be finitely generated over $E(G)$.

From the inclusion $E(G) \subseteq K$ we know that $E(G)$ is a domain. Now let $h \in M(G)$, $f \in E(G)$ be such that $f \neq 0$ and $fh = 0$. Then there exists a nonzero element $k \in K$ such that $f(x) = kx$ for any element $x \in G$. So $kh(x) = 0$ for any $x \in G$. Hence $h = 0$ because G is a torsion-free R-module. This means that $M(G)$ is a torsion-free module over $E(G)$ and from $|E(G)| < |M(G)|$ we have that $M(G)$ is not of finite rank over $E(G)$. $\square$

REFERENCES

[1] P. Fuchs, C. J. Maxson and G. Pilz, *On rings for which homogeneous maps are linear,* Proc. AMS **112** (1991), 1–7.

[2] J. Hausen, *Abelian groups whose semi-endomorphisms form a ring,* Abelian groups,L. Fuchs and R. Gö"below (eds.), Marcel Dekker, New York 1993, 175–180.

[3] J. Hausen, J. A. Johnson, *Centralizer near-rings that are rings*, J. Austral. Math. Soc. (series A) **59** (1995), 173–183.

[4] J. Krempa, *Some examples of indecomposable modules*, in: "Nearrings, Nearfields and K-Loops", G. Saad and M. J. Thomsen (eds), 1997 Kluwer Academic Publishers, 295–299.

[5] C. J. Maxson and A. P. J. van der Walt, *Centralizer near-rings over free ring modules*, J. Austral. Math. Soc. **50** (1991), 279–296.

INSTITUTE OF MATHEMATICS, WARSAW UNIVERSITY, UL. BANACHA 2, 02-097 WARSZAWA, POLAND.
E-MAIL: DOROTAN@MIMUW.EDU.PL

THE NUMBER OF ISOMORPHISM CLASSES OF D. G. NEAR-RINGS ON THE GENERALIZED QUATERNION GROUPS

CHRISTOF NÖBAUER

ABSTRACT. Let Q_n be the generalized quaternion group of order 2^n. In 1972, Malone ([Mal72], Theorem 7) determined that exactly 16 d.g. near-rings can be defined on Q_n and that all of these are in fact distributive. However, as Clay pointed out ([Cla74]), "nothing is said concerning the isomorphism of these 16."

We show in this note that there are exactly 10 non-isomorphic d.g. near-rings on Q_n for $n \geq 4$ and 6 if $n = 3$.

1. GENERALIZED QUATERNION GROUPS

For $n \geq 3$ let $Q_n := \langle \{a, b\} : 2^{n-1}a = 0, 2^{n-2}a = 2b, a + b = b - a \rangle$. Denote the elements of Q_n by $\{0, a, 2a, \ldots, (2^{n-1} - 1)a, b, a + b, 2a + b, \ldots, (2^{n-1} - 1)a + b\}$.

Denote by a' the element $2^{n-2}a$. Then in [Mal72], Theorem 7, it is shown that the d.g. near-rings on Q_n are exactly those with d.g. set $\{0, a, a', b\}$ and multiplication defined by

·	a	b
a	x	y
b	u	v

where $x, y, u, v \in \{0, a'\}$.

Moreover, it is shown that all of these are distributive.

1.1. An automorphism of $Q_n, n \geq 4$.

Lemma 1.1. *If $n \geq 4$ then the mapping $\varphi : Q_n \to Q_n$ which is defined by $a \to a$ and $b \to a + b$ is an automorphism of Q_n.*

Proof. A standard calculation shows that $\varphi(a)$ and $\varphi(b)$ satisfy the defining relations of Q_n; hence φ is a homomorphism $Q_n \to Q_n$. Since a and $a + b$ generate Q_n, φ is an automorphism. □

1.2. An automorphism of $Q_n, n = 3$.

Lemma 1.2. *The mapping $\rho : Q_3 \to Q_3$ which is defined by $a \to b$ and $b \to a$ is an automorphism of Q_3.*

Proof. Analogous to the argument above. □

Note, however, that swapping a and b will no longer be a homomorphism if $n > 3$.

Supported by the Austrian *Fonds zur Förderung der wissenschaftlichen Forschung*, Project P11486-TEC.

Y. Fong et al. (eds.), Near-Rings and Near-Fields, 133–137.
© 2001 Kluwer Academic Publishers. Printed in the Netherlands.

[3] J. Hausen, J. A. Johnson, *Centralizer near-rings that are rings*, J. Austral. Math. Soc. (series A) **59** (1995), 173–183.

[4] J. Krempa, *Some examples of indecomposable modules*, in: "Nearrings, Nearfields and K-Loops", G. Saad and M. J. Thomsen (eds), 1997 Kluwer Academic Publishers, 295–299.

[5] C. J. Maxson and A. P. J. van der Walt, *Centralizer near-rings over free ring modules*, J. Austral. Math. Soc. **50** (1991), 279–296.

INSTITUTE OF MATHEMATICS, WARSAW UNIVERSITY, UL. BANACHA 2, 02-097 WARSZAWA, POLAND.
E-MAIL: DOROTAN@MIMUW.EDU.PL

near-rings in separate classes.

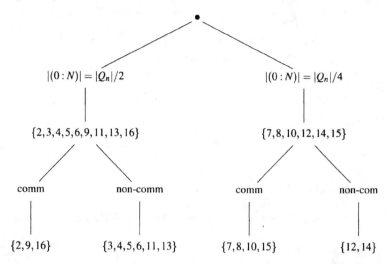

Lemma 3.1. *Near-ring multiplications* 12 *and* 14 *are isomorphic. Hence we have* $C_2 = \{12, 14\}$.

Proof. All we need to show is that group automorphism φ from lemma 1.1 is in fact a near-ring homomorphism. By [Wil70], Theorem 2, it suffices to show the homomorphism property for a and b only. But this is easy:

$$a' = \varphi(a \cdot_{12} a) = \varphi(a) \cdot_{14} \varphi(a) = a \cdot_{14} a = a',$$
$$a' = \varphi(a \cdot_{12} b) = \varphi(a) \cdot_{14} \varphi(b) = a \cdot_{14} (a+b) = a',$$
$$0 = \varphi(b \cdot_{12} a) = \varphi(b) \cdot_{14} \varphi(a) = (b+a) \cdot_{14} a = 0,$$
$$a' = \varphi(b \cdot_{12} b) = \varphi(b) \cdot_{14} \varphi(b) = (a+b) \cdot_{14} (a+b) = a'.$$

$\square$

Lemma 3.2. *Near-rings* 9 *and* 16 *are isomorphic. Near-ring* 2 *is not isomorphic to near-ring* 9; *hence we have* $C_3 = \{2\}$ *and* $C_4 = \{9, 16\}$.

Proof. The first part is shown analogously to the proof of lemma 3.1 by using the same group isomorphism as there.

Now, if near-rings 2 and 9 were isomorphic, we had for some group automorphism ψ:

$$0 = \psi(0) = \psi(a \cdot_2 a) = \psi(a) \cdot_9 \psi(a) = ua \cdot_9 ua$$

for some odd u. But by distributivity,

$$ua \cdot_9 ua = u^2 a' = a',$$

a contradiction.

$\square$

Lemma 3.3. *Near-rings* 10 *and* 15 *are isomorphic. Near-ring* 7 *is not isomorphic to* 10 *and* 8 *and near-ring* 8 *is not isomorphic to* 10, *hence* $C_5 = \{10,15\}, C_6 = \{7\}, C_7 = \{8\}$.

Proof. The isomorphism of 10 and 15 is shown analogously to lemma 3.1 by using the same group isomorphism as there and standard calculation.

Since in near-ring 7 there is no element x with $x^2 \neq 0$, it cannot be isomorphic to 10 or 8.

Near-rings 8 and 10 cannot be isomorphic by the same argument as in the proof of lemma 3.2. ☐

Lemma 3.4. *The remaining near-rings* 3,4,5,6,11,13 *establish the classes* $C_8 = \{3,4\}$, $C_9 = \{5,6\}$, *and* $C_{10} = \{11,13\}$.

Proof. Isomorphism can be shown analogously to lemma 3.1 by using the same group isomorphism as there and standard calculation.

Near-ring 11 cannot be isomorphic to either 3 or 5 by the same argument as in the proof of lemma 3.2.

All that remains to show is that 3 and 5 are non-isomorphic. To this end, suppose there were an isomorphism ψ from 5 onto 3. Now collect the facts:

- $\psi(a \cdot_5 b) = \psi(0) = 0$,
- $\psi(a) = ua$ for some odd $u < 2^{n-1}$ since ψ is a group isomorphism,
- $\psi(b) = ka + b$ for some $0 \leq k < 2^{n-1}$ since ψ is a group isomorphism.

But then, by distributivity,

$$\psi(a \cdot_5 b) = \psi(a) \cdot_3 \psi(b) = ua \cdot_3 (ka + b) = u(a \cdot_3 ka) + u(a \cdot_3 b) =$$

$$u(k(a \cdot_3 a)) + ua' = u0 + ua' \neq 0,$$

a contradiction. ☐

By collecting all the lemmas above, we get

Proposition 3.5. *The* 16 *d.g. near-ring multiplications definable on the generalized quaternion groups* Q_n *for* $n \geq 4$ *are partitioned into* 10 *isomorphism classes.*

4. CLASSES OF D.G. NEAR-RINGS ON Q_n, $n = 3$

In the case of the quaternion group Q of order 8, both, a and b have order 4. Therefore in this case there exists an automorphism of Q which swaps a and b. This is the reason why some of the isomorphism classes collapse.

In this section, we consider the quaternion group Q of order 8 only.

Lemma 4.1. *Near-rings* 2 *and* 9 *are isomorphic. Therefore,* $C_{\{3,4\}} = \{2, 9, 16\}$.

Proof. We show that group isomorphism ρ from lemma 1.2 is in fact a near-ring isomorphism. By [Wil70], Theorem 2, it suffices to show the homomorphism property for a and

b only. (Note that $\rho(a') = \rho(2a) = \rho(2b) = 2a = a'$):

$$0 = \rho(a \cdot_2 a) = \rho(a) \cdot_9 \rho(a) = b \cdot_9 b = 0,$$
$$0 = \rho(a \cdot_2 b) = \rho(a) \cdot_9 \rho(b) = b \cdot_9 a = 0,$$
$$0 = \rho(b \cdot_2 a) = \rho(b) \cdot_9 \rho(a) = a \cdot_9 b = 0,$$
$$a' = \rho(b \cdot_2 b) = \rho(b) \cdot_9 \rho(b) = a \cdot_9 a = a'.$$

$\square$

Lemma 4.2. *Near-rings* 3 *and* 5 *are isomorphic by* ρ *and near-rings* 6 *and* 11 *are isomorphic by* ρ. *Hence* $C_{\{8,9,10\}} = \{3,4,5,6,11,13\}$.

Proof. The proof is analogous to the proof of lemma 4.1. $\square$

Lemma 4.3. *Near-rings* 8 *and* 15 *are isomorphic. Therefore* $C_{\{5,7\}} = \{8,10,15\}$.

Proof. The proof is analogous to the proof of lemma 4.1. $\square$

Class C_6 is unique by the fact that for all x, $x^2 = 0$. No more classes can collapse by considering the size of $(0 : N)$ and commutativity (resp. non-commutativity), therefore we have:

Proposition 4.4. *The number of non-isomorphic d.g. near-rings definable on the quaternion group* Q *is* 6.

REFERENCES

[Cla74] James R. Clay. Review nr. 5059. In *Mathematical Reviews*, volume 47, page 885, 3 - 6 (1974).

[Mal72] J. J. Malone. Generalised quaternion groups and distributively generated near-rings. *Proc. Edinb. Math. Soc.* **18** (1972), 235–238.

[Wil70] M. L. Willhite. Distributively generated near rings on the dihedral group of order eight. Master's thesis, Texas A&M University, December 1970.

INSTITUT FÜR ALGEBRA, STOCHASTIK UND WISSENSBASIERTE MATHEMATISCHE SYSTEME, JOHANNES KEPLER UNIVERSITÄT LINZ, ALTENBERGERSTR. 69, A-4040 LINZ.

E-MAIL: `christof.noebauer@algebra.uni-linz.ac.at`

WHEN IS A CENTRALIZER NEAR-RING ISOMORPHIC TO A MATRIX NEAR-RING? PART 2

ALAN OSWALD, KIRBY C. SMITH, AND LEON VAN WYK

ABSTRACT. Let G be a finite group and A a group of automorphisms of G. It is always the case that for every integer $n \geq 2$ the matrix near-ring $\mathbb{M}_n(M_A(G); G)$ is a subnear-ring of the centralizer near-ring $M_A(G^n)$. We find conditions such that $\mathbb{M}_n(M_A(G); G)$ is a proper subset of $M_A(G^n)$. Assuming both A and G are abelian we find conditions under which $\mathbb{M}_n(M_A(G); G)$ equals $M_A(G^n)$.

1. INTRODUCTION

This work is a continuation of that found in [7]. We will use the notation adopted there.

Let G be a finite group and A a group of automorphisms of G. Using A and G we form the centralizer near-ring

$$M_A(G) = \{r \colon G \to G \mid r(0) = 0 \text{ and } r(\alpha v) = \alpha r(v), \quad \forall \alpha \in A, \forall v \in G\}.$$

The group G forms a module over the near-ring $M_A(G)$ using the operation $r \cdot v = r(v)$ where $r \in M_A(G)$ and $v \in G$. Using $M_A(G)$ and G we form the (generalized) n by n matrix near-ring

$$\mathbb{M}_n(M_A(G); G)$$

which is the subnear-ring of $M_0(G^n)$ generated by the elementary matrices f_{ij}^r where $1 \leq i, j \leq n, r \in M_A(G)$. We recall that the matrix f_{ij}^r is the function $f_{ij}^r \colon G^n \to G^n$ defined by

$$f_{ij}^r(v_1, \ldots, v_n) = (0, \ldots, 0, r(v_j), 0, \ldots, 0)$$

where $r(v_j)$ occupies the ith position.

The automorphism group A may be viewed as a group of automorphisms of G^n as follows: if $\alpha \in A$ and $\vec{v} = (v_1, \ldots, v_n) \in G^n$ then

$$\alpha(\vec{v}) = \alpha(v_1, \ldots, v_n) = (\alpha(v_1), \ldots, \alpha(v_n)).$$

Using A as a group of automorphisms of G^n we have the centralizer near-ring $M_A(G^n)$ and it is easy to verify that

$$\mathbb{M}_n(M_A(G); G) \subseteq M_A(G^n).$$

It is the goal of this work to investigate when we have equality. In particular we seek conditions on A, G and/or $M_A(G)$ such that if $n \geq 2$ is an integer then $\mathbb{M}_n(M_A(G); G) = M_A(G^n)$.

1991 *Mathematics Subject Classification.* 16Y30 .

Y. Fong et al. (eds.), *Near-Rings and Near-Fields*, 138–150.
© 2001 *Kluwer Academic Publishers. Printed in the Netherlands.*

It is instructive to put the above problem in the context of the results in [7]. The following was proven:

Theorem 1.1 ([7, Theorem 2.1]). *Suppose G is a cyclic $M_A(G)$-module. Suppose further that $M_A(G)$ is isomorphic to an $n \times n$ matrix near-ring $\mathbb{M}_n(R;R)$, for some $n \geq 2$, over a near-ring R using R as an R-module. Then there is a subgroup H of G such that $G \cong H^n$, H is A-invariant and A acts on G componentwise. Moreover the near-ring R is isomorphic to the centralizer near-ring $M_A(H)$.*

Our work in this paper addresses the converse of Theorem 1.1.

> Given $M_A(G^n)$ where A acts on G^n componentwise, when is $M_A(G^n)$ the matrix near-ring $\mathbb{M}_n(M_A(G);G)$?

In the next section we find necessary conditions for the equality $\mathbb{M}_n(M_A(G);G) = M_A(G^n)$. In the final section we assume both A and G are abelian and show that when mild conditions are put on A and G, the necessary conditions found in Section 2 are also sufficient. Examples are given in both sections.

2. CONDITIONS FOR PROPER CONTAINMENT

Henceforth G denotes a finite group and A a subgroup of $\text{Aut}(G)$. In this section we present two situations where $\mathbb{M}_n(M_A(G);G)$ is a proper subset of $M_A(G^n)$ for all integers $n \geq 2$.

The action of A on G partitions G into orbits (or A-orbits). If $a \in G$ we denote the orbit of a under A by $\theta(a)$. The stabilizer of $a \in G$ is the set $\text{stab}(a) := \{\alpha \in A \mid \alpha(a) = a\}$, a subgroup of A.

An ordering "$<$" may be put on the set of all A-orbits of G as follows: $\theta(a) < \theta(b)$ if there exists $a' \in \theta(a)$, $b' \in \theta(b)$ with $\text{stab}(a') \supset \text{stab}(b')$. By convention the zero orbit $\{0\}$ is less than every nonzero orbit $\theta(a)$. An orbit $\theta(a)$ is a *minimal orbit* if it is nonzero and not greater than any nonzero A-orbit. An orbit $\theta(a)$ is a *maximal orbit* if it is nonzero and it is not less than any other orbit.

Two nonzero orbits $\theta(a)$, $\theta(b)$ are *equivalent*, or of the *same type*, if there is an $a' \in \theta(a)$, $b' \in \theta(b)$ with $\text{stab}(a') = \text{stab}(b')$. That $\theta(a)$, $\theta(b)$ are equivalent will be denoted by $\theta(a) \sim \theta(b)$. We make the convention that the zero orbit $\{0\}$ is equivalent only to itself. It is easy to see that "$\sim$" is an equivalence relation on the set of A-orbits of G, thus giving rise to equivalence classes of A-orbits. Two orbits are in the same class if they are of the same type.

For $n \geq 2$ we form G^n, the direct sum of n copies of G. The automorphism group A of G becomes an automorphism group of G^n where the action is componentwise. For $a \in G$ let $\bar{a} = (a,0,\ldots,0)$, an element of G^n. We will denote by E_a the function in $M_A(G^n)$ which is the identity on $\theta(\bar{a})$, the A-orbit of $\bar{a}$, and zero on all other A-orbits of G^n.

The following gives our first situation for $\mathbb{M}_n(M_A(G);G) \subset M_A(G^n)$.

Theorem 2.1. *Let $N = M_A(G)$, and let H be a nonzero N-subgroup of G. Suppose K is a proper, nonzero submodule of H. If $h \in H \backslash K$ then $E_h \in M_A(G^n)$, $n \geq 2$, is not a matrix. So $\mathbb{M}_n(M_A(G);G)$ is a proper subset of $M_A(G^n)$ for all $n \geq 2$.*

Proof. For K to be a submodule of H we mean precisely that K is a normal subgroup of H and $r(h+k) - r(h)$ belongs to K for all $h \in H$, $k \in K$ and $r \in M_A(G)$. (Our use of the term "submodule" is not standard. In Pilz [4] it is called an ideal.) Since H is a module over $N = M_A(G)$, then H^n is a module over both $M_A(G^n)$ and $\mathbb{M}_n(M_A(G); G)$. Moreover one verifies that since K is a submodule of H then K^n is a $\mathbb{M}_n(M_A(G); G)$-submodule of H^n.

Select h in $H \backslash K$ and let k be nonzero in K. Then $\vec{k} = (0, k, 0, \ldots, 0)$ belongs to K^n and $\vec{h} = (h, 0, 0, \ldots, 0)$ belongs to $H^n \backslash K^n$. Note that $\vec{h} + \vec{k} = (h, k, 0, \ldots, 0)$ and $\vec{h}$ belong to different A-orbits. So $E_h \in M_A(G^n)$ has the property that

$$E_h(\vec{h} + \vec{k}) - E_h(\vec{h}) = \vec{0} - \vec{h} = -\vec{h}$$

is not in K^n. This shows K^n is not an $M_A(G^n)$-submodule of H^n and in particular E_h is not a matrix. So $\mathbb{M}_n(M_A(G); G)$ is a proper subset of $M_A(G^n)$. $\square$

To get a feeling of how restrictive the conditions of Theorem 2.1 are, we discuss the creation of N-subgroups H of G and N-submodules of H.

It is easy to create N-subgroups of G. For if a is in G then

$$H_a := \{r(a) \mid r \in M_A(G)\}$$

is an N-subgroup of G.

An N-subgroup H of G need not be A-invariant, i.e., H need not be a union of A-orbits of G. What is true is that if a is in H and α is in A such that $\mathrm{stab}(\alpha a) = \mathrm{stab}(a)$ then αa is in H. So H is guaranteed to contain those elements in the orbit of a having the same stabilizer as a. If A is abelian, or more generally if every stabilizer subgroup of A is a normal subgroup, then every N-subgroup of G is A-invariant.

Now let H be a nonzero N-subgroup of G and let K be a proper, nonzero submodule of H. Let a be in $H \backslash K$. Then for every $k \in K$, $r \in M_A(G)$, we must have $r(a+k) - r(a)$ be an element of K. This implies that $a + k$ must be in the orbit of a for every $k \in K$. So in order for K to be a submodule of H it is necessary that for every a in $H \backslash K$, the set $\theta(a) \cap H$ is a union of cosets of K in H. In particular if H is A-invariant then every A-orbit in $H \backslash K$ is a union of cosets of K in H.

Example 2.5 of [7] gives a situation where G has a proper, nonzero submodule H and so for every integer $n \geq 2$, $\mathbb{M}_n(M_A(G); G)$ is a proper subset of $M_A(G^n)$.

We now have our second situation for proper containment.

Theorem 2.2. *Suppose $M_A(G)$ has a ring (with identity) as a homomorphic image. Let L be any faithful (unital) $M_A(G)$-module. Then if $n \geq 2$, $M_A(G^n)$ is not isomorphic to the $n \times n$ generalized matrix near-ring $\mathbb{M}_n(M_A(G); L)$. In particular $\mathbb{M}_n(M_A(G); G)$ is a proper subset of $M_A(G^n)$.*

Proof. Since $M_A(G)$ has a ring as a homomorphic image, then there is an ideal I of $M_A(G)$ such that $M_A(G)/I$ is a ring. Moreover $M_A(G)/I$ is a direct sum of fields ([6]). So without loss of generality we may assume $M_A(G)/I$ is a field.

Let $\theta(a_1), \ldots, \theta(a_k)$ be the nonzero A-orbits of G. For $i = 1, \ldots, k$ let $e_i \in M_A(G)$ be the idempotent corresponding to the ith orbit $\theta(a_i)$. So $e_i(a_i) = a_i$ and e_i is zero off $\theta(a_i)$. We have $1 = e_1 + \cdots + e_k$. Since $I \neq M_A(G)$ then there is at least one index i such that $e_i \notin I$. We may assume $e_1 \notin I$. Let $e = e_1$. But $M_A(G)/I$ is a field and $e = e_1, e_2, \ldots, e_k$ are mutually orthogonal idempotents. This implies that $e_i \in I$ for $i = 2, \ldots, k$, for if $e + I = e_i + I$, then $-e_i + e \in I$, so $(-e_i + e)e = e \in I$, a contradiction.

Since L is a faithful module then there is an m in L such that $e(m) \neq 0$. Let $H = \{r(em) \mid r \in M_A(G)\}$ and $K = \{r(em) \mid r \in I\}$. Then H is a nonzero N-subgroup of L and K is an N-subgroup of H. We claim that K is not all of H. To justify this it suffices to show that em in H is not in K. If em were in K, then there is a function $s \in I$ such that $em = sem$. Then $ese(em) = em \neq 0$, and so ese is a nonzero element of I. As an element of $M_A(G)$, ese takes the orbit $\theta(a_1)$ to itself and all other orbits to 0. Hence there is a function $t \in M_A(G)$ such that $t(ese) = e$. This means e is in I which is a contradiction. We conclude that $em = sem$ is not possible for $s \in I$ and hence $em \notin K$.

We show now that K is a normal subgroup of H. Let h be in H and k in K. There is an r in $M_A(G)$ with $r(em) = h$ and an $s \in I$ with $s(em) = k$. We have $h + k - h = r(em) + s(em) - r(em) = (r + s - r)(em)$, an element of K since $r + s - r$ is in I.

The nonzero factor group H/K is an $M_A(G)$-module under the operation $r(v + K) = r(v) + K$ where $r \in M_A(G)$ and $v \in H$. To show this operation is well defined suppose $v + K = w + K$. Then $-w + v$ belongs to K. So there is an $i \in I$ with $-w + v = i(em)$. We have $v = t(em)$ and $w = s(em)$ for some t, s in $M_A(G)$. This gives $t(em) = s(em) + i(em)$. Since $M_A(G)/I$ is a ring, then $r(f + g) - rg - rf$ belongs to I for all r, f, g in $M_A(G)$. Therefore there is an $i_1 \in I$ such that $r(f + g) = i_1 + rf + rg$. In particular we have, for some $i_2 \in I$, $r(v) = r(t(em)) = r(s(em) + i(em)) = r(s + i)(em) = (i_2 + rs + ri)(em) = i_2(em) + r(s(em)) + r(i(em)) = i_2(em) + r(w) + r(i(em))$. Using the normality of K in H we have $r(v) = r(w) + k$ for some $k \in K$. This implies $r(v) + K = r(w) + K$.

We show now that under this operation r is an endomorphism of H/K. Let v, w be in H with $s(em) = v$ and $t(em) = w$, $s, t \in M_A(G)$. If $r \in M_A(G)$ then $r(v + K + w + K) = r(v + w + K) = r(s(em) + t(em) + K) = r(s(em) + t(em)) + K = r(s + t)(em) + K$. Since $M_A(G)/I$ is a ring then there is a $j \in I$ such that $r(s + t) = j + rs + rt$. So $r(v + K + w + K) = r(s + t)(em) + K = (j + rs + rt)(em) + K = j(em) + r(v) + r(w) + K = r(v + K) + r(w + K)$.

Finally we observe that since $(M_A(G)/I, +)$ is an abelian group then H/K is an abelian group.

We have now shown that every $r \in M_A(G)$ induces an endomorphism on the abelian group H/K. Due to this the elementary matrix f_{ij}^r in $\mathbb{M}_n(M_A(G); L)$ induces an endomorphism on the abelian group $(H/K)^n$. By definition the set of elementary matrices $\{f_{ij}^r \mid r \in M_A(G), 1 \leq i, j \leq n\}$ generates $\mathbb{M}_n(M_A(G); L)$, so every matrix in $\mathbb{M}_n(M_A(G); L)$ induces an endomorphism on $(H/K)^n$. The map $\phi \colon \mathbb{M}_n(M_A(G); L) \to \text{End}(H/K)^n$ defined by $\phi(B) =$ the endomorphism of $(H/K)^n$ induced by B is a nontrivial near-ring homomorphism. Let J be the kernel of ϕ then $\mathbb{M}_n(M_A(G); L)/J$ is a ring. This shows that $\mathbb{M}_n(M_A(G); L)$ has a ring as a homomorphic image.

We now show that this ring is isomorphic to the complete matrix ring $\mathbb{M}_n(F)$, where F is the field $M_A(G)/I$. The module action of $M_A(G)$ on H/K induces a module action of the field $F = M_A(G)/I$ on H/K, namely if $\bar{r}$ is in $M_A(G)/I$, then $\bar{r}(v + K) = r(v) + K$. To show that this is well defined, assume $\bar{r} = \bar{s}$. Then $r - s$ is in I and so $(r - s)(v)$ is in K.

This means $r(v) + K = s(v) + K$, or $\bar{r}(v + K) = \bar{s}(v + K)$. We have now made H/K into a vector space over $F = M_A(G)/I$.

The endomorphism on $(H/K)^n$ induced by the elementary matrix f_{ij}^r is precisely the endomorphism $f_{ij}^{\bar{r}}$ on $(H/K)^n$, where $f_{ij}^{\bar{r}}$ is the usual elementary matrix with $\bar{r}$ in the (i, j)th position and zeroes elsewhere. Since the homomorphic image of ϕ is generated by $\{\phi(f_{ij}^r) \mid r \in M_A(G), 1 \le i, j \le n\}$ and $\phi(f_{ij}^r) = f_{ij}^{\bar{r}}$, then the image of ϕ is the complete matrix ring $\mathbb{M}_n(F)$.

To complete our proof it is enough to show that $M_A(G^n)$ does not have $\mathbb{M}_n(F)$ as a homomorphic image. But this is clear, for $n > 1$ means $\mathbb{M}_n(F)$ is not a direct sum of fields, and the only rings that can be a homomorphic image of a centralizer near-ring are direct sums of fields ([6]). $\square$

We note that in the proof of Theorem 2.2 in order for $M_A(G)$ to have a ring as a homomorphic image G must have an orbit $\theta(v)$ whose type is unique.

In Example 2.4 of [7], G has an orbit $\{3\}$ whose type is unique and further investigation of this example will show that $M_A(G)$ has the field $GF(2)$ as a homomorphic image.

Corollary 2.3. *Suppose I is an ideal of $M_A(G)$ such that $M_A(G)/I$ is a field. Let $e \in M_A(G)$ be the idempotent, with associated orbit $\theta(a)$, such that $e \notin I$. Then for $n \ge 2$, $E_a \in M_A(G^n)$ is not a matrix.*

Proof. The proof of Theorem 2.2 shows there is one and only one orbit $\theta(a)$ whose associated idempotent e is not in I. Then $H = H_a = \{r(a) \mid r \in M_A(G)\}$ and $K = \{r(a) \mid r \in I\} = \{b \mid \text{stab}(b) \supset \text{stab}(a)\} \cup \{0\}$. In the homomorphism $\phi: \mathbb{M}_n(M_A(G); G) \to \text{End}(H/K)^n$ we must have $\phi(f_{11}^e) \ne 0$. We have $f_{11}^e E_a = f_{11}^e$ and if E_a were a matrix then $\phi(f_{11}^e) = \phi(f_{11}^e E_a) = \phi(f_{11}^e)\phi(E_a)$ implies $\phi(E_a) \ne 0$. So E_a is not in I and $\phi(E_a)$ acts nontrivially on $(H/K)^n$. But, as in the proof of Theorem 2.2, $\phi(E_a)$ is the identity on the set of elements of the form $(\alpha a + K, 0, \dots, 0)$ and zero elsewhere. But this is not a linear mapping. So E_a is not a matrix. $\square$

Corollary 2.4. *Let G be the symmetric group S_k where $k \ge 2$. Let A be the group of inner automorphisms of S_n. Then for $n \ge 2$, $\mathbb{M}_n(M_A(G); G)$ is a proper subset of $M_A(G^n)$.*

Proof. It is shown in [6] that $M_A(G)$ always has a ring as a homomorphic image. $\square$

Corollary 2.5. *Let G be the cyclic group $\mathbb{Z}_k$ of order k where $k \ge 2$. Let $A = \text{Aut}(G)$, the complete automorphism group of G. Then for $n \ge 2$, $\mathbb{M}_n(M_A(G); G)$ is a proper subset of $M_A(G^n)$.*

Proof. The near-ring $M_A(G)$ has a field as a homomorphic image as shown following the proof of Proposition 3.6. $\square$

3. CONDITIONS FOR EQUALITY

Throughout this section we assume that $M_A(G)$ is a centralizer near-ring with the following properties:

Property I. $M_A(G)$ does not have a ring as a homomorphic image.

Property II. G has no $M_A(G)$-subgroups that have proper, nonzero submodules.
Property III. Both G and A are abelian.

Our immediate goal is to show that Properties I, II and III imply that $E_a \in M_A(G^n)$ is a matrix for all $\vec{a} = (a, 0, \ldots, 0)$ in G^n. (Note that Theorem 2.1 and Corollary 2.3 imply that Properties I and II are needed.)

Our proof that E_a is a matrix for all $\vec{a}$ will be by induction on m, the number of equivalence classes of A-orbits of G.

If $m = 1$, then $\{0\}$ is the only orbit type. This means $G = \{0\}$ and our result is trivially true.

If $m = 2$ there are two types of A-orbits, namely $\{0\}$ and all nonzero A-orbits. Any two nonzero A-orbits are equivalent, so $M_A(G)$ is simple [1]. Since $M_A(G)$ does not have a ring as a homomorphic image, then $M_A(G)$ is a simple non-ring. Since A is abelian then A is fixed point free [1]. It is known in this case that $M_A(G^n) = \mathbb{M}_n(M_A(G); G)$ [7, Theorem 3.2].

Assume that E_a is a matrix when G has k or fewer equivalence classes of A-orbits. Let $M_A(G)$ be a centralizer near-ring (satisfying Properties I, II and III) such that G has $k+1$ different orbit types.

For $a \in G$ let $H_a = \{r(a) \mid r \in M_A(G)\}$. Then H_a is an N-subgroup of G and $H_a = \{b \in G \mid \text{stab}(b) \supseteq \text{stab}(a)\}$. Since A is abelian then H_a is a union of A-orbits, i.e. H_a is A-invariant.

Select $a \neq 0$ in G such that $H_a \neq G$. Then there is at least one equivalence class of orbits having empty intersection with H_a. Let $A' = A/\text{stab}(a)$. Then A' is an automorphism group on H_a and the number of equivalence classes of A'-orbits of H_a is less than $k+1$.

Define $\phi\colon M_A(G) \to M_{A'}(H_a)$ as follows: if $r \in M_A(G)$ then $\phi(r) = r|_{H_a}$, that is $\phi(r)$ is the function r restricted to H_a. It is easily verified that $\phi(r) \in M_{A'}(H_a)$. Moreover ϕ is a near-ring homomorphism of $M_A(G)$ onto $M_{A'}(H_a)$ with kernel $\{r \mid r(H_a) = \{0\}\}$.

Since $M_A(G)$ does not have a ring as a homomorphic image, neither does $M_{A'}(H_a)$. Suppose H_a were to have an N-submodule H which has a proper, nonzero submodule K. Then one verifies that H is an N-subgroup of $M_A(G)$ and K is in $M_A(G)$-submodule. But this contradicts our assumption on $M_A(G)$. So $M_{A'}(H_a)$ satisfies Properties I, II and III. Our induction hypothesis implies $\overline{E}_a \in \mathbb{M}_n(M_{A'}(H_a); H_a)$ for $n \geq 2$.

Since $\overline{E}_a$ is a matrix then $\overline{E}_a$ is an expression in terms of elementary matrices $f_{ij}^{\overline{r}}$ where $\overline{r}$ is in $M_{A'}(H_a)$. For each $\overline{r}$ in this expression we extend $\overline{r}$ to r in $M_A(G)$ by defining $r(b) = \overline{r}(b)$ if $b \in H_a$ and $r(x) = 0$ if $x \notin H_a$. Using the f_{ij}^r's in place of the $f_{ij}^{\overline{r}}$'s creates a matrix ϕ in $\mathbb{M}_n(M_A(G); G)$. It is possible that ϕ equals E_a, but in any case we will use ϕ to create E_a as a matrix.

Since each r created above is zero on $G\backslash H_a$ then if $\vec{b} = (b_1, \ldots, b_n)$ is such that each b_i is in $G\backslash H_a$ then $\phi(\vec{b}) = \vec{0}$. Also $\phi(\vec{a}) = \vec{a}$ and if $\vec{b}$ is in H_a^n and not in $\theta(\vec{a})$ then $\phi(\vec{b}) = \vec{0}$.

We may assume the range of ϕ is $\theta(\vec{a}) \cup \{\vec{0}\}$. For if not, let $s \in M_A(G)$ be the idempotent associated with $\theta(a)$, then use $f_{11}^s \phi$.

Suppose $\vec{b} = (b_1, \ldots, b_n)$ where, say, $b_i \notin H_a$, and such that $\phi(\vec{b}) \neq \vec{0}$. Since $b_i \notin H_a$ then if $c \in H_a$ we have $c + b_i \notin H_a$. Let $t \in M_A(G)$ be the idempotent associated with $\theta(b_i)$, and let $\psi = f_{1i}^t$.

Consider the matrix $\Omega := \phi(\phi + \psi) - \phi\psi$ where ϕ and ψ are as above. We have Ω agreeing with E_a on H_a^n and if $\phi(\vec{x}) = \vec{0}$ then $\Omega(\vec{x}) = \vec{0}$. In addition

$$
\begin{aligned}
\Omega(\vec{b}) &= \phi(\phi(\vec{b}) + \psi(\vec{b})) - \phi(\psi(\vec{b})) \\
&= \phi(\alpha(\vec{a}) + (b_i, 0, \ldots, 0)) - \phi(b_i, 0, \ldots, 0) \\
&= \phi(\alpha(a) + b_i, 0, \ldots, 0) - \phi(b_i, 0, \ldots, 0) \\
&= \vec{0} - \vec{0} = \vec{0}
\end{aligned}
$$

since $\alpha(a) + b_i \notin H_a$ and $b_i \notin H_a$.

Now we have created a matrix Ω which agrees with E_a on H_a^n, annihilates every orbit that ϕ annihilates and annihilates at least one more orbit, namely $\theta(\vec{b})$. We may repeat the above process until we construct E_a.

The element a in G was chosen such that $H_a \neq G$. If G does not have a unique maximal orbit type, then every $a \in G$ has the above property and we are done.

So we may assume G has a unique maximal orbit type. Let $\theta(a)$ be a maximal orbit. Then $\theta(a)$ is equivalent to every other maximal orbit. The maximality of $\theta(a)$ means either $\theta(b) < \theta(a)$ or $\theta(b)$ is equivalent to $\theta(a)$ for every orbit $\theta(b)$. We must have $\operatorname{stab}(a) = \{1\}$. We show now that E_a is a matrix, where $\vec{a} = (a, 0, 0, \ldots, 0)$, $\operatorname{stab}(a) = \{1\}$, and $E_a \in \mathbb{M}_n(G^n)$ is the idempotent associated with the orbit $\theta(\vec{a})$.

Let $r \in M_A(G)$ be the idempotent associated with a. The matrix $\phi := f_{11}^r$ has the property that $\phi(\vec{a}) = \vec{a}$ and $\phi(x_1, x_2, \ldots, x_n) = \vec{0}$ if $x_1 \notin \theta(a)$. So the only A-orbits of G^n not annihilated by ϕ are those of the form $\theta(a, x_2, \ldots, x_n)$. Moreover the range of ϕ is $\theta(\vec{a}) \cup \{\vec{0}\}$. We now fix $(a, x_2, \ldots, x_n)$ where $(x_2, \ldots, x_n) \neq (0, \ldots, 0)$.

Consider matrices in $\mathbb{M}_n(M_A(G); G)$ of the form $\Omega = \phi(\phi + \psi) - \phi\psi$ where ψ is a matrix such that $\psi(a, x_2, \ldots, x_n) = (z, 0, \ldots, 0)$ for some $z \in G$ and $\psi(\vec{a}) = \vec{0}$. For any such ψ we have $\Omega(\vec{a}) = \vec{a}$ and if $\phi(\vec{b}) = \vec{0}$ then $\Omega(\vec{b}) = \vec{0}$. Our goal is to show that there is a matrix ψ such that $\Omega(a, x_2, \ldots, x_n) = \vec{0}$. Then Ω will agree with ϕ on $\vec{a}$ and on all orbits that ϕ annihilates and Ω will annihilate at least one more orbit. A continuation of this finite process will create E_a as a matrix.

For any ψ we have $\psi(a, x_2, \ldots, x_n) = (z, 0, \ldots, 0)$ for some $z \in G$ and

$$
\begin{aligned}
\Omega(a, x_2, \ldots, x_n) &= \phi(\phi(a, x_2, \ldots, x_n) + \psi(a, x_2, \ldots, x_n)) - \phi(\psi(a, x_2, \ldots, x_n)) \\
&= \phi((\alpha a, 0, \ldots, 0) + (z, 0, \ldots, 0)) - \phi(z, 0, \ldots, 0) \\
&= \phi(\alpha a + z, 0, \ldots, 0) - \phi(z, 0, \ldots, 0).
\end{aligned}
$$

We are done if there is a $z \in G$ such that $z \notin \theta(a)$ and $\alpha a + z \notin \theta(a)$.

Let $K = \{z \in G \mid \psi(a, x_2, \ldots, x_n) = (z, 0, \ldots, 0)$ for some matrix ψ with $\psi(a, 0, \ldots, 0) = \vec{0}\}$. The set K is easily seen to be a subgroup of G, K is not $\{0\}$ since $x_2, \ldots, x_n$ are in K using $\psi = f_{12}^1, f_{13}^1, \ldots, f_{1n}^1$, K is an $M_A(G)$-subgroup of G, and K is A-invariant since A is abelian.

Two cases can occur.

Case 1: Assume $\theta(a)$ is not a unique maximal orbit of G. Then a counting argument shows that there is a z in G such that $z \notin \theta(a)$ and $a + z \notin \theta(a)$, and we are done.

Case 2: Assume $\theta(a)$ is a unique maximal orbit. So $\theta(a) = \{b \in B \mid \text{stab}(b) = \{1\}\}$. Let L be the set of elements y in G such that $\theta(y) < \theta(a)$. Then $L = \{b \in G \mid \text{stab}(b) \neq \{1\}\}$, a nonzero subset of G since $\theta(a)$ is not a minimal orbit (by our induction hypothesis). If L contains an element z such that $a + z \notin \theta(a)$, then we are done. So we may assume $a + L \subseteq \theta(a)$.

We now show that L is a subgroup of G. Since G is finite it is enough to show that L is closed under addition. Suppose b, c are in L such that $b + c \notin L$. Then $b + c \in \theta(a)$ and there is an $\alpha \in A$ with $b + c = \alpha a$. So for every $\gamma \in A$ we have $\gamma b + \gamma c = \gamma \alpha a \in \theta(a)$. This means $a + \gamma \alpha a = a + \gamma b + \gamma c$ is in $\theta(a)$ since $a + \gamma b \in \theta(a)$ and $(a + \gamma b) + \gamma c$ is in $\theta(a)$. Because $G = L \cup \theta(a)$ and $a + L \subseteq \theta(a)$, the above shows $a + G \subseteq \theta(a)$, which is not possible. So L is a group. Since G is abelian L is normal in G.

We show now that in Case 2 the centralizer near-ring $M_A(G)$ has a field as a homomorphic image, which contradicts our assumptions on $M_A(G)$.

Let $I = \{r \in M_A(G) \mid r(G) \subseteq L\}$. We claim that I is an ideal of $M_A(G)$ and $M_A(G)/I$ is a field. Since $a + b$ is in $\theta(a)$ for every $b \in L$, then given $b \in L$ there corresponds a unique $\beta \in A$ with $a + b = \beta a$. Let $B = \{\beta \in A \mid a + b = \beta a$ for some b in $L\}$. It is easy to verify that B is a subgroup of A. This decomposes A into a disjoint union of the cosets of B in A, $A = \gamma_1 B \cup \gamma_2 B \cup \cdots \cup \gamma_m B$. Since $\text{stab}(a) = \{1\}$, an element x in $\theta(a)$ has a unique representation $x = \alpha(a)$ for some α in A. The element α belongs to one and only one coset of B in A, say $\alpha \in \gamma_i B$. So there is a unique β in B such that $\alpha = \gamma_i \beta$. We conclude that given an element x in $\theta(a)$ it may be uniquely represented in the form $x = \gamma_i \beta a$ for some $\beta \in B$. Note that two elements in $\theta(a)$ belong to the same coset of L if and only if they have the form $\gamma_i \beta a$ and $\gamma_i \beta' a$ respectively, where β and β' are in B.

Now let f and g be in $M_A(G)$ and $i \in I$. We need to show that $f(g + i) - fg$ belongs to I. To show this we need only show that $(f(g + i) - fg)(a)$ belongs to L. This is clearly the case if $g(a)$ is in L or if $f(a)$ is in L. So we may assume $f(a)$ and $g(a)$ are both in $\theta(a)$. ¿From the above we may write $f(a) = \gamma_i \beta a$, $g(a) = \gamma_j \beta' a$, uniquely, where $\beta, \beta' \in B$. We have $i(a) = k \in L$ and

$$(f(g + i) - fg)(a) = f(g(a) + i(a)) - f(g(a)) = f(\gamma_j \beta' a + k) - f(\gamma_j \beta' a).$$

Since $\gamma_j \beta' a + k$ and $\gamma_j \beta' a$ are in the same coset of L, then there is a $\beta'' \in B$ such that $\gamma_j \beta' a + k = \gamma_j \beta'' a$. We now have

$$
\begin{aligned}
(f(g + i) - fg)(a) = f(\gamma_j \beta'' a) - f(\gamma_j \beta' a) &= \gamma_j \beta'' f(a) - \gamma_j \beta' f(a) \\
&= \gamma_j \beta'' \gamma_i \beta a - \gamma_j \beta' \gamma_i \beta a \\
&= \gamma_j \gamma_i \beta'' \beta a - \gamma_j \gamma_i \beta' \beta a,
\end{aligned}
$$

an element of L since $\gamma_j \gamma_i \beta'' \beta$ and $\gamma_j \gamma_i \beta' \beta$ belong to the same coset of B in A.

This shows I is an ideal of $M_A(G)$. Since A is abelian, it is easy to verify that $M_A(G)/I$ is a field.

We have now shown that E_a is a matrix for every $\vec{a} = (a, 0, \ldots, 0)$ in G^n. We have proven the following theorem.

Theorem 3.1. *Let $M_A(G)$ satisfy Properties I, II and III. Then E_a is a matrix for every $\vec{a} = (a, 0, \ldots, 0)$ in G^n and all $n \geq 2$.*

We show now that with two additional properties on A and G we have

$$\mathbb{M}_n(M_A(G); G) = M_A(G^n) \quad \text{for all } n \geq 2.$$

The properties are as follows.

Property IV. If a, b are in G then there is a c in G with $\text{stab}(a) \cap \text{stab}(b) = \text{stab}(c)$.

Property V. Suppose $b_1, \ldots, b_n$ are in G with $\text{stab}(b_1) \cap \cdots \cap \text{stab}(b_n) = \text{stab}(a)$. Then there are functions $r_1, \ldots, r_n$ in $M_A(G)$ with at least one r_i invertible such that $\text{stab}(r_1(b_1) + \cdots + r_n(b_n)) = \text{stab}(a)$.

Note that Property IV says that the set of all stabilizer subgroups of A is closed under intersection. Property V is automatically true if $\text{stab}(b_i) = \text{stab}(a)$ for some i. The real need for Property V is the situation where $\text{stab}(b_1) \cap \cdots \cap \text{stab}(b_n) = \text{stab}(a)$ with $\text{stab}(b_i) \supset \text{stab}(a)$ for all i.

Lemma 3.2. *The following matrices in* $\mathbb{M}_n(M_A(G); G)$ *are invertible.*

(i) $Z = f_{11}^{r_1} + f_{21}^{r_2} + \cdots + f_{n1}^{r_n} + f_{22}^{1} + \cdots + f_{nn}^{1}$, *where* r_1 *is invertible.*

(ii) $Z = f_{11}^{r_1} + f_{12}^{r_2} + \cdots + f_{1n}^{r_n} + f_{22}^{1} + \cdots + f_{nn}^{1}$, *where* r_1 *is invertible.*

Proof. (i) We will show Z is 1-1 on G^n. Let $\vec{b} = (b_1, b_2, \ldots, b_n)$, $\vec{c} = (c_1, c_2, \ldots, c_n)$ be in G^n and suppose $Z(\vec{b}) = Z(\vec{c})$. Then

$$(r_1(b_1), r_2(b_1) + b_2, \ldots, r_n(b_1) + b_n) = (r_1(c_1), r_2(c_1) + c_2, \ldots, r_n(c_1) + c_n).$$

This means $r_1(b_1) = r_1(c_1)$ and since r_1 is invertible then $b_1 = c_1$. This implies $b_2 = c_2, \ldots, b_n = c_n$. So $\vec{b} = \vec{c}$ and Z is invertible because G^n is finite.

The proof of (ii) is done similarly. □

We note that if Z is an invertible matrix then Z^{-1} is also a matrix because our near-ring is finite.

For $\vec{x}, \vec{y}$ in G^n with $\text{stab}(\vec{x}) \subseteq \text{stab}(\vec{y})$ let E_{yx} denote that function in $M_A(G^n)$ which takes $\vec{x}$ to $\vec{y}$ and is zero off $\theta(\vec{x})$. To show every function in $M_A(G^n)$ is a matrix, it is enough to show that E_{yx} is a matrix for every such $\vec{x}$ and $\vec{y}$.

Let $\vec{x} = (x_1, \ldots, x_n)$ in G^n be nonzero. We have $\text{stab}\,\vec{x} = \text{stab}(x_1) \cap \cdots \cap \text{stab}(x_n)$. By Property IV there is an $a \in G$ such that if $\vec{a} = (a, 0, \ldots, 0)$ then $\text{stab}(\vec{x}) = \text{stab}(\vec{a})$. By Property V there are functions $r_1, \ldots, r_n$ in $M_A(G)$ with at least one r_i invertible such that $\text{stab}(r_1(x_1) + \cdots + r_n(x_n)) = \text{stab}(a)$. Without loss of generality we may assume r_1 is invertible. By Lemma 3.2 the matrix $X = f_{11}^{r_1} + f_{12}^{r_2} + \cdots + f_{1n}^{r_n} + f_{22}^{1} + \cdots + f_{nn}^{1}$ is invertible. Also, $\vec{y} := X(\vec{x}) = (r_1(x_1) + r_2(x_2) + \cdots + r_n(x_n), x_2, \ldots, x_n)$ where

$$\text{stab}(\vec{y}) = \text{stab}(\vec{a}) = \text{stab}(r_1(x_1) + r_2(x_2) + \cdots + r_n(x_n)).$$

Let $s_1, s_2, \ldots, s_n$ be in $M_A(G)$ defined by $s_1(a) = r_1(x_1) + \cdots + r_n(x_n)$, $s_2(a) = x_2, \ldots$, $s_n(a) := x_n$, and we select s_1 invertible. By Lemma 3.2 the matrix $Z = f_{11}^{s_1} + f_{21}^{s_2} + \cdots + f_{n1}^{s_n} + f_{22}^{1} + \cdots + f_{nn}^{1}$ is invertible. Also $Z(\vec{a}) = \vec{y}$. Note that $X^{-1}Z$ is an invertible matrix such that $X^{-1}Z(\vec{a}) = \vec{x}$, and therefore $Z^{-1}X(\vec{x}) = \vec{a}$. We know E_a is a matrix and hence $E_a Z^{-1} X = E_{ax}$ is a matrix. Also $X^{-1} Z E_a = E_{xa}$ is a matrix. More can be done. If $\vec{y} = (y_1, \ldots, y_n)$ and $\vec{a} = (a, 0, 0, \ldots, 0)$ are in G^n with $\text{stab}(\vec{a}) \subseteq \text{stab}(\vec{y})$ then E_{ya} is a matrix.

For since $\text{stab}(\bar{a}) \subseteq \text{stab}(\bar{y})$ then $\text{stab}(a) \subseteq \text{stab}(y_i)$ for every i. Let $s_i \in M_A(G)$ be such that $s_i(a) = y_i$. Then $E_{ya} = (f_{11}^{s_1} + f_{21}^{s_2} + \cdots + f_{n1}^{s_n})E_a$.

Given $\bar{x} = (x_1, \ldots, x_n)$, $\bar{y} = (y_1, \ldots, y_n)$ in G^n with $\text{stab}(\bar{x}) \subseteq \text{stab}(\bar{y})$ then E_{yx} is a matrix. For by Property V there is an $a \in G$ with $\text{stab}(a) = \text{stab}(\bar{x})$. We know that E_{ya} and E_{ax} are matrices. So $E_{ya}E_{ax} = E_{yx}$ is a matrix.

Theorem 3.3. *Let $M_A(G)$ be a centralizer near-ring satisfying Properties I–V. Then for all integers $n \geq 2$, $\mathbb{M}_n(M_A(G); G) = M_A(G^n)$.*

Corollary 3.4. *Suppose $M_A(G)$ satisfies Properties III, IV and V. Let H be an abelian group with $|H| \geq 3$. Let A act on $G \times H$ by $\alpha(a, h) = (\alpha(a), h)$, $\alpha \in A$. Then for $n \geq 2$, $\mathbb{M}_n(M_A(G \times H); G \times H) = M_A((G \times H)^n)$.*

Proof. It is clear that $M_A(G \times H)$ satisfies Properties III, IV and V. So it is enough to show that $M_A(G \times H)$ satisfies Properties I and II.

Suppose I is an ideal of $M_A(G \times H)$ such that $M_A(G \times H)/I$ is a field. Then, as we have seen, there is a unique orbit $\theta(a)$ and associated idempotent e such that $e + I \neq 0 + I$. The orbit $\theta(a, h)$ is in an equivalence class by itself. But $G \times H$ has no unique orbit types. For $\theta(a, h)$ is equivalent to $\theta(a, h')$ where $h' \in H$ with $h' \neq h$.

Suppose $\{0\} < K < G' \leq G \times H$ where K is a submodule of the N-subgroup G' of $G \times H$. Then K must contain every element of the form $(0, h)$, $h \in H$, and for every $f \in M_A(G \times H)$ we must have

$$f((a, h) + (0, h')) - f(a, h)$$

in K for every (a, h) in G' and $h' \in H$. If $h' \neq 0$ then $(a, h) + (0, h')$ and (a, h) are in different A-orbits. So choose $(a, h) \in G' \backslash K$ and let f be the idempotent associated with $(a, h + h')$. Then $f((a, h) + (0, h')) - f(a, h) = (a, h + h') \notin K$. □

Proposition 3.5. *Let $G = \mathbb{Z}_n$, the cyclic group of integers modulo n where $n \geq 2$, and let $A = \text{Aut}(G)$. Let $\{a_1, a_2, \ldots, a_r\}$ be the complete set of positive integer divisors of n, excluding n itself. Then*

(i) *the nonzero A-orbits of G are $\theta(\bar{a}_1), \theta(\bar{a}_2), \ldots, \theta(\bar{a}_r)$,*

(ii) *$\text{stab}(\bar{a}_i) \cap \text{stab}(\bar{a}_j) = \text{stab}(\bar{a}_k)$ where a_k is the greatest common divisor of a_i and a_j.*

Proof. For positive integers a and b let (a, b) and $[a, b]$ denote the greatest common divisor and the least common multiple of a and b respectively. Recall that the automorphisms of $\mathbb{Z}_n$ have the form $\bar{a} \to k\bar{a}$ where $(k, n) = 1$. We will denote such an automorphism by α_k.

(i) Given a nonzero $\bar{a}$ in $\mathbb{Z}_n$ we will show that $\theta(\bar{a})$ consists of those elements $\bar{b}$ in $\mathbb{Z}_n$ where $(b, n) = (a, n)$. This implies that $\theta(\bar{a}) = \theta(\bar{d})$ where $d = (a, n)$, a divisor of n.

If α_k is an automorphism of $\mathbb{Z}_n$ then $(k, n) = 1$ and $\alpha_k(\bar{a}) = k\bar{a} = \overline{ka}$, where $(ka, n) = (a, n) = d$. So if $\bar{b} \in \theta(\bar{a})$ then $(b, n) = (a, n) = d$.

We now show that if $\bar{b}$ is in $\mathbb{Z}_n$ with $(b, n) = (a, n)$ then there is an automorphism α of $\mathbb{Z}_n$ such that $\alpha(\bar{a}) = \bar{b}$. This is equivalent to finding an integer k with $(k, n) = 1$ such that $k\bar{a} = \bar{b}$. We want $ka \equiv b \pmod{n}$ or equivalently $k\frac{a}{d} \equiv \frac{b}{d} \left(\text{mod } \frac{n}{d}\right)$. Since $d = (a, n) = (b, n)$, then $1 = \left(\frac{a}{d}, \frac{n}{d}\right) = \left(\frac{b}{d}, \frac{n}{d}\right)$. This means both $\frac{a}{d}$ and $\frac{b}{d}$ are units in the ring of integers modulo $\frac{n}{d}$. Because of this there is an integer k' with $\left(k', \frac{n}{d}\right) = 1$ such that $k'\frac{a}{d} \equiv \frac{b}{d} \left(\text{mod } \frac{n}{d}\right)$. Any integer of the form $k' + c\frac{n}{d}$ is also a solution to $x\frac{a}{d} \equiv \frac{b}{d} \left(\text{mod } \frac{n}{d}\right)$, where c is any integer. We want to select c so that $k = k' + c\frac{n}{d}$ has the property that $(k, n) = 1$.

We will choose c as follows. Let $p_1, \ldots, p_l$ be all the prime divisors of n which do not divide $\frac{n}{d}$. (Such primes may not exist in which case we use $c = 0$.) Amongst $p_1, \ldots, p_l$ we distinguish between those p_i's, say $p_1, \ldots, p_{l'}$ (with $l' \leq l$), which divide k', and those p_i's, say $p_{l'+1}, \ldots, p_l$, which do not divide k'. Again, it may happen that all the p_i's divide k' or that none of the p_i's divide k'. If $l' = l$, i.e. if all the p_i's divide k', then choose $c = p_1 \cdots p_{l'} + 1$. Then, for $i = 1, \ldots, l' = l$, p_i does not divide c. If $l' < l$, which includes the possiblity where none of the p_i's divide k', then choose $c = p_{l'+1} \cdots p_l$. Then, for $i = 1, \ldots, l'$, p_i again does not divide c, and, for $i = l' + 1, \ldots, l$, p_i does divide c. We have thus chosen c such that the p_i's which divide k', do not divide c, and that the p_i's which do not divide k', do divide c.

Now, using the fact that $\left(k', \frac{n}{d}\right) = 1$ and our choice of c, one verifies that if p is a prime divisor of n which do not divide $\frac{n}{d}$, then p does not divide $k = k' + c\frac{n}{d}$. This means that $(k, n) = 1$. So $k\frac{a}{d} \equiv \frac{b}{d} \left(\text{mod } \frac{n}{d}\right)$ and hence $ka \equiv b \pmod{n}$ with $(k, n) = 1$. So there is an automorphism α of $\mathbb{Z}_n$ with $\alpha(\bar{a}) = \bar{b}$.

(ii) Let a and b be positive proper integer divisors of n. Then $\text{stab}(\bar{a}) = \{\alpha_k \in A \mid \alpha_k(\bar{a}) = \bar{a}\} = \{\alpha_k \mid ka \equiv a \pmod{n}\} = \{\alpha_k \mid n \text{ divides } (k-1)a\} = \{\alpha_k \mid \frac{n}{a} \text{ divides } k\}$. Likewise $\text{stab}(\bar{b}) = \{\alpha_k \mid \frac{n}{b} \text{ divides } k - 1\}$. We have $\text{stab}(\bar{a}) \cap \text{stab}(\bar{b}) = \{\alpha_k \mid \frac{n}{a} \text{ and } \frac{n}{b}$ divide $k - 1\} = \{\alpha_k \mid [\frac{n}{a}, \frac{n}{b}] \text{ divides } k - 1\} = \text{stab}(\bar{c})$ where $c = \frac{n}{[\frac{n}{a}, \frac{n}{b}]}$.

Using well known identities from elementary number theory we have $ab = (a, b)[a, b]$ and $d[a, b] = [da, db]$ for all positive integers d. So

$$nab = (a, b)n[a, b] = (a, b)[na, nb] = (a, b)ab \left[\frac{n}{b}, \frac{n}{a}\right].$$

This gives $n = (a, b)\left[\frac{n}{a}, \frac{n}{b}\right]$ and therefore $c = (a, b)$. So $\text{stab}(\bar{a}) \cap \text{stab}(\bar{b}) = \text{stab}(\bar{c})$ where $c = (a, b)$.   □

Proposition 3.6. *Let* $G = \mathbb{Z}_n$, $n \geq 2$, *and* $A = \text{Aut}(G)$. *Then* $M_A(G)$ *satisfies Properties III, IV, and V.*

Proof. That $A = \text{Aut}(G)$ is abelian is clear, so $M_A(G)$ satisfies Property III. That Property IV is true follows immediately from Proposition 3.5(ii).

For Property V let a and b be proper divisors of n which are relatively prime. Since $1 = (a, b)$, there are integers c and d such that $1 = ca + db$. We will show that there are integers y and z such that $ya + zb = 1$ and y is relatively prime to n. In order for y to be such that $ya + zb = 1$ for some integer z we must have $ya \equiv 1 \pmod{b}$. Since a is relatively prime to b then $\bar{a}$ has a multiplicative inverse $\bar{a}_1$ in the ring $\mathbb{Z}_b$ where a_1 is relatively prime to b. So $ya \equiv 1 \pmod{b}$ implies $y \equiv a_1 \pmod{b}$ with $1 = (b, a_1)$. By a classical result of Dirichlet on primes in arithmetic progressions ([3], pages 401–402 and the references on page 408) there are infinitely many primes p such that $p \equiv a_1 \pmod{b}$. Let y be such a prime that does not divide n. Then y is relatively prime to n and $ya \equiv 1 \pmod{b}$. Thus there exists an integer z such that $ya + zb = 1$.

Since y is relatively prime to n there is an automorphism α_k of G such that $\alpha_k(\bar{a}) = y\bar{a}$ and $y\bar{a}$ belongs to the orbit $\theta(\bar{a})$. Define $r \in M_A(G)$ by $r(\bar{a}) = y\bar{a}$ and r the identity on all other orbits. Then r is invertible. Define $s \in M_A(G)$ by $s(\bar{b}) = z\bar{b}$ and the identity on all other orbits. We have $\text{stab}(\bar{a}) \cap \text{stab}(\bar{b}) = \text{stab}(\bar{1})$ and $\text{stab}(r(\bar{a}) + s(\bar{b})) = \text{stab}(\bar{1})$ as required for Property V.

Now suppose $\bar{a}_1, \ldots, \bar{a}_m$ are in $\mathbb{Z}_n$ where each a_i is a positive proper divisor of n and such that $\text{stab}(\bar{a}_1) \cap \cdots \cap \text{stab}(\bar{a}_m) = \text{stab}(\bar{1})$. We seek $r_1, \ldots, r_m$ in $M_A(G)$ such that at least one r_i is invertible and $\text{stab}(r_1(\bar{a}_1) + \cdots + r_m(\bar{a}_m)) = \text{stab}(\bar{1})$.

We use induction on m. If $m = 1$, use r_1 to be the identity map. The case $m = 2$ has been established above.

The integers $a_1, \ldots, a_m$ are relatively prime, i.e. $1 = (a_1, \ldots, a_m)$. Let $d = (a_2, \ldots, a_m)$. If $d = 1$, then $\text{stab}(\bar{a}_2) \cap \cdots \cap \text{stab}(\bar{a}_m) = \text{stab}(\bar{1})$. By induction (letting $r_1 = 0$) we are done. So we may assume $d > 1$. Since $1 = (a_1, a_2, \ldots, a_m)$ there are integers $c_1, c_2, \ldots, c_m$ with $1 = c_1 a_1 + c_2 a_2 + \cdots + c_m a_m$ where $1 = (a_1, d)$ and $a_i \equiv 0 \pmod{d}$ for $i \geq 2$. We seek integers $y_1, y_2, \ldots, y_m$ such that y_1 is relatively prime to n and $1 = y_1 a_1 + y_2 a_2 + \cdots + y_m a_m$.

Toward this goal we note that $y_1 a_1 \equiv 1 \pmod{d}$ with $1 = (d, a_1)$. As above there is an integer b relatively prime to d with $y_1 \equiv b \pmod{d}$. By Dirichlet's theorem there is a prime p, p not dividing n, such that $p \equiv b \pmod{d}$. Let $y_1 = p$ and then $1 = y_1 a_1 + sd$ for some integer s. Now write d in terms of $a_2, \ldots, a_m$ to obtain an equation of the form $1 = y_1 a_1 + y_2 a_2 + \cdots + y_m a_m$. For $i = 1, \ldots, m$ define r_i in $M_A(G)$ by $r_i(a_i) = y_i a_i$, and extend r_1 to all of G so that r_1 is invertible.

Finally, suppose $\bar{a}_1, \ldots, \bar{a}_m$ are in $\mathbb{Z}_n$ with each a_i being a proper divisor of n and $\text{stab}(\bar{a}_1) \cap \cdots \cap \text{stab}(\bar{a}_m) = \text{stab}(\bar{c})$. By Proposition 3.5(ii), $c = (a_1, \ldots, a_m)$. So $1 = \left(\frac{a_1}{c}, \ldots, \frac{a_m}{c}\right)$. By the above, there are integers $y_1, \ldots, y_m$ with at least one y_j relatively prime to n such that $1 = y_1\left(\frac{a_1}{c}\right) + \cdots + y_m\left(\frac{a_m}{c}\right)$. This gives $c = y_1 a_1 + \cdots + y_m a_m$ where $(y_j, n) = 1$. Now define $r_i \in M_A(G)$ such that $r_i(a_i) = y_i a_i$ with r_j extended to G so that it is invertible. $\qquad \square$

The near-ring $M_A(\mathbb{Z}_n)$ with $A = \text{Aut}(\mathbb{Z}_n)$ does not satisfy Property I. We will show that it has a field as a homomorphic image. Let p be the smallest prime divisor of n. Let $a = \frac{n}{p}$. We will show that $\theta(\bar{a})$ is a minimal orbit of unique type. Since $\bar{a}$ has prime order p in $\mathbb{Z}_n$, then it is easy to see that $|\theta(a)| = p - 1$. In fact $\theta(\bar{a}) = \{\alpha_1(\bar{a}), \alpha_2(\bar{a}), \ldots, \alpha_{p-1}(\bar{a})\}$, the set of all elements in $\mathbb{Z}_n$ with order p.

Now let $\theta(\bar{b})$ be a nonzero orbit different from $\theta(\bar{a})$. We will show that $|\theta(\bar{b})| > |\theta(\bar{a})|$ which implies $\theta(\bar{b})$ has different type than $\theta(\bar{a})$. Since p is the smallest prime divisor of n then $\alpha_1(\bar{b}), \alpha_2(\bar{b}), \ldots, \alpha_{p-1}(\bar{b})$ are all different. This means $|\theta(\bar{b})| \geq |\theta(\bar{a})|$.

Assume $p \neq 2$. Then $p + 1$ is relatively prime to n and α_{p+1} is an automorphism of $\mathbb{Z}_n$. Suppose $\alpha_{p+1}(\bar{b}) = \alpha_i(\bar{b})$ for some $i \in \{1, 2, \ldots, p-1\}$. Then $(p+1)\bar{b} = i\bar{b}$, or equivalently $(p + 1 - i)\bar{b} = \bar{0}$. This means $\bar{b}$ has order less than p in $\mathbb{Z}_n$ which is not possible due to the minimality of p. So $|\theta(\bar{b})| > |\theta(\bar{a})|$.

Assume $p = 2$. Then $\bar{a}$ is the unique element of $\mathbb{Z}_n$ of order 2 and $\theta(\bar{a}) = \{\bar{a}\}$. Let $\bar{b}$ be a nonzero element in $\mathbb{Z}_n$ different from $\bar{a}$. Then $\bar{b}$ has order $m > 2$ and $(m-1)\bar{b}$ is an element different from $\bar{b}$ also having order m. One verifies that $(b, n) = ((m-1)b, n)$. Because of this $\bar{b}$ and $\overline{(m-1)b} = (m-1)\bar{b}$ are in the same orbit. This shows $|\theta(\bar{b})| > |\theta(\bar{a})|$.

Let $H = \theta(\bar{a}) \cup \{0\}$, the cyclic subgroup of $\mathbb{Z}_n$ of order p. The subgroup H is both A-invariant and $M_A(\mathbb{Z}_n)$-invariant. Let B be the automorphism group which is A restricted to

H. Then $B = \text{Aut}(H)$. Let $\Phi\colon M_A(\mathbb{Z}_n) \to M_B(H)$ be the restriction map: if f is a function on $\mathbb{Z}_n$, then $\Phi(f)$ is f restricted to H. One verifies that Φ is a surjective homomorphism and that $M_B(H)$ is isomorphic to the field $GF(p)$.

Even though $M_A(\mathbb{Z}_n)$ does not satisfy Property I, Corollary 3.4 may be used to construct centralizer near-rings $M_A(G \times H)$ such that $\mathbb{M}_n(M_A(G \times H); G \times H) = M_A((G \times H)^n)$ for all $n \geq 2$.

REFERENCES

[1] C. J. Maxson and K. C. Smith, *The centralizer of a set of group automorphisms*, Comm. Alg. **8** (1980), 211–230.

[2] J. D. P. Meldrum and A. P. J. van der Walt, *Matrix near-rings*, Arch. Math. **47** (1986), 312–319.

[3] I. Niven, H. Zuckerman and H. Montgomery, An introduction to the theory of numbers. John Wiley & Sons, New York (1991).

[4] G. Pilz, Near-rings (second edition), North Holland/American Elsevier, Amsterdam (1983).

[5] K.C. Smith, *Generalized matrix near-rings*, Comm. Alg. **24** (1996), 2065–2077.

[6] K.C. Smith, *Rings which are a homomorphic image of a centralizer near-ring*, in Proc. of the 1993 Fredericton Conference on near-rings and near-fields, (eds. Y. Fong et al). Kluwer Academic Publishers (1995), 257–270.

[7] K.C. Smith and L. van Wyk, *When is a centralizer near-ring isomorphic to a matrix near-ring?* Comm. Alg. **24** (1996), 4549–4562.

SCHOOL OF COMPUTING AND MATHEMATICS, UNIVERSITY OF TEESSIDE, MIDDLESBROUGH, CLEVELAND, TS1 3BA ENGLAND. E-MAIL: a.oswaldtees.ac.uk

DEPARTMENT OF MATHEMATICS, TEXAS A & M UNIVERSITY, COLLEGE STATION, TEXAS 77843-3368, U.S.A. E-MAIL: ksmithmath.tamu.edu

DEPARTMENT OF MATHEMATICS, UNIVERSITY OF STELLENBOSCH, PRIVATE BAG XI, MATIELAND 7602, SOUTH AFRICA. E-MAIL: lvwmaties.sun.ac.za

TOPOLOGY AND PRIMARY N-GROUPS

S. D. SCOTT

ABSTRACT. A primary N-group is a compatible N-group without ring submodules and two non-zero submodules with zero intersection. A very natural zero set topology arises and an adaptation of this is used throughout the paper. Topological features are studied and these are related to algebraic properties of the nearring. Many surprising results are obtained. The last part of the paper is concerned with showing, that with connectedness, direct decomposition implies we are dealing with the reals. Furthermore, local compactness implies the nearring is a subnearring of the nearring of all zero-fixing continuous self maps on the reals.

Primary N-groups will shortly be defined. They arise frequently in many mathematical contexts. Also, they can be studied by powerful topological methods. This is a very pleasing development providing much hope that their study will, for some time to come, continue to furnish new horizons.

INTRODUCTION

Throughout this paper all near-rings are left distributive, zero-symmetric and have an identity. Also, all N-groups will be unitary. The near-ring N is called a non-ring if it is not a ring. Any subset S of a N-group V for which $SN \subseteq S$ will be called an N-subset. The N-group V is called *minimal* if $V \neq \{0\}$ and V has no non-zero proper N-subgroups. This is equivalent to requiring V is of type 2, but since we are mainly concerned with tame situations, the term 'minimal N-group' is adopted. The near-ring N is said to be *primitive* (on V), if it has a faithful minimal N-group (i.e., V). The *radical* $J(N)$, which in other contexts is often denoted $J_2(N)$, is simply the intersection of the annihilators of all minimal N-groups.

A considerable number of results of this paper are related to direct decomposition. A *direct decomposition* of a near-ring N is an expression of N as a direct sum $R_1 \oplus R_2$, of two right ideals R_i, $i = 1, 2$. This direct decomposition is called *non-trivial* if $R_i \neq \{0\}$, for $i = 1$ and 2. If such an expression of N as a direct sum exists, then N is said to have *non-trivial direct decomposition*. One very substantial result of this paper (see §§11–13) concerns implications of such decompositions.

Almost at the opposite extreme from direct decomposition is uniformity. An N-subgroup U of an N-group V is said to be *uniform* in V if no two non-zero submodules of V contained in U, have zero intersection. If U is not uniform in V, we call it *non-uniform*

1991 *Mathematics Subject Classification.* 16Y30 .

Y. Fong et al. (eds.), Near-Rings and Near-Fields, 151–197.
© 2001 *Kluwer Academic Publishers. Printed in the Netherlands.*

in V. In the case $U = V$ we omit the term '*in* V' from 'V is uniform (non-uniform) in V'. Clearly these definitions apply to N-subgroups of the N-group N and to N itself.

An N-group V has been called *tame* if every N-subgroup is a submodule. If the near-ring N has a faithful tame N-group (V say), then it will be called *tame* (*tame on* V). The N-group V is in fact tame if, and only if, for any given v and w in V and α in N we can find β in N such that $(v + w)\alpha - v\alpha = w\beta$. This leads to the notion of *n-tame*. If $n \geq 1$ is an integer, then the N-group V is called *n-tame* if, for any given v in V, α in N and n-tuple $(w_1, \ldots, w_n)$ of elements of V, we can find β in N such that $(v + w_i)\alpha - v\alpha = w_i\beta$, for $i = 1, \ldots, n$. From above, an N-group is 1-tame if, and only if, it is tame. If the near-ring N has a faithful n-tame N-group (V say) it will be called *n-tame* (*n-tame on* V).

The notion of an *n*-tame N-group ($n \geq 1$ an integer) has a further specialisation, which is still very general and covers many important examples from quite diverse mathematical directions. An N-group V will be called *compatible*, if for any given v in V and α in N there exists β in N such that $(v + w)\alpha - v\alpha = w\beta$, for all w in V. The near-ring N is said to be *compatible* (*compatible on* V), if it has a faithful compatible N-group (i.e., V). Clearly a compatible N-group is *n*-tame for $n = 1, 2, \ldots$. Primitive compatible near-rings have been the object of considerable study (see [5]). A near-ring N is a *primitive compatible near-ring* (primitive and compatible on V), if it is faithful on a compatible minimal N-group (i.e., V). Although examples of such near-rings given in [5] are, in some ways, only a small selection, they suffice to show these near-rings arise in many many different contexts. However, in some respects theory can be greatly enriched by dispensing with the minimality of V. Before considering the condition (being primary) on which these developments rest, important examples of compatibility are given.

If V is a topological group (written additively) and $C_0(V)$ the set of all continuous zero fixing functions of V into V, then under pointwise addition and composition $C_0(V)$ is a near-ring and $C_0(V)$ is compatible on the $C_0(V)$-group V. Another important example is obtained for any group V (written additively) and semigroup S of endomorphisms of V into V containing the inner automorphisms of V. In this case, if N is the near-ring of maps of V into V generated by S, then N is compatible on the N-group V. Another important example comes from considering an Ω-group V (written additively). The set $V_0[x]$ of all zero-symmetric polynomials over V in the indeterminate x, is a near-ring where, the operations in $V_0[x]$ are addition and substitution for x. Now, the near-ring $P_0(V)$ of zero fixing polynomial maps of V into V is easily defined. Elements of $P_0(V)$ are all maps of V into V taking v in V to $(v)\varphi$ where $(x)\varphi$ is some element of $V_0[x]$. The operations of $P_0(V)$ are composition and pointwise addition. The fact that $P_0(V)$ is now compatible on the $P_0(V)$-group V is highly useful. Indeed much advanced Ω-group theory can be obtained from investigations that stem from such basic observations (see [6]). In several places in this paper attention focuses on this important example.

Fairly well known background definitions are behind us and, before discussing more fully the contents of this paper, it is desirable to know exactly what is meant by semiprimary and primary N-groups. In [4] a *semiprimary* N-group was defined as one with no non-zero ring submodules (i.e., no non-zero submodules which as N-groups are ring modules). Also in [4] a semiprimary N-group which was uniform was called *primary*. A central result of [4] was that any semiprimary N-group is a subdirect product of primary N-groups. The importance of primary N-groups (as in [4]) is that because the definition is only weakly restrictive, they can be expected to occur very frequently. As far as this paper is concerned it will be convenient to require that, in the definition of a *primary* or *semiprimary* N-group this N-group is also compatible. With this definition primary N-groups are certainly still very plentiful and, now that this terminology is fixed, examples are given.

Any compatible N-group V can be taken, and the lower radical U of V determined by the ring submodules factored out. Now V/U is semiprimary and, from above, it is a subdirect product of primary N-groups. More specialised examples can be supplied. If V is any group without non-zero normal abelian subgroups, then it is a subdirect of $I(V)$-groups V_i, $i \in I$, each having the property, that they have no non-zero normal abelian subgroups and no two non-zero normal subgroups have zero intersection (here $I(V)$ is the nearring generated by the inner automorphisms of V). It follows now that the V_i are primary $I(V_i)$-groups. Examples do not stop at this stage. The wreath product V of a non-abelian simple group with any group has a unique minimal normal subgroup which is non-abelian. Thus V is a primary $I(V)$-group which is not subdirectly irreducible. Finally, primary N-groups can readily be obtained from nearrings of all zero-fixing continuous self maps on certain topological groups. These examples illustrate just how wide the possibilities are.

Discussion of the contents of this paper is now given. The next section is concerned with preliminaries. The one that follows shows how a very important topology Z^t can be imposed on a primary N-group in such a way that it reflects much of the N-group structure. In this way all the power of topology becomes available for the study of such N-groups. The third section covers connectedness. Connectedness in primary N-groups has quite a number of rather unique aspects. In this context reasonable restrictions on the $P_0(V)$-group V (V an Ω-group) yield interesting consequences. There follows in the fourth a discussion of discreteness. For primary N-groups this is a surprisingly general notion. The remainder of four covers an assortment of material. Conditions similar to the existence or non-existence of small subgroups (arising in topological groups) are considered. In the fifth section the Hausdorff assumption is examined. A Hausdorff primary N-group is shown to be regular and this property (viz: being Hausdorff) can be lifted from an arbitrarily small non-zero N-subgroup. Also being Hausdorff implies the non-uniformity of every non-zero right ideal of N. The section that follows shows, amongst other things, that closures are well behaved. For example, the closure of a subgroup (N-subgroup) of a primary N-group is a subgroup (N-subgroup).

In section seven we look at the material related to sparseness. This is a property (defined later) at the opposite extreme from being Hausdorff. It is shown that the uniformity of N and sparseness are equivalent. Also meaningful investigation of the important subset of V, consisting of all points of V unseparated from 0, takes place. Eight starts by showing prime compatible non-rings have a faithful primary N-group. It is shown that such an N-group has no proper closed N-subgroup, although in the Hausdorff case it necessarily has a minimal N-subgroup. Gradings, introduced in nine, may be described as localized idempotents. Their existence implies the openness of certain N-subgroups and has implications for the situation where there exists direct decomposition. For a compact primary N-group a graded N-subgroup is shown to be compact. In ten we investigate chain conditions. This investigation covers precise conditions for N to have no infinite direct sums. Questions related to the nilpotency of $J(N)$ are also covered. The next three sections are concerned with the proof of a remarkable theorem. This theorem states that connectedness and direct decomposition imply the near-ring involved is very similar to certain subnear-rings of the near-ring of zero fixing continuous functions on the reals. The statement of the theorem has four parts to it. The proof of the first two takes place in section eleven. The third part requires a section of greater length (viz: twelve). Local compactness is a condition employed in part four. This part is fully proved in section thirteen. Section thirteen also looks very briefly at certain other issues. For example, the question as to when a primary N-group is a topological group, is touched on.

Finally section fourteen has been added to tie up loose ends. There are a number of results taken from unpublished manuscripts that need proof. These proofs are supplied in that section.

1. PRELIMINARIES

This section loosely covers material required later and not dealt with in the introduction. When proofs are required, they may be found in the final section.

If V is an N-group, then an N-subgroup U of V is called *central* if $(v+u)\alpha = v\alpha + u\alpha$ for all v in V, u in U and α in N. Alternatively, since U is necessarily a submodule, it is called a *central submodule*. N-subgroups of central submodules are central (i.e., central submodules) and N-homomorphic images of such are central submodules of the image. Also any sum of such submodules is again central. The centre $z(V)$ of the N-group V is, accordingly, taken as the sum of all central submodules of V. Although the notation $Z(V)$ is used elsewhere it is not adopted here since this paper makes extensive use of Z for other purposes.

The next result is an obvious extension of 2.1 of [2]. A proof may be found in the final section.

Proposition 1.1. *If V is a compatible N-group and $U_1 \le U_2$ submodules of V such that $(U_1 : V) = (U_2 : V)$, then $U_2/U_1 \le z(V/U_1)$.*

Later, in relation to 1.1, we shall be using z-constraint. This is now defined. If V is an N-group, then a *factor* of V is an N-group of the form U_2/U_1 where $U_1 \le U_2$ are submodules of V. The factor U_2/U_1 is *non-zero* if $U_1 < U_2$. An N-group V is now called z-*constrained*, if there are no non-zero factors U_2/U_1 of V such that $U_2/U_1 \le z(V/U_1)$.

We come now to central sums. If the N-group V is a sum of the submodules $V_i, i = 1, 2$, in such a way that $(v_1 + v_2)\alpha = v_1\alpha + v_2\alpha$ for all v_i in V and α in N, then V is said to be a *central sum* of V_1 and V_2. An elementary property of central sums used quite frequently later is the following:-

Proposition 1.2. *If the N-group V is a central sum of the submodules V_i, $i = 1, 2$, then $V_1 \cap V_2 \le z(V)$.*

It should be noted that in many situations the converse of 1.2 fails to hold. Now another elementary result which is also used later follows.

Proposition 1.3. *If the N-group V is a central sum of V_i, $i = 1, 2$, and μ an N-homomorphism on V, then $V\mu$ is a central sum of $V_i\mu$, $i = 1, 2$.*

Proposition 1.3 is often used in the context of V being a direct sum of the V_i, $i = 1, 2$. Here $V\mu$ is, if not a direct sum of $V_i\mu$, $i = 1, 2$, at least a central sum and in particular $V_1\mu \cap V_2\mu \leq z(V\mu)$.

If V is an N-group and there exists an integer $k \geq 1$ and a sequence of submodules

$$\{0\} = V_0 \leq V_1 \leq \cdots \leq V_k = V$$

of V with $V_{i+1}/V_i \leq z(V/V_i)$, for $i = 0, \ldots, k-1$, then we call V, N-nilpotent. Clearly a submodule of the N-group V is N-nilpotent, if it is N-nilpotent as an N-group. With $L(V)$ taken as the sum of all N-nilpotent submodules of V, we have the remarkable result stating that:-

Theorem 1.4. *If V is a compatible N-group, where N has descending chain condition on right ideals (i.e., DCCR), then $L(V)$ is N-nilpotent and $V/L(V)$ is finite.*

The proof of 1.4 along with much supplementary material can be found in [3].

We now look at centralizers. Definitions and elementary theory can be found in §5 of [3], but for completeness, some coverage is given here. If V is an N-group and S a non-empty subset of V, then an N-subset P of V *centralizes* S if $(h + u)\alpha = h\alpha + u\alpha$, for all h in P, u in S and α in N. This expression is equivalent to requiring $(u + h)\alpha = u\alpha + h\alpha$, for all u in S, h in P and α in N. Now the union of all such P is an N-subset of V centralizing S which is called the *centralizer* of S in V and denoted by $C_V(S)$. In the particular case where $S = \{u\}, u \in V$, we denote $C_V(S)$ by $C_V(u)$. The importance of centralizers in many tame and compatible situations rests on the following result (see 5.3 of [3]).

Theorem 1.5. *If V is a 3-tame N-group and S a non-empty subset of V, then $C_V(S)$ is an N-subgroup of V.*

Theorem 1.5 means that centralizers in compatible situations are a very valuable tool. Indeed the use of such N-subgroups appears essential for the proof of 1.4. However, they also allow an alternative characterisation of a primary N-group. The next result follows from 3.10 of [4]. The proof is given in the final section.

Proposition 1.6. *A compatible N-group V is primary if, and only if, $C_V(u) = \{0\}$ for all $u \neq 0$ in V.*

There is at least one other condition that a primary N-group satisfies which is close to being a characterisation. This follows from 4.6 of [4] and the proof may be found in the final section.

Theorem 1.7. *If V is a primary N-group and v_i, $i = 1, 2$, non-zero elements of V such that $(0 : v_1) \leq (0 : v_2)$, then $v_1 = v_2$.*

For a compatible N-group V an approximate converse of 1.7 is possible. Thus, if for all pairs (v_1, v_2) of non-zero elements of V with $(0 : v_1) \leq (0 : v_2)$ we have $v_1 = v_2$, then V is very close to being primary. However, in this paper we shall not pursue this matter further. What is now required is some basic coverage of polynomials.

If V is an Ω-group, then $V_0[x]$ and $P_0(V)$ are defined in the previous section. We may readily extend the definition of polynomials and in particular zero-symmetric polynomials, to more than one indeterminate. Accordingly, $V_0[x, y]$ is taken as the set of all polynomials over V in the indeterminates x and y, that are zero-symmetric in both x and y. This simply means that if $(x,y)\varphi$ is in $V_0[x,y]$, then

$$(0,y)\varphi = (x,0)\varphi = 0.$$

Now many Ω-groups have additional properties that can be specified in terms of such polynomials. For example, any ring (associative or otherwise) is full if it has an identity. Here the *fullness* of an Ω-group V (see §20 of [6]) is defined by requiring that for any ideal A of V the ideal generated by all $(a,v)\varphi, a \in A, v \in V$ and $(x,y)\varphi \in V_0[x,y]$, coincides with A. This is equivalent to the requirement that all $(v,a)\varphi, v \in V, a \in A$ and $(x,y)\varphi \in V_0[x,y]$ generates A. Another example of such a property which is frequently fulfilled in more elementary contexts (e.g. when some type of semisimplicity holds) is that of rigidity. The Ω-group V is called *rigid* (see §§21 and 22 of [6]), if for any ideal A of V the ideal generated by all $(a,b)\varphi, a \in A, b \in A$ and $(x,y)\varphi \in V_0[x,y]$ coincides with A.

Underlying the above notions and many developments on polynomials is G. Pilz's lovely theorem stating:-

Theorem 1.8. *If V is an Ω-group, then U is a $P_0(V)$-subgroup of V if, and only if, U is an ideal.*

All preliminaries are now behind us and the next requirement is to lay topological foundations.

2. INTRODUCING TOPOLOGY

This section is concerned with covering elementary topological requirements on which the remainder of the paper will rest. Certain features of the exposition given here are similar to those of [5]. However, much of it represents a rather different approach to Z and Z' (c.f. §1 of [5]).

If V is a compatible N-group and S a non-empty subset of N, then $Z(S)$ is taken as the set of all v in V such that $v\sigma = 0$ for all σ in S. A subset $\triangle$ of V is said to be Z-closed, if either $\triangle = \emptyset$ or there exists a non-empty subset S of N such that $Z(S) = \triangle$. Clearly $Z(0) = V$ and V is Z-closed. Also $Z(1) = \{0\}$ so that $\{0\}$ is also Z-closed. The collection of all Z-closed subsets of V is denoted by Z. It is worth noting that Z-closed subsets are determined by their annihilators.

Proposition 2.1. *If V is a compatible N-group and Z_1, a non-empty element of Z, then $Z(0 : Z_1) = Z_1$.*

It is easily verified that an arbitrary intersection of elements of Z is again in Z. We would like Z in fact to form the closed subsets of a topology on V. A very meaningful condition ensuring this is the case is that of requiring V to be primary. Indeed theory developed throughout this paper rests heavily on this basic requirement.

Theorem 2.2. *If V is a primary N-group then the elements of Z form the closed subsets of a topology on V.*

Proof. As the proof that arbitrary intersections of elements of Z are in Z is somewhat elementary we move to showing that if Z_i, $i = 1, 2$, are in Z, then so is $Z_1 \cup Z_2$. Clearly we may assume $Z_i \neq \emptyset$ for $i = 1$ and 2. The result will follow if it is shown that

$$Z[(0 : Z_1) \cap (0 : Z_2)] = Z_1 \cup Z_2.$$

Obviously elements of $(0 : Z_1) \cap (0 : Z_2)$ annihilate $Z_1 \cup Z_2$ and

$$Z[(0 : Z_1) \cap (0 : Z_2)] \geq Z_1 \cup Z_2.$$

Now the N-group

$$[(0 : Z_1) + (0 : Z_2)] / [(0 : Z_1) \cap (0 : Z_2)] (= A \text{ say})$$

is a direct sum of

$$(0 : Z_i) / [(0 : Z_1) \cap (0 : Z_2)] \text{ for } i = 1, 2.$$

Thus for v in $Z[(0 : Z_1) \cap (0 : Z_2)]$,

$$[v(0 : Z_1) + v(0 : Z_2)] / v[(0 : Z_1) \cap (0 : Z_2)],$$

being an N-homomorphic image of A, is a central sum of

$$v(0 : Z_i) / v[(0 : Z_1) \cap (0 : Z_2)] \text{ for } i = 1, 2.$$

However, $v[(0 : Z_1) \cap (0 : Z_2)] = \{0\}$ and this N-group is N-isomorphic to $v(0 : Z_1) + v(0 : Z_2)$, which is consequently a central sum of the $v(0 : Z_i)$, $i = 1, 2$. The semiprimary nature of V now implies this sum is direct and the primary nature of V means $v(0 : Z_1) = \{0\}$ or $v(0 : Z_2) = \{0\}$. By 2.1, $Z(0 : Z_i) = Z_i$, for $i = 1, 2$, and v is in Z_1 or Z_2. Thus v is in $Z_1 \cup Z_2$ and the theorem is completely proved. $\square$

In order to make more substantial steps forward we need information about certain Z-closed subsets of V.

Proposition 2.3. *If V is a primary N-group and $v \neq 0$ in V, then $\{0, v\}$ is Z-closed.*

Proof. If $u \neq 0$ is in $Z[(0 : v)]$, then $u(0 : v) = 0$ and $(0 : v) \leq (0 : u)$. Now, by 1.7, $u = v$ and since $Z[(0 : v)]$ contains 0, we see $Z[(0 : v)] = \{0, v\}$. Thus $\{0, v\}$ is Z-closed and the proposition holds. $\square$

Topological investigation of a primary N-group V will on occasions make use of Z. However, there is an adaptation of this topology which provides the backbone of all serious developments. In order to introduce this topology we first consider a drawback to using Z. Clearly from the way in which Z is defined all non-empty Z-closed subsets contain 0. Consequently, the only Z-open subset containing 0 is V. This illustrates that in using Z one would expect complications arising from the exceptional nature of 0. Certainly such problems occur. However, there is another reason to extend Z. Since V is a group, it is very desirable that the topology used is such that translations of closed (open) subsets of V are again closed (open). The Z'-topology will meet above requirements, not be far removed from Z and be much more useful.

For a compatible N-group V we define Z' as the set of all left translations of Z-closed subsets of V. The definition of Z' is in fact symmetric.

Proposition 2.4. *For a compatible N-group V, Z' is the set of all right translations of Z-closed subsets of V.*

Proof. A typical element of Z^t is of the form $v + Z_1$, where v is in V and Z_1 is in Z. This may be written as $v + Z_1 - v + v$. Now let α in N be such that $w\alpha = -v + w + v$ for all w in V. We shall show

$$Z[\alpha(0 : Z_1)] = v + Z_1 - v.$$

Clearly if u is in $v + Z_1 - v$, then $u\sigma = 0$ for all σ in $\alpha(0 : Z_1)$ and u is in $Z[\alpha(0 : Z_1)]$. Also if u is in $Z[\alpha(0 : Z_1)]$, then $-v + u + v$ is in Z_1 and u is in $v + Z_1 - v$. Thus $v + Z_1 - v$ is Z-closed and $v + Z_1$ is of the form $Z_2 + v$, where Z_2 is in Z. The proposition is proved. □

It is clear that $Z \subseteq Z^t$ but a very useful type of reverse inclusion holds.

Proposition 2.5. *If V is a compatible N-group, then a subset of Z^t containing 0 is in Z.*

Proof. Suppose $v + Z_1$ is in Z^t, where v is in V, Z_1 in Z and $v + Z_1$ contains 0. We shall show $v + Z_1 = Z(S)$, where S is the set of all β in N such that, for some α in $(0 : Z_1)$

$$w\beta = (-v + w)\alpha - (-v)\alpha,$$

for all w in V. Now $-v$ is in Z_1 and $(-v)\alpha = 0$, for all α in $(0 : Z_1)$. Consequently β is in $(0 : v + Z_1)$ and $v + Z_1 \subseteq Z(S)$. If on the other hand w is such that $w\beta = 0$ for all β in S, then clearly $-v + w$ is in Z_1 and w in $v + Z_1$. The proof of the proposition is complete. □

The last two propositions provide very useful information but the most important property of Z^t is that:-

Theorem 2.6. *If V is a primary N-group, then elements of Z^t are the closed subsets of a topology on V.*

Proof. Suppose $K_i, i \in I$, is a family of subsets of Z^t. If the intersection K of the K_i is $\emptyset$, then K is in Z^t. Otherwise there exists v in K and, by 2.5, $-v + K_i$ is in Z for each i in I. Thus the intersection L of all the $-v + K_i$, $i \in I$, is in Z (see 2.2). However, $L \supseteq v + K$ and if u is in L, then $u = -v + w_i$ where w_i is in K_i so that for all i in I, $w_i = a$ with a in K. Thus $L = -v + K$, from which it readily follows that K is in Z^t.

Now, suppose H_i, $i = 1, 2$, are in Z^t. To complete the proof we show $H_1 \cup H_2$ is in Z^t. Clearly we may assume $H_i \neq \emptyset$ for $i = 1, 2$. Translating by an element $-v$ with v in H_1, yields $(-v + H_1) \cup (-v + H_2)$ and, if this union is in Z^t, then so is $H_1 \cup H_2$. Thus, we may assume H_1 contains 0 and is, by 2.5, in Z. Let b be in H_2. By 2.2 and 2.3, $H_1 \cup \{b\}$ is in Z. Now $H_1 \cup H_2 = H_1 \cup \{b\} \cup H_2$ and translation of $H_1 \cup H_2$ by $-b$ yields the union

$$(-b + H_1 \cup \{b\}) \cup (-b + H_2)(= -b + H_1 \cup H_2)$$

of two Z^t-closed subsets of V containing 0. By 2.2 and 2.5, this subset of V is Z-closed and consequently $H_1 \cup H_2$ is in Z^t. The proof of the theorem is complete. □

It is useful to keep in mind several examples of the topology of 2.6. For the nearring of all zero-fixing self maps of a group V, we have that if V has more than two elements, then it is primary and Z^t is the discrete topology. The nearring of all zero-fixing continuous self maps on the reals, involves a primary situation where Z^t is the classical topology. In the case of the nearring of all zero-fixing polynomial self maps on the reals, we are again in a primary situation. Here Z^t is just the cofinite topology. All these are examples where the nearring, is in fact primitive on the nearring group. Non-primitive examples are not difficult to supply. The wreath product V of a non-abelian simple group U and a non-zero group, is a primary $I(V)$-group ($I(V)$ the nearring generated by the inner automorphisms of V), which is not of type 2. In this case Z' exists and if U is finite, then Z' is often discrete. More complex topological features become possible when U is infinite.

Consideration of primary N-groups involves extensive use of Z' (see comments above). The contents of the remainder of this paper follow this course. Therefore, if V is a primary N-group it is convenient to call a subset of V *closed* if it is in Z' and *open* if it is of the form $V \backslash Z_1$ where Z_1 is in Z'. Also, it is worth noting that translation of $\{0\}$ yields the fact that one element subsets of V are closed and consequently all finite subsets of V are closed. Theorem 2.6 has another aspect. In a sense it is close to being optimal.

Theorem 2.7. *If V is a compatible N-group for which elements of Z' form the closed subsets of a topology on V, then either V is primary or there exists a unique minimal N-subgroup U of V with $|U| = 2$ and $C_V(d) = \{0\}$ for all d in $V \backslash U$ (c.f. 1.6).*

Proof. Suppose $a \neq 0$ is in V and $b \neq 0$ is in $C_V(a)$. Since $\{0\}$ is in Z' all one element subsets of V are in Z'. Because Z' is a topology, all finite subsets of V are in Z'. In particular $\{0, a, b\}$ is in Z' and, by 2.5, this finite set is in Z. Now suppose $b \neq -a$ so that $a + b$ is not in $\{0, a, b\}$. In this case there exists α in N such that $\{0, a, b\}\alpha = \{0\}$ but $(a + b)\alpha \neq 0$. However, b is in $C_V(a)$ and we have the contradiction that $(a + b)\alpha = a\alpha + b\alpha$. This means $C_V(a) = \{0\}$ for all $a \neq 0$ in V or there exists $a \neq 0$ in V, such that the only non-zero element of $C_V(a)$ is $-a$. In the first case V is primary by 1.6. In the second case we take $U = C_V(a)$. Now, since $U = \{0, -a\}$, we have $|U| = 2$ and U is a minimal N-subgroup. If U_1 is some minimal N-subgroup distinct from U, then since a is in $U, U_1 \leq C_V(a)$. This yields the contradiction that $U_1 = U$ and implies U is unique. If d is in $V \backslash U$, then from above $C_V(d)$ is either $\{0\}$ or of the form $\{0, -d\}$. Thus, if $C_V(d) \neq \{0\}$, it is a minimal N-subgroup and coincides with U. This means $-d$ and therefore d is in U. This case does not occur so $C_V(d) = \{0\}$. The theorem is completely proved. $\qquad\square$

If V is a compatible N-group, S a subset of V and α in N, then the *inverse image* (of S under α) denoted $S\alpha^{-1}$ is the set of all v in V such that $v\alpha$ is in S. Suppose $S_i, i \in I$, is a family of subsets of V with union and intersection respectively H and K. It is easy to verify that $H\alpha^{-1}$ and $K\alpha^{-1}$, respectively, coincide with the union and intersection of all the $S_i\alpha^{-1}, i \in I$. In particular, if $S_j, j = 1, 2$, are subsets of V such that $S_1 \cup S_2 = V$, then $S_1\alpha^{-1} \cup S_2\alpha^{-1} = V$ and a similar statement can be made when $S_1 \cap S_2 = \emptyset$. Furthermore for subsets in Z' we have:

Theorem 2.8. *If V is a compatible N-group, α in N and Z_1 in Z', then $Z_1\alpha^{-1}$ is in Z'.*

Proof. It is first shown that if S is in Z, then $S\alpha^{-1}$ is in Z. Clearly we can assume $S \neq \emptyset$. If it is shown that $S\alpha^{-1} = Z[\alpha(0 : S)]$, then this result will follow. If v is in $Z[\alpha(0 : S)]$, then $v\alpha(0 : S) = \{0\}, v\alpha$ is, by 2.1, in S and v is in $S\alpha^{-1}$. On the other hand, if v is in $S\alpha^{-1}$, then $v\alpha$ is in S, $v\alpha(0 : S) = \{0\}$ and v is in $Z[\alpha(0 : S)]$. The conclusion therefore holds for subsets in Z. Now $Z_1 = b + Z_2$, where Z_2 is in Z. Also we can assume $Z_1\alpha^{-1} \neq \emptyset$. Let a in V be such that $a\alpha$ is in Z_1. Now $a\alpha = b + c$, where c is in Z_2 and $Z_1 = a\alpha + (-c + Z_2)$. Since c is in $Z_2, -c + Z_2$ is, by 2.5, in Z. Now $Z_1\alpha^{-1}$ is the set of all w in V such that $w\alpha - a\alpha$ is in $-c + Z_2$. However, there exists β in N such that $(a + u)\alpha - a\alpha = u\beta$ for all

u in V. Thus $Z_1\alpha^{-1}$ is precisely the set of all w in V such that $(-a+w)\beta$ is in $-c+Z_2$. This means $Z_1\alpha^{-1}$ is the set of all w in V such that $-a+w$ is in $(-c+Z_2)\beta^{-1}$. So we have shown

$$Z_1\alpha^{-1} = a+(-c+Z_2)\beta^{-1},$$

where from the first part, $(-c+Z_2)\beta^{-1}$ is in Z. It clearly follows $Z_1\alpha^{-1}$ is in Z' and the proof is complete.   □

Let V be a compatible N-group and α in N. The *function induced by* α is simply the map of V into V taking w in V to $w\alpha$. It is now easy to verify from 2.8 (see previous comments) that:-

Corollary 2.9. *Let V be a primary N-group S a subset of V and α in N. If S is open (closed), then $S\alpha^{-1}$ is open (closed). Furthermore, the function induced by α is continuous (with respect to Z').*

In the sense explained in 2.9 elements of N are continuous. However, left (or right) translation by an element of V provides not just a continuous map but a homeomorphism. Thus, under translation, topological properties of subspaces of V are preserved. In particular, if v is in V and S is a connected, Hausdorff or sparse (see §7) subspace of V, then the same is true for $v+S$.

We now have behind us all material (topological and algebraic) enabling advancement of impressive theory.

3. CONNECTEDNESS

In a topological space S a *disconnection* is a pair (D_1, D_2) of non-empty open and closed subsets of S with $D_1 \cup D_2 = S$ and $D_1 \cap D_2 = \emptyset$. If S has no such disconnection, it is said to be *connected*. The *connected component* of a point b of S is the union of all connected subspaces of S containing b. This subspace of S is connected and closed (see §7 of [1]). If, for all b in S, the connected component of b is $\{b\}$, then S is said to be *totally disconnected*. All these definitions immediately translate by means of the Z'-topology of §2 to primary N-groups. Such use of Z' will now be standard. Also in this context detailed study of the above concepts proves very profitable.

If V is a primary N-group, then the connected component of 0 is denoted by $c(V)$ and the intersection of all open and closed subsets of V containing 0 is denoted by $i(V)$. In the case of a compact Hausdorff space S, the connected component of a point b of S and the intersection of all open and closed subsets of S containing b in fact coincide (see 14 of [1]). However, generally such equality does not hold and in the case of V, $c(V)$ and $i(V)$ often differ. It is therefore somewhat surprising that quite conservative algebraic conditions ensure $i(V) = c(V)$. This fact will be the content of the final theorem of this section. Our immediate goal is to gain information about $c(V)$ and $i(V)$.

Theorem 3.1. *If V is a primary N-group, then $i(V)$ and $c(V)$ are closed N-subgroups of V.*

Proof. If v is in $c(V)$, then it is easily seen that $-v+c(V)$ is a connected subspace of V containing 0. Thus $-v+c(V) \subseteq c(V)$ and for any b in $c(V), -v+b$ is in $c(V)$. Thus $c(V)$ is a subgroup of V. It is now shown $c(V) \cdot N \subseteq c(V)$. If for some α in $N, c(V)\alpha$ is not contained in $c(V)$, then the definition of $c(V)$ yields the fact that $c(V)\alpha$ is not a connected subspace of V. Thus there exist non-empty open and closed subsets D_i, $i = 1,2$ of $c(V)\alpha$ with $D_1 \cup D_2 = c(V)\alpha$ and $D_1 \cap D_2 = \emptyset$. Now, by 2.9, the inverse images $D_i\alpha^{-1}$, $i = 1,2$, of the D_i in $c(V)$, are non-empty open and closed subsets of $c(V)$ with $D_1\alpha^{-1} \cup D_2\alpha^{-1} = c(V)$ and $D_1\alpha^{-1} \cap D_2\alpha^{-1} = \emptyset$. This contradiction to the connectedness of $c(V)$ means $c(V)\cdot N \subseteq c(V)$ and $c(V)$ is an N-subgroup of V.

Now, if D is any open and closed subset of V containing 0, then since $D(-1)$ is another such subset, $i(V)(-1) = i(V)$. Thus for v in $i(V)$, we have $-v$ is in $i(V)$ and $v + D$ contains 0. Clearly $v + D$ is an open and closed subset of V containing $\{0\}$ and

$$-v + i(V) \subseteq -v + (v + D) = D,$$

so that $-v + i(V) \subseteq i(V)$. Thus for all b in $i(V)$, $-v + b$ is in $i(V)$ and $i(V)$ is a subgroup of V. If for some α in N, $i(V)\alpha$ is not contained in $i(V)$, then there exists an open and closed subset D_1 of V containing 0 with $i(V)\alpha$ not contained in D_1. Now $D_1\alpha^{-1}$ is, by 2.9, an open and closed subset of V containing 0. This would mean $i(V) \subseteq D_1\alpha^{-1}$, thus yielding the contradiction that $i(V)\alpha \subseteq D_1$. It follows that $i(V)$ is an N-subgroup of V and theorem 3.1 is completely proved. $\qquad\square$

The fact that connected components are just cosets of $c(V)$ will be of use later.

Proposition 3.2. *If V is a primary N-group, then the connected component of some element b of V is the coset $b + c(V)$.*

Proof. Translation of connected subspaces of V yields connected subspace and $b + c(V)$ is therefore a connected subspace of V containing b. However, if S is any connected subspace of V containing b, then $-b + S \subseteq c(V)$. Clearly the proposition now follows. $\qquad\square$

The next question that we address is that of the connectedness of N-subgroups of a connected primary N-group. It is an interesting fact that connectedness is quite frequently inherited by such subobjects. Since a primary N-group V is connected precisely when $c(V) = V$ our study again involves consideration of connected components of 0.

Theorem 3.3. *If V is a connected primary N-group and H an N-subgroup of V, then $(H : V) = (c(H) : V)$.*

Proof. We first note that, being an N-subgroup of V, H is primary and the resulting Z^t topology (see 2.6) is simply that obtained by viewing H as a subspace of V. Now clearly $c(H)$ is an N-subgroup of V, $c(H) \leq H$ and

$$(c(H) : V) \leq (H : V).$$

If this inclusion is proper, then there exists α in $(H : V)$ such that $V\alpha$ is not contained in $c(H)$. Now $V\alpha$ is contained in H and, consequently, the subspace $V\alpha$ of H has a disconnection $((D_1, D_2)$ say$)$. The $D_i\alpha^{-1}$, $i = 1,2$, are, by 2.9, non-empty open and closed subsets of V. Clearly, also $D_1\alpha^{-1} \cup D_2\alpha^{-1} = V$ and $D_1\alpha^{-1} \cap D_2\alpha^{-1} = \emptyset$. This is a contradiction to the connectedness of V. Thus $V\alpha \subseteq c(H)$ and α is in $(c(H) : V)$. The theorem is entirely proved. $\qquad\square$

Two examples of connectedness being inherited by N-subgroups can now be given.

Corollary 3.4. *If the connected primary N-group V is z-constrained, then every N-subgroup of V is connected.*

Proof. If H is an N-subgroup of V and $c(H) < H$, then by 3.3 and 1.1,

$$H/c(H) \leq z(V/c(H)).$$

By z-constraint we conclude $c(H) = H$ and H is connected. The corollary is proved. □

The second example shows that for a large selection of situations involving polynomials over Ω-groups connectedness is similarly inherited.

Corollary 3.5. *Suppose V is a full Ω-group that as a $P_0(V)$-group is primary. If V is connected, then so is every $P_0(V)$-subgroup of V.*

Proof. The $P_0(V)$-subgroups of V are, by 1.8, precisely the ideals of V. If A is an ideal of V, then fullness implies A is generated, as an ideal, by all $(v,a)\varphi$ with v in V, a in A and $(x,y)\varphi$ in $V_0[x,y]$. Thus, if $c(A) < A$, we can find v_1 in V, a_1 in A and $(x,y)\psi$ in $V_0[x,y]$ such that $(v_1,a_1)\psi$ is not in $c(A)$ (see §1). This means the map λ taking w in V to $(w,a)\psi$ is in $P_0(V)$ but not in $(c(A) : V)$. However, with w fixed we see that the map taking u in V to $(w,u)\psi$ is in $P_0(V)$ and consequently for all w in $V, (w,a_1)\psi$ is in A. Thus λ is in $(A : V)$ which contradicts theorem 3.3. We conclude $A = c(A)$ and thereby show that the corollary holds. □

Although fullness allows the proof of 3.5, rigidity often has further implications. The final theorem of this section (mentioned above) relates to such Ω-groups.

Theorem 3.6. *Suppose V is a rigid Ω-group. If V is a primary $P_0(V)$-group, then $c(V) = i(V)$.*

Proof. Suppose (D_1,D_2) is a disconnection of V with 0 in D_1. Either $(D_1 \cap c(V), D_2 \cap c(V))$ is a disconnection of $c(V)$ or, since $D_1 \cap c(V)$ contains 0, $D_2 \cap c(V) = \emptyset$. However, $c(V)$ is connected and therefore $D_2 \cap c(V) = \emptyset$ implying $c(V) \leq D_1$. Now D_1 was an arbitrary open and closed subset of V containing 0 and we have $c(V) \leq i(V)$. Suppose $c(V) < i(V)$. In this case $i(V)$ has a disconnection (E_1, E_2) with 0 in E_1. Let v be in E_2. Because V is rigid, the ideal $vP_0(V)$ of V (see 1.8) is generated as an ideal by all $(u_1,u_2)\varphi$ with $(x,y)\varphi$ in $V_0[x,y]$ and $u_i, i = 1,2$, in $vP_0(V)$. Any sum

$$(v\alpha_1, v\beta_1)\varphi_1 + \cdots + (v\alpha_k, v\beta_k)\varphi_k$$

with $k \geq 1$ an integer, α_i and β_i, $i = 1,\ldots,k$, in $P_0(V)$ and $(x,y)\varphi_i$, $i = 1,\ldots,k$, in $V_0[x,y]$, is of the form $(v,v)\theta$ with $(x,y)\theta$ in $V_0[x,y]$. The collection S of all such $(v,v)\theta$ is therefore a $P_0(V)$-subgroup of V. Thus, being an ideal of V (see 1.8), S coincides with $vP_0(V)$. This means there exists $(x,y)\psi$ in $V_0[x,y]$ such that $(v,v)\psi = v$. Now let λ in $P_0(V)$ be given by $w\lambda = (v,w)\psi$ for all w in V. With w fixed the map taking u in V to $(u,w)\psi$ is in $P_0(V)$ and thus for any w in V, $(v,w)\psi$ is in $vP_0(V)$. We see, $V\lambda \subseteq vP_0(V)$ and $v\lambda = v$. However, $i(V)$ is, by 3.1, a $P_0(V)$-subgroup of V and $vP_0(V) \leq i(V)$. Hence $V\lambda \subseteq i(V)$ and $E_1 \cap (V\lambda)$ contains 0 and $E_2 \cap (V\lambda)$ contains v. Now $(E_1 \cap V\lambda, E_2 \cap V\lambda)$ is clearly a disconnection of $V\lambda$. Thus, by 2.9, $(E_i \cap V\lambda)\lambda^{-1}$ are for $i = 1,2$, open and closed subsets of V. They are also disjoint with union equal to V. Now 0 is in $(E_1 \cap V\lambda)\lambda^{-1}$ and this would mean

$i(V) \leq (E_1 \cap V\lambda)\lambda^{-1}$. However, v in $i(V)$ is in $(E_2 \cap V\lambda)\lambda^{-1}$ and we have a contradiction. It has been shown that $i(V) = c(V)$ and theorem 3.6 is completely proved. □

Frequently restrictions on Z^t still encompass wide classes of primary N-groups. Even discreteness and related conditions are surprisingly inclusive.

4. DISCRETENESS, ETC.

In this section we discuss not only discreteness but several other loosely related topics.

If V is a primary N-group in which some subset $\{a\}$ where a is an element of V, is open, then translation yields the fact that any such singleton subset is open. Any non-empty subset is a union of such subsets and is also open. Thus V is discrete. It follows that:-

Proposition 4.1. *The primary N-group V is discrete if, and only if, for some a in V, $\{a\}$ is open.*

Discreteness is, however, a more subtle property than might at first be imagined. As a preliminary to giving a more significant characterisation we first prove:-

Proposition 4.2. *A non-zero discrete primary N-group contains a minimal N-subgroup.*

Proof. Let $V \neq \{0\}$ be a discrete primary N-group and $a \neq 0$ an element of V. Since $V \backslash \{a\}$ is Z-closed, there exist α in N such that $(V \backslash \{a\})\alpha = \{0\}$ but $a\alpha \neq 0$. Taking β in N such that $(a+w)\alpha - a\alpha = w\beta$, for all w in V, we see $w\beta = -a\alpha$, for all $w \neq 0$ in V. Now $H\beta \subseteq H$, for any non-zero N-subgroup H of V. It follows that any such N-subgroup of V contains $-a\alpha$ and the N-subgroup of V generated by $-a\alpha$ is non-zero and minimal. The proof of 4.2 is complete. □

We now give the characterisation mentioned above.

Theorem 4.3. *A faithful primary N-group $V \neq \{0\}$ is discrete if, and only if, N has a minimal right ideal.*

Proof. Suppose V is discrete and H the minimal N-subgroup of V of 4.2. If $h \neq 0$ is in H, then since hN is N-isomorphic to $N/(0:h)$, $(0:h)$ is a maximal right ideal of N. However, discreteness means $(0 : V \backslash \{h\}) \neq \{0\}$. Also elements of the intersection $(0 : V \backslash \{h\}) \cap (0 : h)$ annihilate V and faithfulness implies

$$(0 : V \backslash \{h\}) \cap (0 : h) = \{0\}.$$

Thus the natural N-isomorphism of

$$[(0 : V \backslash \{h\}) + (0 : h)]/(0 : h)(= N/(0 : h))$$

onto $(0 : V \backslash \{h\})/\{0\}$ yields the fact that $(0 : V \backslash \{h\})$ is minimal.

Now suppose N has a minimal right ideal R. Clearly $vR \neq \{0\}$ for some $v \neq 0$ in V. Now $R \leq (0 : u)$ or $R \cap (0 : u) = \{0\}$ for each $u \neq 0$ in V. Suppose $u \neq 0$ in V is such that the sum $R + (0 : u)$ is direct. In this case $vR + v(0 : u)$ is a central sum and the semiprimary nature of V implies this sum is direct. Also the primary nature of V implies $vR = \{0\}$ or $v(0 : u) = \{0\}$. Since we are assuming $vR \neq \{0\}$, we see $v(0 : u) = \{0\}$ and $(0 : u) \leq (0 : v)$. Now, by 1.7, we have $v = u$ and $R \leq (0 : a)$ for all $a \neq v$ in V. Thus $(0 : V \backslash \{v\})$ is non-zero and, by faithfulness, $Z(\alpha) = V\backslash\{v\}$ for any $\alpha \neq 0$ in $(0 : V\backslash\{v\})$. Thus $\{v\}$ is open and the theorem follows, by 4.1. $\qquad\square$

If $V \neq \{0\}$ is a discrete primary N-group, then for $v \neq 0$ in V, $V\backslash\{v\}$ is Z-closed and there exists α in N such that $v\alpha \neq 0$ and $u\alpha = 0$, for all $u \neq v$ in V. This brings us to the notion of pre-images which were first defined in §3 of [5]. If V is an N-group then a non-empty subset S of V excluding 0 will be called a *pre-image*, if there exists γ in N such that $v\gamma = w\gamma \neq 0$ for all v and w in S and $u\gamma = 0$ for all u in $V\backslash S$. In the primary N-group situation pre-images do not play quite the same role as that occurring with minimal compatible N-groups. However, they gain in importance when direct decomposition is under consideration. Quite basic use of pre-images will be made in §11.

Suppose S is a pre-image of the primary N-group V. By definition there exists $a \neq 0$ in V and γ in N such that $w\gamma = a$ for all w in S and $w\gamma = 0$ for all w in $V\backslash S$. Now $\{a\}\gamma^{-1} = S$, $\{0\}\gamma^{-1} = V\backslash S$ and by 2.9, $\{a\}\gamma^{-1}$ and $\{0\}^{-1}$ are both closed. It follows that S and $V\backslash S$ are both open and closed. This is covered by the following statement (c.f. 3.8 of [5]).

Proposition 4.4. *If V is a primary N-group with a pre-image S, then $(S, V\backslash S)$ is a disconnection of V.*

Another feature of pre-images is that they are always associated with a certain amount of discreteness.

Theorem 4.5. *A primary N-group V has a pre-image if, and only if, there exists α in N such that $V\alpha$ is non-zero and discrete.*

Proof. If V has a pre-image, then clearly there exists α in N such that $V\alpha = \{0, u\}$, where u is a non-zero element of V. Since $V\alpha$ is discrete the 'only if' part of the proof holds. Now suppose for some α in N, $V\alpha$ is non-zero and discrete. If $w\alpha, w \in V$, is a non-zero element of $V\alpha$, then there exists a Z-closed subset Z_1 of V such that $Z_1 \cap V\alpha = V\alpha\backslash\{w\alpha\}$. Take β in $(0 : Z_1)$ such that $w\alpha\beta \neq 0$. We have for all v in V, $v\alpha\beta = 0$ if $v\alpha \neq w\alpha$ and $v\alpha\beta = w\alpha\beta$ if $v\alpha = w\alpha$. Thus the set S of all v in V such that $v\alpha = w\alpha$ is a pre-image and the theorem holds. $\qquad\square$

It is well known that a topological group may be totally disconnected without necessarily being discrete (see §7 of [1]). The same is certainly true for primary N-groups. Also non-discrete primary N-groups may contain non-zero N-subgroups which are discrete. On the other hand total disconnection can be lifted from an arbitrarily small non-zero N-subgroup.

Theorem 4.6. *If V is a primary N-group and H a non-zero totally disconnected N-subgroup of V, then V is totally disconnected.*

Proof. It is necessary to show that the connected component of any b of V is $\{b\}$. This connected component is, by 3.2, a left coset of the right N-subgroup $c(V)$ (see §3). We need only show $c(V) = \{0\}$ and, in order to do this, the contrary is assumed. Now since V is primary $c(V) \cap H \neq \{0\}$. If

$$(c(V) \cap H : c(V)) = \{0\}$$

then, by 1.1, $c(V) \cap H \leq z(c(V))$ and $c(V) \cap H$ is a ring module. This contradicts the semiprimary nature of V. Thus there exists α in N such that $c(V)\alpha \subseteq c(V) \cap H$ and $c(V)\alpha \neq \{0\}$. Now $c(V)\alpha$ cannot be a connected subspace of V, otherwise $c(H)$ would contain $c(V)\alpha$ contrary to H being totally disconnected. Let (D_1, D_2) be a disconnection of $c(V)\alpha$. The inverse images $D_i\alpha^{-1}$ of D_i, $i = 1, 2$, in $c(V)$ are, by 2.9, disjoint open and closed subsets of $c(V)$. They are also non-empty and such that $D_1\alpha^{-1} \cup D_2\alpha^{-1} = c(V)$. Since $c(V)$ is connected, this contradiction can only mean $c(V) = \{0\}$ and V is totally disconnected. The theorem is completely proved. $\qquad\square$

Loosely connected with discreteness is the question of the existence of small N-subgroups (defined shortly). According to 4.2, a non-zero discrete primary N-group contains a minimal N-subgroup. An interesting fact is that the existence of such N-subgroups is equivalent to the non-existence of small N-subgroups. A primary N-group V is said to have *small N-subgroups* (have *SNS*), if every open subset of V containing 0 contains a non-zero N-subgroup. Otherwise V is said to have *no small N-subgroups* (have *NSNS*). A further assumption of this type also occurs. A primary N-group V is said to have *small closed N-subgroups* (have *SCNS*), if every open subset of V containing 0 contains a non-zero closed N-subgroup. Otherwise V is said to have *no small closed N-subgroups* (have *NSCNS*).

Theorem 4.7. *A primary N-group V has NSNS if, and only if, V has a minimal N-subgroup.*

Proof. If $V \neq \{0\}$ has *NSNS*, then there exists an open subset O of V containing 0 and such that O does not contain any non-zero N-subgroup of V. Now $O \neq V$ since otherwise O contains V. Consequently, there exists a in $V \backslash O$ and $-a + (V \backslash O)$ is, by 2.5, a non-empty Z-closed subset of V. Thus there exists α in N such that $(-a + (V \backslash O))\alpha = \{0\}$ and $(-a)\alpha \neq \{0\}$ (clearly 0 is not in $V \backslash O$ and $-a$ is not in $-a + (V \backslash O)$). If β in N is such that $(-a + w)\alpha - (-a)\alpha = w\beta$ for all w in V, then $w\beta = (-a)\alpha$ for all w in $V \backslash O$. Thus if H is any non-zero N-subgroup of V, then since $H \subseteq O$, there exists h in H such that $h\beta = (-a)\alpha$ and $(-a)\alpha$ is in H. It follows that, since $(-a)\alpha$ is contained in all such N-subgroups, the N-subgroup of V generated by $(-a)\alpha$ is minimal.

Now suppose $V \neq \{0\}$ has a minimal N-subgroup U. If $u \neq 0$ is in U, then since $\{u\}$ is closed, $V \backslash \{u\}$ is an open subset of V containing 0. If V has *SNS*, then there exists a non-zero N-subgroup U_1 of V contained in $V \backslash \{u\}$. The primary nature of V implies $U \cap U_1 \neq \{0\}$ and $U \leq U_1$. This is a contradiction, since u is not in U_1. The proof of the theorem is complete. $\qquad\square$

An obvious but interesting topological consequence is that:-

Corollary 4.8. *A non-zero primary N-group without a minimal N-subgroup has SNS.*

Another interesting fact is that the proof of 4.7 goes through with only superficial changes when *NSNS* is replaced by *NCSNS* and 'minimal N-subgroup' by 'minimal closed N-subgroup'. This gives:-

Proposition 4.9. *A primary N-group $V \neq \{0\}$ has NSCNS if, and only if, V has a minimal closed N-subgroup.*

The obvious but interesting topological consequence is that:

Corollary 4.10. *A non-zero primary N-group without a minimal closed N-subgroup has SCNS.*

A more well known property now occupies our attention.

5. THE HAUSDORFF PROPERTY

If V is a primary N-group, then as indicated in §2, any subset of V of the form $\{a\}$, with a in V, is closed. This means that if b and c are two distinct points of V, then we can find an open set containing b but excluding c. This is also expressed by saying V is a T_1-*space* (see §5 of [1]).

A stronger separation axiom that will be of use is the Hausdorff property. However, we first define what it means for points to be separated or unseparated. Here a subspace of a topological space (possibly V) is just a subset, with the topology of the space restricted to this subset. If V is a primary N-group and S a subspace of V, then two points a and b of S are said to be *separated in S* if there exist disjoint open subsets O_i, $i = 1, 2$, of S with a in O_1 and b in O_2. If no such open subsets of S exist, then a and b are said to be *unseparated in S*. If $S = V$ or it is clear what is implied, then the words 'in S' are omitted from the expressions 'separated in S' and 'unseparated in S'. The subspace S of V will be called *Hausdorff* if any two distinct points of S are separated in S.

Yet another useful property of this kind is regularity. The subspace S of V will be called *regular*, if for any a in S and closed subset S_1 of S excluding a we can find disjoint open subsets O_i, $i = 1, 2$, of S with a in O_1 and $S_1 \subseteq O_2$. In general for T_1-spaces the Hausdorff property is much weaker than regularity. The surprising fact that for primary N-groups the Hausdorff requirement implies regularity will now be proved. This has already been shown to hold for the primitive compatible case (see 2.1 of [5]) but, to make our present treatment reasonably self contained, full coverage is desirable.

Theorem 5.1. *Suppose V is a primary N-group. If V is Hausdorff, then it is regular.*

Proof. Let S be a closed subset of V and b a point of V not in S. If S is empty, then $\emptyset$ and V are disjoint open subsets of V with $S \subseteq \emptyset$ and b in V. Assume $S \neq \emptyset$ and let c be in S. Clearly $-c + S$ is a closed subset of V containing 0. Thus, by 2.5, $-c + S$ is Z-closed and of the form $Z(H)$, where H is a non-empty subset of N. Now $-c + b$ is not in $-c + S$ and there exists α in H such that $(-c + S)\alpha = \{0\}$ but $(-c + b)\alpha \neq 0$. Now, since V is Hausdorff, there exist disjoint open subsets O_i, $i = 1, 2$, of V with 0 in O_1 and $(-c + b)\alpha$ in O_2. Since $(-c + S)\alpha$ is contained in O_1, we see $O_1\alpha^{-1}$ contains $-c + S$ and, since $(-c + b)\alpha$ is in O_2, $-c + b$ is in $O_2\alpha^{-1}$. By 2.9, $c + O_1^{-1}$ and $c + O_2\alpha^{-1}$ are disjoint open subsets of V, respectively, containing S and b. The proof of the theorem is therefore complete. $\qquad\square$

The requirement that a faithful primary N-group is Hausdorff is intimately connected with the existence of direct sums of right ideals. The exposition of this fact is best given in terms of uniformity. According to the definition given in the introduction a non-zero right ideal of N is non-uniform in N, if it contains two non-zero right ideals of N with zero intersection. Apart from discreteness the Hausdorff assumption implies this is always the case.

Theorem 5.2. *Let N be a faithful primary N-group. If V is Hausdorff, then either V is discrete or every non-zero right ideal of N is non-uniform in N.*

Proof. Let $R \neq \{0\}$ be a right ideal of N. Now $Z(R)$ is properly contained in V and, if $V \backslash Z(R)$ consists of one element only, then by 4.1, V is discrete. Thus we may assume that a_i, $i = 1, 2$, are two distinct elements of $V \backslash Z(R)$. The Hausdorff property ensures the existence of open sets L_i, $i = 1, 2$, with a_i in L_i and $L_1 \cap L_2 = \emptyset$. Thus, with O_i, $i = 1, 2$, taken as $(V \backslash Z(R)) \cap L_i$, we see O_i are open, a_i is in O_i, $O_i \cap Z(R) = \emptyset$ and $O_1 \cap O_2 = \emptyset$. Let Z_i, $i = 1, 2$, be $V \backslash O_i$. Clearly Z_i, $i = 1, 2$, are proper closed subsets of V, $Z_1 \cup Z_2 = V$ and $Z(R) \subseteq Z_i$ for $i = 1, 2$. It will be shown that for $i = 1, 2, R \cap (0 : Z_i) \neq \{0\}$. If $R \cap (0 : Z_1) = \{0\}$, then the sum $R + (0 : Z_1)$ is direct and for all v in V the sum $vR + v(0 : Z_1)$ is central. The semiprimary nature of V implies this sum is direct and the primary nature of V implies $vR = \{0\}$ or $v(0 : Z_1) = \{0\}$. Thus for all v in V, v is in $Z(R)$ or, because Z_1 is Z-closed, v is in Z_1. However, $Z_1 \supseteq Z(R)$ and Z_1 is a proper subset of V. This contradiction can only mean $(0 : Z_1) \cap R \neq \{0\}$ and similarly $(0 : Z_2) \cap R \neq \{0\}$. Now the sum of the $(0 : Z_i) \cap R, i = 1, 2$, is direct, since an element of the intersection annihilates $Z_1 \cup Z_2$ and, by faithfulness, is zero. Thus the right ideal R contains the direct sum

$$(0 : Z_1) \cap R \oplus (0 : Z_2) \cap R,$$

and, since both right ideals of N are non-zero, the theorem is completely proved. $\square$

It appears theorem 5.2 does not have a converse. However, the opposite extreme from being Hausdorff is sparseness (see §7) and a corresponding result (see theorem 7.3) has a two way statement.

A further feature of being Hausdorff is that, in the case of primary N-groups, this property can be lifted from an arbitrarily non-zero N-subgroup.

Theorem 5.3. *If a primary N-group V has a non-zero N-subgroup H which is Hausdorff, then V is Hausdorff.*

Proof. Let $u \neq 0$ be an element of V. By 1.6, we have $C_V(u) = \{0\}$ and there exists h in H and α in N such that

$$(u + h)\alpha - h\alpha - u\alpha (= a \text{ say})$$

is non-zero. Let β in N be such that

$$(w + h)\alpha - h\alpha - w\alpha = w\beta,$$

for all w in V. Clearly β is in $(H : V)$ and since H is Hausdorff we can find disjoint open subsets L_i, $i = 1, 2$, of H such that 0 is in L_1 and a in L_2. By 2.9, $L_1\beta^{-1}$ and $L_2\beta^{-1}$ are disjoint open subsets of V. Now since $0\beta = 0$ we have 0 is in $L_1\beta^{-1}$ and since $u\beta = a, u$ is in $L_2\beta^{-1}$. Thus if u_i, $i = 1, 2$, are any two distinct points of V, then there exist disjoint open subsets O_i, $i = 1, 2$, of V with 0 in O_1 and $-u_1 + u_2$ in O_2. Since $u_1 + O_1$ and $u_1 + O_2$ are disjoint open subsets of V, respectively containing u_1 and u_2, the theorem is completely proved. $\square$

In most of this paper progress depends on the interplay of topology and algebra and, in this regard, the next section is certainly of quite real help.

6. CLOSURES

Several of the results of this section are not unlike those existing in other contexts. In general a primary N-group equipped with the Z' topology will not be a topological group. Primary N-group topology is often a lot more rudimentary. However, the proofs of some elementary topological group theory and primary N-group results are often similar (see 6.1 and 6.2). In order to give adequate coverage, some of the results of this section are of this type.

Let S be a subspace of the primary N-group V. The *closure in S of a subset S_1 of S* is the intersection of all closed subsets of S containing S_1. This subset of V is clearly a closed subset of S. In the case where $S = V$, or where it is evident as to what subspace S is being considered, the expression 'in S' is omitted from the term 'the closure in S of S_1'. An important aspect of closures is that underlying algebraic structure is often preserved.

Theorem 6.1. *If V is a primary N-group, then the closure K of a subgroup H of V is a subgroup of V.*

Proof. Now $V\backslash K$ is open and $H + (V\backslash K)$ being a union of the open sets $h + (V\backslash K), h \in H$, is open. Clearly $V\backslash K$ avoids H so the right cosets $H + v, v \in V\backslash K$, of H in V avoid H. Thus $H + (V\backslash K)$ avoids H and contains $V\backslash K$. If $K \cap [H + (V\backslash K)]$ is non-empty, then $V\backslash [H + (V\backslash K)]$ is a closed subset of V properly contained in K and containing H. This cannot happen since K is the closure of H and therefore $H + (V\backslash K) = V\backslash K$. Thus $V\backslash K$ is a union of right cosets of H and it follows that K is also. Now $H(-1) = H$ and, since -1 is a unit of N, $K(-1)$ is closed and $K \cap K(-1)$ contains H. This can only mean $K \cap K(-1) = K$ and $K \subseteq K(-1)$. Clearly this implies $K(-1) \subseteq K$ and $K = K(-1)$. Now take v in K and consider $K - v$. This is a union of right cosets of H and, since it contains $v - v$, it contains H. Since $K - v$ is a right translation of K, it is closed. Because $H \subseteq (K - v) \cap K$, we must therefore have $(K - v) \cap K = K$. It follows that $K \subseteq K - v$ and $K + v \subseteq K$. If a is in K, then $-a$ is in K and $-a + v$ is in K. Because a and v where arbitrary elements of K, it follows that K is a subgroup of V. The theorem is completely proved.   □

After necessary preliminaries we shall make an interesting application of 6.1. The intermediate proposition, which we now prove, is well known in the context of topological groups. It is an important result which will be used on several occasions in the remainder of this paper.

Proposition 6.2. *If V is a primary N-group then a subgroup H of V containing a non-empty open subset is both open and closed.*

Proof. Let O be a non-empty open subset of V contained in H. It is easily verified that $H + O = H$ and, since $H + O$ is a union of the open subsets $h + O, h \in H$, it follows that H is open. Now $H + [V\backslash H]$ is the union of all right cosets $H + v, v$ in $V\backslash H$ and coincides with $V\backslash H$. Also, since H is open, the union of all such $H + v$ is open. Thus H and $V\backslash H$ are both open and the proposition clearly holds.   □

We come now to the use of 6.1 which was hinted at above. The result is concerned with an interesting instance of the existence of essential right ideals. A right ideal R of a near-ring N is called *essential*, if there does not exist a right ideal $R_1 \neq \{0\}$ of N such that $R \cap R_1 = \{0\}$. A right ideal of N which is not essential is called *non-essential*.

Theorem 6.3. *If V is a faithful connected primary N-group and H a subgroup of V, then either $(0 : H) = \{0\}$ or $(0 : H)$ is essential.*

Proof. If $(0 : H) = \{0\}$, then the closure of H in V is V. Thus, by 6.1, we may assume that the closure K of H in V is a proper subgroup of V. Now, if $H\alpha = \{0\}, \alpha \in N$, then $K\alpha = \{0\}$ and consequently $(0 : K) = (0 : H)$. If $(0 : K)$ is non-essential, then there exists a right ideal $R \neq \{0\}$ of N such that $(0 : K) \cap R = \{0\}$. For all v in $V, v(0 : K) + vR$ is a central sum. Since V is semiprimary, this sum is direct. Also V is primary and either $v(0 : K) = \{0\}$ or $vR = \{0\}$. Thus v is in K or $Z(R)$ and $K \cup Z(R) = V$. However, faithfulness implies $Z(R) \neq V$ and $V \backslash Z(R)$ is a non-empty open subset of V contained in K. By 6.2, this would imply K is both open and closed contrary to V being connected. Thus $(0 : K)(= (0 : H))$ is essential and the theorem holds. $\qquad\square$

The above theorem made quite real use of 6.1. It is certainly nice to know that closures of subgroups are subgroups, but primary N-groups have more than just group structure. The important corollary of the next result will find use later.

Proposition 6.4. *If V is a primary N-group and H an N-subset of V, then the closure K of H in V is an N-subset of V.*

Proof. If α is in N then, by 2.9, $K\alpha^{-1}$ is closed. Since $H\alpha \subseteq H \subseteq K$, we have $K\alpha^{-1} \supseteq H$. Thus $K \cap K\alpha^{-1}$ is a closed subset of V containing H. It follows that $K \cap K\alpha^{-1} = K$ and $K\alpha^{-1} \supseteq K$. Thus for all a in $K, a\alpha$ is in K and K is an N-subset of V. The proof is complete. $\qquad\square$

Corollary 6.5. *If V is a primary N-group and H an N-subgroup of V, then the closure K of H is an N-subgroup of V.*

Proof. This follows by 6.1 and 6.4. $\qquad\square$

We now turn our attention to a rather different notion.

7. SPARSENESS

This section covers certain properties of sparse subspaces of a primary N-group. Some of these will be used later while others are very interesting in their own right.

A subspace S of a primary N-group V will be called *sparse* if any two points of S are unseparated in S. This definition differs from that given in [7] or §2 of [5]. However the following proposition removes this anomaly.

Proposition 7.1. *A subspace S of a primary N-group V is sparse if, and only if, the closure (in S) of any non-empty open subset of S is S.*

Proof. Suppose S is sparse and O is a non-empty open subset of S. If K is the closure in S of O and $S \backslash K$ is non-empty, then since $S \backslash K$ is open, points of $S \backslash K$ are separated (in S) from those of O. Thus $K = S$ and the 'only if' part follows. Now if the closure (in S) of any non-empty open subset of S is S, then clearly there cannot exist two non-empty disjoint open subsets of S. Thus all points are unseparated and the proposition is completely proved. □

When dealing with sparseness, closures tend to be well behaved.

Proposition 7.2. *The closure of a sparse subspace of a primary N-group V is sparse.*

Proof. Suppose K is the closure of a sparse subspace S of V and two points of K are separated in K. In this case there exist open subsets O_i, $i = 1,2$, of V with $O_i \cap K \neq \emptyset$ for $i = 1,2$, and $O_1 \cap K$ and $O_2 \cap K$ disjoint. Clearly if $O_i \cap S \neq \emptyset$ for $i = 1,2$, then points of $O_1 \cap S$ are separated (in S) from those of $O_2 \cap S$. We may assume $O_1 \cap S = \emptyset$, so that $K \setminus O_1$ is a closed subset of V containing S and strictly contained in K. This contradiction to the nature of K establishes the result. □

It is clear from the definition that the Hausdorff property and the sparse property represent opposite extremes. In between these extremes there exists a whole range of possibilities. Furthermore, the Hausdorff property has been seen (see 5.2) to be associated with an abundance of direct sums. Thus the absence in the near-ring of non-zero right ideals with zero intersection might be expected to indicate sparseness. Not only is this so but, unlike the Hausdorff case, the condition is a characterization.

Theorem 7.3. *A faithful primary N-group V is sparse if, and only if, N is uniform.*

Proof. Suppose V is sparse but N is not uniform. In this case there exists two non-zero right ideals R_i, $i = 1,2$, of N such that the sum $R_1 + R_2$ is direct. Now for v in V, $vR_1 + vR_2$ is a central sum. The semiprimary nature of V implies this sum is direct and the primary nature of V implies $vR_1 = \{0\}$ or $vR_2 = \{0\}$. It follows that $V = Z(R_1) \cup Z(R_2)$ and since $R_i \neq \{0\}$ for $i = 1,2, V \setminus Z(R_i)$ is non-empty. Clearly

$$(V \setminus Z(R_1)) \cap (V \setminus Z(R_2)) = \emptyset$$

and the $V \setminus Z(R_i)$, $i = 1,2$, are non-empty open disjoint subsets of V. This would clearly mean points of $V \setminus Z(R_1)$ are separated from those of $V \setminus Z(R_2)$ contrary to V being sparse. The 'only if' part of the proof is established.

Now suppose N is uniform, but V is not sparse. There exists two non-empty disjoint open subsets O_i, $i = 1,2$, of V. Set $Z_i = (V \setminus O_i) \cup \{0\}$, for $i = 1,2$. It is clear from 2.5, that the Z_i being closed subsets containing 0, are Z-closed. Also if Z_i, $i = 1,2$, are proper subsets of V, then $(0 : Z_i)$ are non-zero right ideals of N. However, an element of $(0 : Z_1) \cap (0 : Z_2)$ acts as zero on Z_1 and Z_2 and therefore on $Z_1 \cup Z_2$. Since $Z_1 \cup Z_2 = V$, faithfulness implies $(0 : Z_1) \cap (0 : Z_2)$ is zero contrary to N being uniform. The only other possibility is that Z_1 or Z_2 (Z_1 say) is V. Now $O_1 \neq \emptyset$ so, if $Z_1 = V$ we have $O_1 = \{0\}$ and V is discrete (see 4.1). Since V is semiprimary, $|V| > 2$ and for $v \neq 0$ in V, we have $(0 : V \setminus \{v\})$ and $(0 : v)(= (0 : \{0, v\}))$ are non-zero. In this case faithfulness implies the sum $(0 : V \setminus \{v\}) + (0 : v)$ is direct, contrary to N being uniform. We have shown that if N is uniform, then V must be sparse and the theorem is completely proved. □

For a primary N-group V, deeper study of sparseness rests on properties of the subset of V consisting of all v in V, which are unseparated from 0. This very useful subset of V

will be denoted by $u(V)$. Even in entirely general situations $u(V)$ enjoys the properties set forth in the next theorem.

Theorem 7.4. *If V is a primary N-group then $u(V)$ is a closed N-subset of V.*

Proof. If a in V is seperated from 0, then there exists disjoint open subsets O containing 0 and O_a containing a. Clearly all points of O_a are separated from 0 and the union of all the O_a is an open subset H of V consisting precisely of all points of V separated from 0. Thus $V \backslash H$ is a closed subset of V coinciding with $u(V)$. It remains to show $u(V)$ is an N-subset of V. Suppose there exists v in $u(V)$ and α in N such that $v\alpha$ is in H. Clearly, since $v\alpha$ is separated from $0, v\alpha \neq 0$ and there exist open disjoint subsets O_1 containing 0 and O_2 containing $v\alpha$. Now $O_i\alpha^{-1}$ are disjoint open subsets of V and, since $0\alpha = 0, O_1\alpha^{-1}$ contains 0. Also since O_2 contains $v\alpha, O_2\alpha^{-1}$ contains v. This would mean 0 and v are separated. We conclude that $v\alpha$ is in $u(V)$. This holds for all v in $u(V)$ and α in N and $u(V)$ is an N-subset of V. The theorem is proved. $\qquad\square$

We shall see shortly (in 7.6) that there exist relatively general situations in which an N-subgroup of V contained in $u(V)$ is sparse. However, much more information can be obtained, if $u(V)$ contains a non-empty open subset of V (c.f. 6.2).

Theorem 7.5. *If V is a primary N-group in which $u(V)$ contains a non-empty open subset, then $u(V)$ is a sparse open and closed N-subgroup coinciding with $c(V)$.*

Proof. Suppose O_1 is a non-empty open subset of V contained in $u(V)$. Since $u(V)$ is an N-subset of $V, O_1(-1)$ is another such subset. Thus $O = O_1 \cup O_1(-1)$ is a non-empty open subset of V contained in $u(V)$ and such that $O(-1) = O$. It will now be shown that the subgroup K of V generated by O is contained in $u(V)$. Since $O(-1) = O$, it is required that we show that if $k \geq 1$ is an integer and $h_i, i = 1, \ldots, k$, are in O, then $h_1 + \cdots + h_k$ is in $u(V)$. This will be proved by induction on k (assumed minimal). Clearly, if $k = 1$, the result holds since $O \subseteq u(V)$. Inductively it follows that for $k \geq 2, h_2 + \cdots + h_k$ is unseparated from 0 and translation means $h_1 + \cdots + h_k$ is unseparated from h_1. Since O is an open subset of V containing h_1, any open subset O_2 of V containing $h_1 + \cdots + h_k$ is such that $O \cap O_2 \neq \emptyset$. However, points of $O \cap O_2$ are unseparated from 0 and any open subset O_3 of V containing 0 must be such that $O \cap O_2 \cap O_3 \neq \emptyset$. Clearly this means $O_2 \cap O_3$ is non-empty and $h_1 + \cdots + h_k$ is unseparated from 0. We have shown the subgroup K of V is contained in $u(V)$.

Now K contains the non-empty open subset O and is therefore open and closed in V by 6.2. Also, if K is properly contained in $u(V)$ and b is in $u(V)\backslash K$, then $V\backslash K$ and K are disjoint open subsets of V containing b and 0 respectively. This is contrary to the nature of $u(V)$ and $u(V)$ is, therefore, an open and closed N-subgroup of V. Also, using the fact that $u(V)$ is open, two points of $u(V)$ are separated (in V) precisely when they are separated in $u(V)$. If g_1 and g_2 are two such points, then under translation, we see $g_1 - g_2$ is separated from 0. This cannot happen because $u(V)$ is a group. Thus $u(V)$ is sparse. It remains to show $u(V)$ coincides with $c(V)$. Since, if $u(V) < V, (u(V), V\backslash u(V))$ is a disconnection and 0 is in $u(V)$, it clearly follows $c(V) \leq u(V)$. However, $u(V)$ is connected, since a disconnection would imply the existence of points of $u(V)$ unseperated from 0. Thus $u(V) \leq c(V)$ and the theorem is entirely proved. $\qquad\square$

Previous results illustrate the fact that sparseness is very useful in allowing development of quite deep theoretical material. However, it is a concept which also makes itself felt in more concrete settings. The final theorem of this section shows that serious study of Ω-groups and their polynomial maps can also involve the use of sparseness.

Theorem 7.6. *Let V be a rigid Ω-group. If the $P_0(V)$-group V is primary, then a $P_0(V)$-subgroup H of V contained in $u(V)$ is sparse and if $u(V) \neq \{0\}$, then V contains non-zero sparse closed $P_0(V)$-subgroups.*

Proof. Suppose v in H is separated in H from 0. The rigidity of V means the ideal of V generated by all $(u_1, u_2)\varphi$ with $(x, y)\varphi$ in $V_0[x, y]$ and u_i, $i = 1, 2$ in the ideal $vP_0(V)$ (see 1.8) coincides with $vP_0(V)$. Any sum

$$(v\alpha_1, v\beta_1)\varphi_1 + \cdots + (v\alpha_k, v\beta_k)\varphi_k,$$

with $k \geq 1$ an integer, α_i and β_i, $i = 1, \ldots k$, in $P_0(V)$ and $(x, y)\varphi_i$, $i = 1, \ldots, k$, in $V_0[x, y]$ is of the form $(v, v)\theta$ with $(x, y)\theta$ in $V_0[x, y]$. The collection S of all such $(v, v)\theta$ is therefore a $P_0(V)$-subgroup of V. Thus, being an ideal of V (see 1.8) S coincides with $vP_0(V)$. This means there exists $(x, y)\psi$ in $V_0[x, y]$ such that $(v, v)\psi = v$. Now let λ in $P_0(V)$ be given by $w\lambda = (v, w)\psi$ for all w in V. For the moment consider w to be fixed. In this situation the map taking u in V to $(u, w)\psi$ is in $P_0(V)$. Returning to the definition of λ we see $w\lambda$ must be in $vP_0(V)$ for all w in V and λ maps V into H and is such that $v\lambda = v$. Also there exist disjoint open subsets O_1 and O_2 of H such that 0 is in O_1 and v in O_2. By 2.9, $O_1\lambda^{-1}$ and $O_2\lambda^{-1}$ are disjoint open subsets of V respectively containing 0 and v. This contradiction can only mean 0 and v are unseperated in H. However, if g_1 and g_2 are any two points of H, then the unseperated nature of 0 and $g_1 - g_2$ implies that of g_1 and g_2. We have shown H is sparse. Now, if $a \neq 0$ is in $u(V)$, then from above (see also 7.4) $aP_0(V)$ is a non-zero sparse $P_0(V)$-subgroup of V. By 6.5, the closure of $aP_0(V)$ is a non-zero $P_0(V)$-subgroup of V and, by 7.2, this $P_0(V)$-subgroup is sparse. The theorem is completely proved. $\square$

In view of 7.6 the following question certainly appears very meaningful:- When do rigid Ω-groups as in 7.6, possess the property that $u(V)$ is a $P_0(V)$-subgroup? Further questions can be raised but 7.6 signals the completion of this section and time to consider other specific uses of Z^t.

8. PRIMENESS

A near-ring N has been called *prime*, if $AB \neq \{0\}$, whenever A and B are non-zero ideals of V. This section is particularly concerned with prime compatible non-rings. The first theorem shows that the study of such non-rings finds its rightful place within the study of primary N-groups.

Theorem 8.1. *If the prime non-ring N is compatible on V, then $V/z(V)$ is a faithful primary N-group.*

Proof. Since

$$V(z(V) : V)(0 : z(V)) \subseteq z(V)(0 : z(V)) = \{0\}$$

we have

$$(z(V) : V)(0 : z(V)) = \{0\}.$$

If $(0 : z(V)) = \{0\}$, then N is faithful on the ring module $z(V)$. The fact that N is a non-ring implies $(0 : z(V)) \neq \{0\}$ and the primeness of N means $(z(V) : V) = \{0\}$. Thus N is faithful on the N-group $V/z(V)$.   □

Let $V_1 = V/z(V)$. We show that if U_1 is a submodule of V_1 such that $(U_1 : V_1) = \{0\}$, then $U_1 = \{0\}$. Now there exists a submodule $U \geq z(V)$ of V such that $U/z(V) = U_1$. Also

$$(U_1 : V_1) = (U : V)/(z(V) : V)$$

and, since $(U_1 : V_1) = \{0\}$, we have $(U : V) = \{0\}$. By 1.1, this implies $U \leq z(V)$ and $U_1 = \{0\}$. Now suppose H_1 is a ring submodule of V_1. Since V_1 is faithful it follows that

$$V_1(H_1 : V_1)(0 : H_1) = \{0\}$$

and $(H_1 : V_1)(0 : H_1) = \{0\}$. If $H_1 \neq \{0\}$, then $(H_1 : V_1) \neq \{0\}$ and primeness implies $(0 : H_1) = \{0\}$. This would mean N is faithful on the ring module H_1 and is a ring. The only conclusion is that $H_1 = \{0\}$ and V_1 is semiprimary. Now, if X_i, $i = 1, 2$, are two submodules of V_1 such that $X_1 \cap X_2 = \{0\}$, then

$$V_1(X_1 : V_1)(X_2 : V_1) \subseteq X_1 \cap X_2 = \{0\}$$

and $(X_1 : V_1)(X_2 : V_1) = \{0\}$. Primeness implies one of the $(X_i : V_1)$ is zero and, by what has been proved above, this X_i is $\{0\}$. Thus V_1 is primary and the proof of 8.1 is complete.

Faithful primary N-groups of a prime near-ring N enjoy a property similar in some respects to minimality.

Theorem 8.2. *If V is a faithful primary N-group of the prime near-ring N, then every non-zero N-subgroup of V is faithful.*

Proof. If H is a non-zero N-subgroup of V and $h \neq 0$ in H, then since $z(V) = \{0\}$, there exists u in V and α in N such that $(h + u)\alpha - u\alpha - h\alpha$ is non-zero. Let β in N be such that

$$(h + w)\alpha - w\alpha - h\alpha = w\beta,$$

for all w in V. Because $u\beta \neq 0$ we have $\beta \neq 0$ and since h is in H, β is in $(H : V)$. Now

$$V \cdot (H : V)(0 : H) = \{0\},$$

and $(H : V)(0 : H) = \{0\}$ so, by the primeness of N, $(0 : H) = \{0\}$. Thus N is faithful on H and the theorem is entirely proved.   □

Theorem 8.2 tends to indicate that there may be relatively general circumstances under which N is faithful on a minimal N-subgroup of V and therefore primitive. It will be seen in 8.7 that this, is indeed the case. First, another general result indicating another aspect of the 'minimality' of V is given.

Theorem 8.3. *Let V be a faithful primary N-group. The near-ring N is prime if, and only if, V has no non-zero proper closed N-subgroups.*

Proof. Suppose V has no non-zero proper closed N-subgroups but N is not prime. In this case there exists two non-zero ideals A and B of N such that $AB = \{0\}$. Thus $VA \subseteq Z(B)$ and since $B \neq \{0\}$, $Z(B) \neq V$. Now there exists v in V such that $vA \neq \{0\}$. It follows that vA is contained in the proper closed subset $Z(B)$ of V and, by 6.5, the closure of vA is a non-zero proper closed N-subgroup of V. This contradiction can only mean N is prime.

Now suppose N is prime and U is a non-zero proper closed N-subgroup of V. Since V is primary we have, by 1.6, that $C_V(u) = \{0\}$ for some non-zero element u of U. Thus there exists $v \neq 0$ in V and α in N such that $(u+v)\alpha - v\alpha - u\alpha$ is non-zero. By compatibility there exists β in N such that

$$(u+w)\alpha - w\alpha - u\alpha = w\beta$$

for all w in V. Since u is in U, $w\beta$ is in U, for all w in V. Thus β is in $(U:V)$ and since $v\beta \neq 0, (U:V) \neq \{0\}$. Now $Z[(0:U)] = U$ and $(0:U) \neq \{0\}$. However,

$$V(U:V)(0:U) \subseteq U(0:U) = \{0\}$$

and $(U:V)(0:U) = \{0\}$. This contradicts the fact that N is prime. Thus V contains no non-zero proper closed N-subgroups and the theorem is proved.   □

Theorem 8.3 allows us to deduce a corollary that will be of value in §9.

Corollary 8.4. *Let V be a faithful primary N-group and H a proper non-zero N-subgroup of V. If N is prime, then a non-empty open subset O of V has non-empty intersection with H and $V \backslash H$.*

Proof. If $O \subseteq H$, then by 6.2, H is closed contrary to 8.3. If $O \subseteq V \backslash H$, then the closure of H in V is, by 6.5, a closed N-subgroup contained in $V \backslash O$. Again this contradicts 8.3 and the corollary is proved.   □

We now look at further properties of primary N-groups as in 8.3.

Proposition 8.5. *If V is a primary N-group with no non-zero proper closed N-subgroups, then V is connected or totally disconnected. Furthermore, if V is connected, then every N-subgroup of V is also.*

Proof. By 3.1, $c(V)$ is either V or $\{0\}$. The first part of the result now follows, by 3.2. Suppose V is connected and H a non-zero N-subgroup of V. By 3.1, $c(H)$ is a closed N-subgroup of H. The closure L of $c(H)$ in V is, by 6.5, a closed N-subgroup of V and clearly also such that $L \cap H = c(H)$. Thus $L = V$ or $L = \{0\}$. If $L = \{0\}, c(H) = \{0\}$ and, by 3.2, H is totally disconnected. This contradiction to 4.6 means $L = V, H = c(H)$ and H is connected. The proof is complete.   □

There is also a clear distinction with N-groups of 8.3 between those which are or are not Hausdorff.

Proposition 8.6. *If V is a primary N-group with no non-zero proper closed N-subgroup, then V is either sparse or Hausdorff.*

Proof. If V is not Hausdorff, then there exist two distinct unseperated points of V. Translation gives a point $a \neq 0$ of V unseperated from 0. Now, by 7.4, all points of V unseperated from 0 form a closed N-subset $u(N)$ of V. The closure of aN is, by 6.5, a closed N-subgroup of V contained in $u(V)$. Since this closure must be $V, u(V) = V$ and all points of V are unseperated from 0. Thus, if $g_i, i = 1, 2$, are two points of V, then using translation and the unseperated nature of $g_2 - g_1$ from 0, we see g_1 and g_2 are unseperated. The proof is complete.   □

By 8.1, a prime compatible non-ring N has a faithful primary N-group. By 8.3, such an N-group V has no non-zero proper closed N-subgroups. Now, by 8.6, V is either sparse or Hausdorff and this respectively occurs (see 7.3) when N is uniform or not uniform. It is therefore consistent with previous developments to call such an N *sparse* if it is uniform and *Hausdorff* otherwise. The important fact that prime compatible near-rings which are Hausdorff are primitive compatible near-rings, follows from our next result.

Theorem 8.7. *Suppose $V \neq \{0\}$ is a primary N-group without non-zero proper closed N-subgroups. If V is Hausdorff, then it contains a minimal N-subgroup.*

Proof. Clearly we may assume V contains non-zero proper N-subgroups. Let H be such an N-group. Now if O is a non-empty open subset of V, then $O + H$ being a union of the $O + h, h \in H$, is open. Also $O + H$ is a union of left cosets of H and if $O + H \neq V$, then translation of $V \backslash [O + H]$ by $-b$, with b in $V \backslash [O + H]$, yields a proper closed subset $-b + [V \backslash [O + H]]$ containing H. The closure of H is however a closed N-subgroup of V (see 6.5) and coincides with V. This yields the contradiction that $-b + [V \backslash [O + H]]$ is not proper and implies $O + H = V$. If the closure K of O in V does not contain $V \backslash \{0\}$, then $K \cup \{0\}$ is Z-closed and there exists $\alpha \neq 0$ in N such that $K\alpha = \{0\}$ but $V\alpha \neq \{0\}$. Thus

$$V\alpha \subseteq O\alpha + H \subseteq H$$

and this holds for all non-zero proper N-subgroups H, so that the N-subgroup of V generated by $V\alpha$ is necessarily minimal. The only other possibility is that for all non-empty open subsets O of V, the closure of O is V or $V \backslash \{0\}$. If for some such O this closure is $V \backslash \{0\}$, then V is discrete and since $|V| \neq 2$, some O exists with closure a proper subset of $V \backslash \{0\}$. In this case V has a minimal N-subgroup from above. Thus we may assume that all non-empty open subsets of V have closure coinciding with V. This situation is excluded (see 7.1) since V is Hausdorff. The theorem is completely proved. $\square$

Corollary 8.8. *A prime compatible non-ring which is Hausdorff is a primitive compatible near-ring.*

Proof. This follows from 8.1, 8.2, 8.3 and 8.7. $\square$

It might be infered from 8.1, 8.2, 8.3 and 8.7 that in the Hausdorff case the study of a prime compatible non-ring N depends entirely on the use of Z^t within a faithful compatible minimal N-group. This is certainly not the case. The fact that faithful primary N-groups exist which may not be minimal raises many questions and allows fuller use of the Z^t topology. However, it is true, in a number of important situations, that such an N-group is minimal. Results of this kind are contained within the contents of the next section (see 9.3 and 9.5).

9. GRADINGS

The existence of two mutually orthogonal idempotents e_i, $i = 1, 2$, with $e_1 + e_2 = 1$ in some near-ring N is often of interest. For example, one question relating to direct decomposition, is whether or not the $e_i N, i = 1, 2$, are right ideals. However, for primary N-groups the near-ring N may have very useful elements exhibiting much weaker orthogonality properties.

If V is a primary N-group, then a *grading* on V is a triple (e, O_1, O_2), where O_i, $i = 1, 2$, are non-empty open subsets of V and e an element of N, such that $we = w$, for all w in O_1 and $we = 0$, for all w in O_2. It is clearly true that (e, O_1, O_2) is such a grading if, and only if, $(1 - e, O_2, O_1)$ is also. The primary N-group V will be called *totally graded* if for any two distinct points a_i, $i = 1, 2$, in V we can find a grading (e, O_1, O_2) with a_i in O_i. The fact that gradings allow us to establish the existence of open N-subgroups is of very real interest.

Theorem 9.1. *If V is a primary N-group and (e, O_1, O_2) a grading on V, then any N-subgroup H of V such that $H \cap O_i \neq \emptyset$ for $i = 1, 2$, is both open and closed in V.*

Proof. Clearly $H + H \cap O_1$ coincides with H and there exists h in H such that $h + O_1$ intersects O_2 non-trivially. Now there exists β in N such that

$$(-h + w)e - we - (-h)e = w\beta$$

for all w in V. Clearly since $-h$ is in H, $w\beta$ is in H for all w in V. However, if w is in $(h + O_1) \cap O_2$, then $-h + w$ is in O_1 and $w\beta = -h + w - (-h)e$.
  Thus

$$[(h + O_1) \cap O_2]\beta = -h + (h + O_1) \cap O_2 - (-h)e (= X \text{ say})$$

is contained in H. Now X is clearly a non-empty open subset of V, and, by 6.2, H is both open and closed. The proof is complete $\square$

An easy consequence of 9.1 follows:-

Corollary 9.2. *If the primary N-group V is totally graded, then every non-zero N-subgroup of V is both open and closed in V.*

Certainly an N-subgroup of a primary N-group is primary. Such a subgroup H is called *graded* if there exists a grading (e, O_1, O_2) of H. It should be noted that the O_i, $i = 1, 2$, are open in H but may not be open in V. It is clear that saying V is graded simply implies the existence of a grading.
  It has been shown (see 8.1) that if the compatible non-ring N is prime, then it has a faithful primary N-group. Also such an N-group has no non-zero proper closed N-subgroups (see 8.3). Furthermore, in the case where V is Hausdorff, it contains a minimal N-subgroup H (see 8.7). Since, by 8.2, N is faithful on H it is an interesting question as to when $H = V$.

Theorem 9.3. *Suppose the prime non-ring N is faithful on the primary N-group V (see 8.1). If V is graded, then V is a minimal N-group.*

Proof. Let (e, O_1, O_2) be a grading on V. Suppose H is a non-zero proper N-subgroup of V. By 8.4, $H \cap O_i \neq \emptyset$ for $i = 1, 2$, and by theorem 9.1, H is open and closed in V. Thus $V \backslash H$ is a non-empty open subset of V avoiding H. This contradiction to 8.4 means the theorem has been proved. $\square$

We now examine an important example of a grading.

Lemma 9.4. *Suppose V is a faithful primary N-group and N has a non-trivial direct decomposition $R_1 \oplus R_2$. If $1 = e_1 + e_2$, where e_i is in $R_i, i = 1, 2$, then $Ve_1 \cap Ve_2 = \{0\}, Ve_1 \cup Ve_2 = V$ and*

$$(e_1, Ve_1 \backslash \{0\}, Ve_2 \backslash \{0\})$$

is a grading.

Proof. Suppose v in V is such that $ve_i \neq 0$ for $i = 1, 2$. We have vN is a central sum $vR_1 + vR_2$, where $vR_i \neq \{0\}$ for $i = 1, 2$. Since vN is semiprimary $vR_1 \cap vR_2 = \{0\}$ and, since vN is primary we have a contradiction. Thus for all v in $V, ve_1 = 0$ or $ve_2 = 0$. Now $v = v1 = ve_1 + ve_2$ and each v in V is in Ve_1 or Ve_2. Also $e_i, i = 1, 2$ are orthogonal idempotents so if $ue_1 = we_2$ (u and w in V), then $ue_1 = we_2 = 0$. We have shown $Ve_1 \cup Ve_2 = V$ and $Ve_1 \cap Ve_2 = \{0\}$. Again using the fact that the $e_i, i = 1, 2$, are orthogonal idempotents we see $Z(e_1) = Ve_2$ and $Z(e_2) = Ve_1$. Therefore $Ve_1 \backslash \{0\}$ and $Ve_2 \backslash \{0\}$ are open. Now e_1 clearly acts as the identity on $Ve_1 \backslash \{0\}$ and as zero on $Ve_2 \backslash \{0\}$ so the lemma is proved. $\square$

A corollary again addressing the question, of when certain primary N-groups are minimal, follows.

Corollary 9.5. *Suppose the prime non-ring N is faithful on the primary N-group V (see 8.1). If N has non-trivial direct decomposition, then V is a minimal N-group.*

Proof. This is a direct consequence of 9.3 and 9.4. $\square$

If in accordance with 9.4, we have direct decomposition, then the indicated grading provides important consequences.

Theorem 9.6. *Suppose V is a faithful primary N-group and N has a non-trivial direct decomposition $R_1 \oplus R_2$. If H is an N-subgroup of V such that $HR_i \neq \{0\}$ for $i = 1, 2$, then H is both open and closed in V.*

Proof. Since $HR_i \neq \{0\}$ and $R_i = e_i N$ ($e_i, i = 1, 2$, as in 9.4) we have $He_i \neq \{0\}$ for $i = 1, 2$, and $H \cap (Ve_i \backslash \{0\}) \neq \emptyset$. Thus, by 9.1 and 9.4, H is both open and closed. The proof of 9.6 is complete. $\square$

Gradings are also of considerable interest when compactness is considered. The next theorem deals with an important situation where the compact assumption implies the compactness of certain N-subgroups.

Theorem 9.7. *If V is a compact primary N-group, then a graded N-subgroup H of V is compact.*

Proof. (the main point of this theorem is that H is graded as a primary N-group although V may have no such grading). Let K be the closure of H and O a non-empty open subset of K. Now $H + O$ being a union of the open sets $h + O, h \in H$, is open and, by 6.1, contained in K. Also it is a union of right cosets of H and if it avoids H, then $K \backslash (H + O)$ is a proper closed subset of K containing H. Clearly this cannot happen and $H + O$ contains H. Also if $H + O \neq K$, then $K \backslash (H + O)$ is a closed non-empty subset of K and a union of right cosets of H in K. Thus if b is in $K \backslash (H + O)$, then $[K \backslash (H + O)] - b$ is a union of right cosets of H and contains 0. It follows that $[K\backslash(H + O)] - b$ is a closed subset of K containing H and $[K\backslash(H + O)] - b = K$. Clearly this contradicts the assumption that $K\backslash(H + O)$ is a proper subset of K. We have shown that any non-empty open subset O of K is such that $H + O = K$.

Now suppose (e_1, O_1, O_2) is a grading of H. There exists h in H such that $(h + O_1) \cap O_2 \neq \emptyset$. Let β in N be such that

$$(-h + w)e - we - (-h)e = w\beta,$$

for all w in V. Clearly $V\beta \subseteq H$ and for all w in $(h + O_1) \cap O_2$, we have $w\beta = -h + w - (-h)e$. Thus $V\beta$ contains the non-empty open subset

$$-h + (h + O_1) \cap O_2 - (-h)e(= X \text{ say})$$

of H. By 2.9, the function of V into V induced by β is a continuous map. Now the image of a compact space under a continuous map is compact (see §8 of [1]) and consequently $V\beta$ is compact. Since X is open in H, there exists an open subset X_1 of K such that $H \cap X_1 = X$. Now from the first part of the proof $H + X_1 = K$. Thus K is covered by open sets $h + X_1, h$ coming from H. Now K is closed in V and a closed subspace of a compact space is compact (see §8 of [1]). Thus there exists an integer $k \geq 1$ and $h_1, \ldots, h_k$, in H such that the union of the $h_i + X_1$, $i = 1, \ldots, k$, covers K. However, $h_i + (X_1 \backslash X)$ is contained in $K\backslash H$ for $i = 1, \ldots, k$, so that the union of the $h_i + X$ covers H. Also we have

$$h_i + X \subseteq h_i + V\beta \subseteq H$$

for $i = 1, \ldots, k$. Thus the union of the $h_i + V\beta$, $i = 1, \ldots, k$, covers H and since these subsets of V are in H this union coincides with H. However, each $h_i + V\beta$ is a translation of a compact subset of V and therefore compact. Thus H is a union of a finite number of compact subsets of V. From §8 of [1] the theorem now follows. □

Theorem 9.7 allows the proof of more results along the lines of 9.3 or 9.5.

Theorem 9.8. *Suppose the prime non-ring N is faithful on the primary N-group V (see 8.1). If V is compact, then it has no non-zero proper graded N-subgroups.*

Proof. If H is a proper graded N-subgroup of V with grading (e, O_1, O_2), then the nature of e implies $O_1 \cap O_2 = \{0\}$ or $O_1 \cap O_2 = \emptyset$. In the first case $\{0\}$ is open in H and, by 4.1, H is discrete. In the second case, we have, by 8.2, N is faithful on H and, by 8.3 and 8.6, H is Hausdorff. Thus in either case H is Hausdorff and, by 5.3, V is Hausdorff. Since a compact subspace of a Hausdorff space is closed (see §8 of [1]) theorem 9.7 implies H is closed in V. This, however, is a contradiction to 8.3 and the theorem is proved. □

Before leaving this section we briefly look at the situation in which every N-subgroup of a primary N-group is graded.

Theorem 9.9. *If V is a primary N-group in which every N-subgroup is graded, then every N-subgroup of V is closed.*

Proof. Let H be an N-subgroup of V and K the closure of H. Now, by 6.5, K is an N-subgroup of V and is graded. Let (e, O_1, O_2) be a grading of K. If $H \cap O_i \neq \emptyset$ for $i = 1, 2$, then theorem 9.1 implies H is closed in K and $H = K$. If on the other hand one of the O_i,

$i = 1, 2, O_1$ say, is disjoint from H, then $K \backslash O_1$ is a proper closed subset of K containing H. This clearly contradicts the fact that K is the closure of H and implies $H = K$. Thus H is closed and the theorem is proved.   □

10. USE OF CHAIN CONDITIONS

In this section we make a reasonably systematic study of how chain conditions, taken in conjunction with the Z^t topology, can supply valuable information. Descending chain condition on right ideals (i.e., $DCCR$) may be thought of as the strongest assumption of this type. According to 1.4, a compatible N-group V where N has $DCCR$ is such that $V/L(V)$ is finite. Since, by 1.4, $L(V)$ is an N-nilpotent N-group this means that if V is primary, then it is finite. Consequently we have:

Proposition 10.1. *If V is a primary N-group and N has DCCR, then V is finite.*

An assumption that may be considered a weaker form of chain condition on N, is that obtained by requiring N to have no infinite direct sums. If N is a near-ring with an infinite family R_i, $i \in I$, of non-zero right ideals in which the sum $\sum R_i$, $i \in I$, is direct, then we say N has infinite *direct sums*. Otherwise we say N *has no infinite sums*. When N has a faithful primary N-group a basic characterisation of the no infinite direct sums requirement can be given. Before presenting this interesting theorem a lemma is in order.

Lemma 10.2. *If V is a faithful primary N-group, then N has an infinite direct sum if, and only if, there exists a sequence O_i, $i = 1, 2, \ldots$, of mutually disjoint non-empty open subsets of V.*

Proof. First we show that if such a sequence O_i, $1 = 1, 2, \ldots$, exists, then N has an infinite direct sum. By renaming the O_i, if necessary, we may assume that no $O_j (j$ in $\{1, 2, \ldots\})$ is $\{0\}$. Let $\triangle_i$, $i = 1, 2, \ldots$, be the union of all O_j with $j = i+1, i+2, \ldots$. Clearly O_i and $\triangle_i$ are non-empty disjoint open subsets of V and $(V \backslash O_i) \cup (V \backslash \triangle_i) = V$. Now, by 2.5, $(V \backslash O_i) \cup \{0\}$ and $(V \backslash \triangle_i) \cup \{0\}$ are non-zero proper Z-closed subsets of V and their annihilators $(0 : V \backslash O_i)$ and $(0 : V \backslash \triangle_i)$ are non-zero (otherwise these subsets of V would coincide with V). Also an element of the intersection of these annihilators acts as the zero on $V \backslash O_i$ and $V \backslash \triangle_i$ and, by faithfulness, is 0. Thus for $i = 1, 2, \ldots$, the sum

$$(0 : V \backslash O_i) + (0 : V \backslash \triangle_i),$$

is direct and in particular

$$(0 : V \backslash O_1) + (0 : V \backslash \triangle_1)$$

is direct. However, $(0 : V \backslash \triangle_1)$ contains $(0 : V \backslash O_2)$ and $(0 : V \backslash \triangle_2)$ and the sum

$$(0 : V \backslash O_1) + (0 : V \backslash O_2) + (0 : V \backslash \triangle_2),$$

is direct. Since this process can be continued we obtain an infinite direct sum of the non-zero right ideals $(0 : V \backslash O_i)$, $i = 1, 2, \ldots$.

Now suppose N contains an infinite direct sum. In this case it is clear that we can find non-zero right ideals R_i, $i = 1, 2, \ldots$, of N such that the sum $\sum R_i$, $i = 1, 2, \ldots$, is direct. Now define the O_i, $i = 1, 2, \ldots$, to be $V \backslash Z(R_i)$. Clearly the O_i are open and, since $R_i \neq \{0\}, Z(R_i) \neq V$ and the O_i are non-empty. Now if j and k are distinct integers in $\{1, 2, \ldots\}$, then the sum $R_j + R_k$ is direct. Thus if v is in V, then by 1.3, $vR_j + vR_k$ is a central sum. The semiprimary nature of V implies this sum is direct and the primary nature of V implies $vR_j = \{0\}$ or $vR_k = \{0\}$. Thus v is in $Z(R_j) \cup Z(R_k)$ and $V = Z(R_j) \cup Z(R_k)$. It therefore follows that

$$O_j \cap O_k = (V \backslash Z(R_j)) \cap (V \backslash Z(R_k)) = V \backslash [Z(R_j) \cup Z(R_k)],$$

must be empty and O_i, $i = 1, 2, \ldots$, is a sequence of mutually disjoint non-empty open subsets of V. The lemma is completely proved.   □

We now come to the theorem mentioned above:-

Theorem 10.3. *Suppose V is a faithful primary N-group. The near-ring N has no infinite direct sums if, and only if, $c(V)$ is sparse and such that $V/c(V)$ is finite.*

Proof. We shall show that there exists an open subspace O of V containing 0 and such that all points of O are unseperated from 0. If V has this property, take $O = V$. Otherwise we can find non-empty disjoint open subsets O_1 and H_1 of V with 0 contained in H_1. If all points of H_1 are unseperated from 0 take $H_1 = O$. If this is not the case, then it is clear that there exists non-empty disjoint subspaces O_2 and H_2 of H_1 with 0 contained in H_2. Thus O_1, O_2 and H_2 are disjoint and either all points of H_2 are unseperated from 0 or the process can be continued. This process cannot be continued indefinitely otherwise O_i, $i = 1, 2, \ldots$, would be a sequence of non-empty disjoint open subsets of V contrary to lemma 10.2. Thus the subspace O exists and, by 7.5, $c(V)$ is sparse and open. Now the distinct left cosets of $c(V)$ in V are clearly non-empty open and disjoint. It follows readily from 10.2 that, the number of distinct cosets of $c(V)$ in V is finite and $V/c(V)$ is finite.

Now suppose $c(V)$ is sparse and $V/c(V)$ is finite. Since $c(V)$ is closed, $V \backslash c(V)$ being a union of a finite number of left cosets of $c(V)$, is clearly closed. Thus $c(V)$ and every coset of $c(V)$ is open. If $O_i, i = 1, 2, \ldots$, were a sequence of non-empty disjoint open subsets of V, then the finiteness of the number of distinct left cosets of $c(V)$ in V implies the existence of distinct j and k in $\{1, 2, \ldots\}$, such that $O_j \cap \Gamma \neq \emptyset$ and $O_k \cap \Gamma \neq \emptyset$, for some such coset Γ. Now Γ is open and, being a left translation of the sparse subspace $c(V)$, is also sparse. Thus points of $O_j \cap \Gamma$ and $O_k \cap \Gamma$ must be unseperated in Γ and clearly $O_j \cap O_k \cap \Gamma \neq \emptyset$. This contradicts the fact that $O_j \cap O_k = \emptyset$ and means no such sequence O_i, $i = 1, 2, \ldots$, can exist. By 10.2, the theorem is fully proved.   □

Theorem 10.3 has implications for the case where V is Hausdorff.

Corollary 10.4. *Suppose V is a faithful primary N-group. If V is Hausdorff, then either N contains an infinite direct sum or V is finite.*

Proof. If N does not contain infinite direct sums and a_i, $i = 1, 2$, are two distinct points in $c(V)$ then, since H is Hausdorff we can find disjoint open subsets O_i, $i = 1, 2$, of V such that a_i is in O_i. Clearly $O_i \cap c(V)$, $i = 1, 2$, are disjoint open subsets of $c(V)$ containing a_i and $c(V)$ cannot be sparse. By 10.3, $c(V)$ contains only one element (i.e., 0) and $V/\{0\}$ is finite. The corollary holds.   □

A chain condition that has proved to be of considerable use in other contexts is *ACC* on right ideals (i.e., *ACCR*). Clearly the presence of this condition rules out the possibility of infinite direct sums and we have:-

Corollary 10.5. *If V is a faithful primary N-group and N has ACCR, then $c(V)$ is sparse and $V/c(V)$ is finite.*

In the context of primary N-groups more general chain conditions are often of interest. One such condition on a near-ring N is DCC on ideals (i.e., *DCCI*). Here we have:-

Proposition 10.6. *If V is a primary N-group and N has DCCI, then V has ACC on closed N-subgroups.*

Proof. Suppose that

$$H_1 \leq H_2 \leq \ldots$$

is an ascending chain of closed N-subgroups of V. Now $(0:H_i) \geq (0:H_{i+1})$ for $i = 1, 2, \ldots$, and by *DCCI*, there exists an integer $n \geq 1$ such that $(0:H_n) = (0:H_m)$, for all integers $m \geq n$. However, by 2.1

$$H_n = Z((0:H_n)) = Z((0:H_m)) = H_m,$$

and $H_n = H_m$ for all integers $m \geq n$. The proposition, therefore, clearly holds.   □

In a number of important situations (see 9.9 and §4) all N-subgroups of V are closed. It is an interesting fact that in this case $J(N)$ is often nilpotent.

Theorem 10.7. *Suppose N is a faithful primary N-group in which all N-subgroups are closed. If N has DCCI, then $J(N)$ is nilpotent.*

Proof. If $J(N)$ is not nilpotent, then by *DCCI*, we can find a non-zero ideal $B \leq J(N)$ of N minimal for being non-nilpotent. If the ideal $Id(B^2)$ generated by B^2 is properly contained in B, then B is clearly nilpotent. Thus $Id(B^2) = B$. Suppose v in V is such that $vB \neq \{0\}$. Since, by 10.6, the N-subgroups of V satisfy *ACC*, vB has a maximal N-subgroup W say. Since $B \leq J(N)$, $vB^2 \subseteq W$ and $B^2 \subseteq (W:v)$. Now the right ideal $R(B^2)$ generated by B^2 is contained in $(W:v)$. However, $N \cdot B^2 \subseteq R(B^2)$ and $N.R(B^2) \subseteq R(B^2)$. Thus $R(B^2)$ is an ideal and consequently $R(B^2) = Id(B^2) = B$. It follows that $B \leq (W:v)$ and $vB \leq W$. This is a contradiction and all v in V must be such that $vB = \{0\}$. Thus B cannot be non-zero and the theorem holds.   □

Corollary 10.8. *Suppose V is a faithful primary N-group and N has DCCI. If V is such that every N-subgroup is graded, then $J(N)$ is nilpotent.*

The corollary follows from 9.9. However, a very much more surprising consequence of 4.3 follows.

Corollary 10.9. *Suppose N is a faithful primary N-group and N has DCCI. If N has a minimal right ideal, then $J(N)$ is nilpotent.*

This follows from 4.3. By that result, V is discrete and all N-subgroups of V are necessarily closed.

The final condition of this section that merits brief consideration is ACC on ideals (i.e., *ACCI*). In a manner completely similar to that of 10.6 it can be proved that:-

Proposition 10.10. *If V is a primary N-group and N has ACCI, then V has DCC on closed N-subgroups.*

Unlike *DCCI* we have the following interesting consequence of assuming V is not Hausdorff (c.f. 5.3).

Proposition 10.11. *Suppose V is a primary N-group and N has ACCI. If V is not Hausdorff, then it contains a non-zero sparse closed N-subgroup.*

Proof. Clearly since V is not Hausdorff $V \neq \{0\}$ and, by 10.10, it contains a minimal closed N-subgroup H. If H where Hausdorff then, by 5.3, V would also be. Thus, by 8.6, H is sparse and the proposition holds.   □

The following three sections cover the proof of a very unexpected result.

11. THE UNEXPECTED

The next three sections will be taken up with proving a remarkable theorem. For faithful primary N-groups assuming connectedness cannot be considered to be particularly restrictive. If, however, N has non-trivial direct decomposition, then surprising things happen. The near-ring N frequently coincides with one of far more familiar character. Statement of this result makes use of the topological group $\mathbb{R}$ (the reals) for which the additive structure is denoted by $(\mathbb{R}, +)$. Also in the interests of completeness some terms involving direct decomposition need defining. A non-trivial direct decomposition $R_1 \oplus R_2$ of a near-ring N is called *singular* (see §6 of [5]) if any non-trivial direct decomposition $H_1 \oplus H_2$ of N is such that $R_i = H_i$ for $i = 1, 2$, or $R_1 = H_2$ and $R_2 = H_1$ (i.e., $R_1 \oplus R_2$ is the unique non-trivial direct decomposition). If non-trivial direct decomposition exists which is not singular it is called *non-singular*. In the situation where N has no non-trivial direct decomposition it is called *indecomposable*. Presentation of the theorem can now be made.

Theorem 11.1. *Suppose V is a connected faithful primary N-group. If N has non-trivial direct decomposition, then*

    (i) *N has singular direct decomposition,*
    (ii) *V is a minimal N-group,*
    (iii) *there exists a group isomorphism μ of V onto $(\mathbb{R}, +)$ and*
    (iv) *if V is locally compact, then the μ of (iii) can be taken as a homeomorphism of V onto $\mathbb{R}$*

Some brief explanation of the significance of (iv) seems desirable. Two faithful N_i-groups V_i, $i = 1, 2$, may be considered algebraically equivalent if there is a group isomorphism τ of V_1 onto V_2 and a near-ring isomorphism λ of N_1 onto N_2, such that $(v\alpha)\tau = (v\tau)(\alpha\lambda)$, for all v in V_1 and α in N_1. This can be expressed by saying the isomorphisms τ and λ faithfully transfer the action of N_1 on V_1 to that of N_2 on V_2. Now, with V as in 11.1, and N having non-trivial direct decomposition the meaning of (iv) becomes more apparent. The local compactness requirement ensures the existence of μ as in (iv). For α in N we can define a function $(\alpha)\lambda$ of $\mathbb{R}$ into $\mathbb{R}$ by $(v\mu)(\alpha\lambda) = (v\alpha)\mu$, for all v in V. Certainly $(\alpha)\lambda$ is well defined. However, if β is in N, then

$$(v\mu)(\alpha\beta)\lambda = ((v\alpha)\beta)\mu = (v\alpha)\mu(\beta)\lambda = (v\mu)(\alpha\lambda)(\beta\lambda)$$

and $(\alpha\beta)\lambda = (\alpha\lambda)(\beta\lambda)$. Similarly

$$(\alpha + \beta)\lambda = (\alpha)\lambda + (\beta)\lambda$$

and λ is a near-ring homomorphism of N into the near-ring $M_0(\mathbb{R})$ of all zero-fixing functions of $\mathbb{R}$ into $\mathbb{R}$. Now the fact that μ is a homeomorphism ensures λ is a near-ring
monomorphism and that $(\alpha)\lambda, \alpha \in N$, is in the near-ring $C_0(\mathbb{R})$ of all zero-fixing continuous functions of $\mathbb{R}$ into $\mathbb{R}$. The N-group V and $N\lambda$-group $\mathbb{R}$ are in fact algebraically
equivalent. Indeed the isomorphism μ of V onto $\mathbb{R}$ and near-ring isomorphism λ of N onto
$N\lambda$ satisfy $(v\alpha)\mu = (v\mu)(\alpha\lambda)$, for all v in V and α in N. The compatibility of $N\lambda$ on $\mathbb{R}$ also
follows and what we have shown is that local compactness (provided the other conditions
hold) means the N-group V is equivalent to a faithful compatible subnear-ring of $C_0(\mathbb{R})$
acting on $\mathbb{R}$. Thus confinement to the general conditions of 11.1(iv), yields concrete well
known near-rings.

The above explanation still begs a question. Can the situation outlined above arise at
all? This amounts to asking if subnon-rings of $C_0(\mathbb{R})$ which are primitive and compatible
on $\mathbb{R}$ can have non-trivial direct decomposition. There is in fact a respectable class of such
non-rings and $C_0(\mathbb{R})$ is itself a member. The near-ring $C_0(\mathbb{R})$ contains the function e of $\mathbb{R}$
into $\mathbb{R}$ given by $ve = 0$ for all $v \le 0$ in $\mathbb{R}$ and $ve = v$, for all $v \ge 0$ in $\mathbb{R}$. Now $eC_0(\mathbb{R})$ and
$(1 - e)C_0(\mathbb{R})$ are non-zero right ideals of $C_0(\mathbb{R})$ such that

$$C_0(\mathbb{R}) = eC_0(\mathbb{R}) \oplus (1 - e)C_0(\mathbb{R}).$$

Indeed, any subnon-ring N of $C_0(\mathbb{R})$ that is primitive and compatible on $\mathbb{R}$ has non-trivial
direct decomposition if, and only if, e is in N. The point to be taken is that many respectable
subnon-rings of $C_0(\mathbb{R})$ satisfy theorem 11.1(iv).

The remainder of this section is essentially concerned with the proof of (i) and (ii)
although material needed in proving (iii) and (iv) is also touched on. This undertaking
will draw on substantial results from [5]. In the interests of completeness these results (or
obvious modifications) will be stated when required.

For an additive group V a subset S of V is *symmetric* if $S(-1) = S$. If S is a subset of V
containing 0, then the notation $S^\perp$ is used for $(V\backslash S) \cup \{0\}$. A subset S of V containing 0 is
said to be *antisymmetric* if $S(-1) = S^\perp$. Now suppose V is an N-group. We call a subset
P of V containing 0 an *OP-subset* (*orthogonally projective subset*) of V if there exists e in
N such that $ve = v$, for all v in P and $ve = 0$, for all v in $V\backslash P$. If P is distinct from $\{0\}$ or
V, then it is called a *proper OP*-subset. Combining 5.2 and 6.1 of [5] we obtain.

Proposition 11.2. *If V is an N-group and P_i, $i = 1, 2$, OP-subsets of V, then $P_1(-1)$, $P_1^\perp$,
$P_1 \cap P_2$ and $P_1 \cup P_2$ are OP-subsets of V.*

The existence of proper OP-subsets is often related to the existence of non-trivial direct
decomposition. In the situation with which we are dealing such a subset can be found.

Proposition 11.3. *If V and N are as in the statement of theorem 11.1, then V has a proper
OP-subset.*

Proof. Let $R_1 \oplus R_2$ be a non-trivial direct decompositon of N and $1 = e_1 + e_2$, where e_i is
in R_i. Clearly the e_i, $i = 1, 2$, are mutually orthogonal non-zero idempotents. Now $e_1 \ne 0$
and, by faithfulness $Ve_1 \ne \{0\}$. Also $Ve_1 \ne V$ since otherwise e_1 acts as the identity on V
and, by faithfulness $e_1 = 1$. Clearly $ue_1 = u$ for all u in Ve_1 and because $Ve_1 \cup Ve_2 = V$
(see 9.4) we have $ue_1 = 0$ for all u in $V\backslash Ve_1$. Thus Ve_1 is a proper OP-subset of V and 11.3
holds. $\square$

Theorem 5.3 of [5] gives us the following:-

Theorem 11.4. *If V is a compatible N-group with a proper symmetric OP-subset, then V has a pre-image.*

This result can in fact be considerably extended.

Proposition 11.5. *If V is a compatible N-group with a proper OP-subset which is not antisymmetric, then V has a pre-image.*

Proof. Let P be a proper OP-subset of V which is not antisymmetric. Since $P \cup P(-1)$ is, by 11.2, a symmetric OP-subset of V we have, from 11.4, that either V has a pre-image or $P \cup P(-1) = V$. If $P \cup P(-1) = V$, then $P \cap P(-1) \neq \{0\}$ otherwise $P(-1) = P^{\perp}$ (i.e., P would be antisymmetric). Now, by 11.2, $P \cap P(-1)$ is an OP-subset of V which is proper. However, $P \cap P(-1)$ is clearly symmetric and the result follows by 11.4.   □

Corollary 11.6. *If V and N are as in theorem 11.1, then any proper OP-subset of V is antisymmetric.*

Proof. If this were not so, then by 11.5, V would have a pre-image. Since, by 4.4, this yields a disconnection of V, the corollary holds.   □

We are now ready to prove (ii) of theorem 11.1.

Proposition 11.7. *If V and N are as in 11.1, then V is a minimal N-group.*

Proof. Suppose V has a proper N-subgroup H. Let $R_1 \oplus R_2$ be a non-trivial direct decomposition of N. If $HR_i \neq \{0\}$ for $i = 1, 2$, then by 9.6, H is open and closed in V and V has a disconnection. Clearly $HR_i \neq \{0\}$, for $i = 1$ or 2, and to obtain a contradiction we may assume $HR_1 \neq \{0\}$ but $HR_2 = \{0\}$. With $1 = e_1 + e_2$, where e_i is in R_i the proof of 11.3 yields the fact that Ve_1 is an OP-subset of V. However, by corollary 11.6, this OP-subset is antisymmetric. Since $He_1 \neq \{0\}$ (otherwise $HR_1 = \{0\}$) we have the existence of h in H such that $he_1 \neq 0$. Because Ve_1 is an antisymmetric OP-subset of V, $-he_1$ is not in Ve_1. Now, by 9.4, $Ve_1 \cup Ve_2 = V$, $-he_1$ is in Ve_2 and $(-he_1)e_2 \neq 0$. However, since $-he_1$ is in H we have the contradiction that $HR_2 \neq \{0\}$. The proof of 11.7 is complete.   □

Certainly from 11.7 we see that if V and N are as in 11.1, then N is primitive and compatible on V. The semiprimary nature of V means N is a non-ring. A powerful theorem about direct decomposition of such non-rings is 8.1 of [5].

Theorem 11.8. *If the non-ring N is primitive and compatible on V, then either*

    (a) *N is indecomposible,*

    (b) *N has direct decomposition which is singular,*

    (c) *such non-singular decomposing and minimal right N-sugroups exist, or*

    (d) *$(V, +)$ is a group of order 3.*

Thus it is possible to deduce (i) of theorem 11.1 if in theorem 11.8 we can eliminate (a), (c) and (d). Clearly (a) does not apply and in case (d), V has a disconnection. To deal with (c) we need a further result from [5]. In that paper the Z' topology was defined for V as in 11.7. This is a special case of Z' as used in this paper. According to 3.9 of [5] we have:-

Proposition 11.9. *If the non-ring N is primitive and compatible on V and N has a minimal right N-subgroup, then V is totally disconnected.*

Returning now to our use of 11.8. With V and N as in 11.1 we have $N \neq \{0\}$ so that (c) cannot hold. The only possibility existing is (b) and therefore:-

Proposition 11.10. *If V and N are as in 11.1, then N has singular direct decomposition.*

The main object of this section has been reached and (i) and (ii) of 11.1 are proved. However, a little more information about proper OP-subsets of the V of 11.1 is needed. The main ingredients of one such result is covered by 5.4 of [5] which states:-

Theorem 11.11. *Let V be a compatible N-group with an antisymmetric OP-subset P. If either P or $P^{\perp}$ is not an additive subsemigroup of V, then V has a pre-image.*

Now if V and N are as in 11.1, then by 11.3 and 11.6, V has proper OP-subsets and any such subset is antisymmetric. Since V is connected it follows from 11.11 and 4.4 that such an OP-subset is an additive subsemigroup of V. Thus we have:-

Proposition 11.12. *If V and N are as in 11.1, then V has proper OP-subsets and any such subset is antisymmetric and an additive subsemigroup of V.*

The final result of this section is about the uniqueness of such OP-subsets.

Proposition 11.13. *If V and N are as in 11.1 and P a proper OP-subset of V, then P and $P^{\perp}$ are the only such subsets.*

Proof. If P_1 is a proper OP-subset of V and $P_1 \cap P \neq \{0\}$, then by 11.2 and 11.12 and obvious manipulation we see

$$P_1^{\perp} \cup P^{\perp} = (P_1 \cap P)^{\perp} = (P_1 \cap P)(-1) \subseteq P^{\perp}$$

so that $P_1^{\perp} \subseteq P^{\perp}$ and $P \subseteq P_1$. Clearly $P_1 \cap P^{\perp} \neq \{0\}, P^{\perp} \subseteq P_1$ and $P_1 = V$. Thus either $P \subseteq P_1$ and $P_1 \cap P^{\perp} = \{0\}$ implying $P_1 = P$ or with $P^{\perp}$ replacing P we see $P_1 = P^{\perp}$. The proposition is proved.   □

12. SUPPLY OF PROOF

This section is concerned with the proof of (iii) of 11.1. Accordingly, throughout this section V is a connected faithful primary N-group and N has a non-trivial direct decomposition. Also, in order that the proof be understandable, a respectable amount of preliminaries are now given.

By 11.3, V has a proper OP-subset. One such subset is now fixed and denoted by V^{+}. Also we denote the OP-subset $V^{+}(-1)$ (see 11.2) by V^{-}. From 11.6 it follows that $V^{-} \cup V^{+} = V$ and $V^{-} \cap V^{+} = \{0\}$. Now for a and b in V we write $a \leq b$ if $b - a$ is in V^{+}. Since V^{+} contains 0 the relationship $\leq$ is reflexive. Because $V^{+} \cap V^{-} = \{0\}$, antisymmetry can also be checked. Furthermore, from the semigroup nature of V^{+} (see 11.12) one proves that transitivity holds. If $a \leq b$ and $a \neq b$ we write $a < b$.

To understand such relationships better we note that if v is in V and γ in N is such that $w\gamma = -v + w + v$ for all w in V, then the unit γ is such that $V^{+}\gamma = V^{+}$. Indeed there exists e in N such that $we = w$ for all w in V^{+} and $we = 0$ for all w in V^{-}. Now $(w\gamma)(\gamma^{-1}e\gamma) = w\gamma$, for all w in V^{+} and equals 0 otherwise. Thus $V^{+}\gamma$ is a proper OP-subset of V and, by 11.13, $V^{+}\gamma = V^{+}$ or $V^{+}\gamma = V^{-}$. The first statement holds if $v = 0$. If $v \neq 0$ is in V^{+}, then $v\gamma = v$ and $V^+\gamma = V^+$. If $v \neq 0$ is in V^-, then $V^-\gamma = V^-$ which fairly readily yields $V^+\gamma = V^+$. We conclude that if a and b are in V and $b - a$ is in V^+, then $-a + b$ is in V^+. This means if $a \leq b$, then $-b \leq -a$. However, above explanation has another readily deduced consequence.

Proposition 12.1. *If a, b, c and d are in V and $a \leq b$ and $c \leq d$, then $a + c \leq b + d$.*

The reverse relationships $\geq$ or $>$ can be defined for any a and b in V by respectively requiring $a < b$ does not hold, or $a \leq b$ does not hold. Clearly $a \geq b$ precisely when $b - a$ is in V^- and $a > b$ is the same as $a \geq b$ but $a \neq b$. Above properties that hold for $\leq$ hold for $\geq$. The relationship $>$ tends to be similar.

The *minimum* of two elements a and b of V will be a, if $a \leq b$ and b otherwise. The *maximum* of a and b is b, if $a \leq b$ and a otherwise. The *absolute value* $|c|$ of c in V is the maximum of c and $-c$.

If a and b are in V, then (a, b) is all c in V such that $c > a$ and $c < b$. With $<$ replaced by $\leq$, we obtain $[a, b]$. The definition of $[a, b)$ and $(a, b]$ should be clear. All c in V with $c < b$ is denoted by $(-\infty, b)$ and with $<$ replaced by $\leq$ we obtain $(-\infty, b]$. All c in V with $c > b$ (i.e., $b < c$) is denoted (b, ∞) and with $>$ replaced by $\geq$ we obtain $[b, \infty)$. Now V^- is the zero set of some e in N such that $we = w$, for all w in V^+ and $we = 0$, for w in V^-. Thus V^- is closed and since $V^- + b = (-\infty, b]$ this subset of V is closed. Similarly $[b, \infty)$ is closed and considering intersections it follows that for a in $V, [a, b]$ is closed. However,

$$(V \backslash V^+) + b = (V^- \backslash \{0\}) + b = (-\infty, b)$$

is clearly open. Similarly (b, ∞) is open and with a in $V, (a, b)$ is open. Preliminary requirements have now been covered. The remainder of this section is divided into four parts. The first consists of gaining some understanding of V as a $\mathbb{Z}$-group (here $\mathbb{Z}$ is the integers). The second uses the first to establish V as a certain $\mathbb{Q}$-group (here $\mathbb{Q}$ is the rationals). Thirdly, we establish an 'action' of $\mathbb{R}$ on V. Although this makes V an $\mathbb{R}$-group this fact is not proved. However, what is necessary for the fourth part is obtained. The final part establishes that the μ of 11.1 is an isomorphism.

We start with considerations relating to the action of $\mathbb{Z}$ on V. As now explained, it is convenient to take a particular copy of $\mathbb{Z}$. If 1 is the identity of N, then for n in $\mathbb{Z}$, $1n$ is taken as the 0 of N if $n = 0$, as $1 + 1 + \cdots + 1$ (n-times) if $n > 0$ and as $-1 - 1 - \cdots - 1$ (($-n$)-times) if $n < 0$. The value of $v(1n), v \in V$ is now apparent. Also, it can be checked that for n and m in $\mathbb{Z}$, $(1n)(1m) = 1(nm)$ and $1n + 1m = 1(n + m)$. Thus the additive subgroup A of N generated by 1 is a subnear-ring of N. In fact the map taking n in $\mathbb{Z}$ to $1n$ is a homomorphism onto A. Thus A is a ring but, as will be shown, this homomorphism is an isomorphism. If for $n \neq 0$ in $\mathbb{Z}$, $1n = 0$, then V has an element $a \neq 0$ of finite order. If necessary taking $-a$ we can assume $a > 0$. By 12.1, we have the contradiction that

$$0 < a \leq a + a \leq a + a + a \leq \ldots$$

Thus the above homomorphism is an isomorphism. From now on we consider $\mathbb{Z}$ to coincide with A and let $1n$, $n \in \mathbb{Z}$ be simply denoted by n. Clearly V is a faithful unitary $\mathbb{Z}$-group.

Two straightforward propositions follow.

Proposition 12.2. *If $a < b$ are in V and $n \geq 1$ an integer, then $an < bn$.*

Proof. This is proved inductively. For $n \geq 2$ we have

$$an = a(n-1)+a \leq b(n-1)+a < bn$$

The proposition holds. $\qquad\qquad\qquad\qquad\qquad\qquad\qquad\qquad\qquad\qquad\qquad\qquad\quad\square$

If in 12.2, $n \leq -1$, then as is easily seen $an > bn$. Another easily deduced proposition is:-

Proposition 12.3. *If $a > 0$ is in V and $n > m$ in $\mathbb{Z}$, then $am < an$.*

Proposition 12.3 gives some information on how elements of $a\mathbb{Z}$ are distributed in V. However, a more significant aspect of such distribution is the following:-

Proposition 12.4. *If $a > 0$ and h are in V, then there exists n in $\mathbb{Z}$ such that $an > h$.*

Proof. Suppose $am = h$ for some m in $\mathbb{Z}$. In this case $a(m+1) = h+a > h$. Thus in order to obtain a contradiction we may assume $an < h$ for all n in $\mathbb{Z}$. Let H be the set of all b in V such that $|b| < ar$ for some r in $\mathbb{Z}$. If b_i, $i = 1,2$ are in H, then $|b_i| < an_i$, $i = 1,2$, and

$$-b_1 + b_2 < -b_1 + an_2 < an_1 + an_2 = a(n_1 + n_2).$$

Similarly $-b_2 + b_1 < a(n_1 + n_2)$ so that $-b_1 + b_2$ is in H. Thus H is a subgroup of V. However, b is in H precisely when b is in $(-ar, ar)$ for some integer $r \geq 1$. Thus H is a union over all $r \geq 1$ of the open sets $(-ar, ar)$ of V and is open. By 6.2, this means H is open and closed in V. Now V is connected and since a is in H, $H = V$. This is a contradiction since h is not in H. The proposition is proved. $\qquad\qquad\qquad\quad\square$

In moving toward a sensible presentation of V as a $\mathbb{Q}$-group, the obtaining of a solution b of $bm = a$, with m in $\mathbb{Z}\backslash\{0\}$ and a in V, becomes important.

Proposition 12.5. *If a is in V and $m \neq 0$ in $\mathbb{Z}$, then there exists a unique b in V such that $bm = a$.*

Proof. Clearly we may assume $a \neq 0, m \neq 1$ and $m \neq -1$. To establish the existence of b adopt the contrary position that a is not in Vm. Our assumptions imply (see 12.3) that either am is in $(-\infty, a) \cap Vm$ and $a(-m)$ in $(a, \infty) \cap Vm$ or am is in $(a, \infty) \cap Vm$ and $a(-m)$ in $(-\infty, a) \cap Vm$. Thus the pair of subsets $(-\infty, a) \cap Vm$ and $(a, \infty) \cap Vm$ of Vm are non-empty and, since a is not in Vm, form a disconnection (D_1, D_2) of Vm. Now, by 2.9, $D_i m^{-1}$, $i = 1,2$, are non-empty open and closed subsets of V with empty intersection and union V. This contradiction means that b exists. If b_1 is such that $b_1 m = a$ and $b_1 \neq b$, then supposing $b_1 < b$, it follows readily from 12.3 that $b_1 m < bm$ or $bm < b_1 m$. The only possibility is that $b_1 = b$ and the proposition holds. $\qquad\qquad\qquad\qquad\quad\square$

We now move to the second phase of the proof of (iii) of 11.1. Consequently $\mathbb{Z}$ (as above) is considered embedded in $\mathbb{Q}$. Although $\mathbb{Z} \leq N$, there is no guarantee that such an embedding holds for $\mathbb{Q}$. However, we can define an 'action' of $\mathbb{Q}$ on V that agrees with that of $\mathbb{Z}$. Also, as will be seen shortly, this action makes V into a $\mathbb{Q}$-group.

For $s \neq 0$ in $\mathbb{Z}$, the element $1/s$ of $\mathbb{Q}$ acts on V as follows. For b in $V, b(1/s)$ is the unique c of V for which $cs = b$ (see 12.5). It follows from this definition that $(bs)(1/s) = b$ and, as is easily proved

$$(b(1/s_1))(1/s_2) = b(1/s_1 s_2)$$

for all s_i, $i = 1, 2$, in $\mathbb{Z}\backslash\{0\}$. Now, if n/m is in $\mathbb{Q}$ where n and $m \neq 0$ are in $\mathbb{Z}$, then $b(n/m)$ is defined as $(bn)(1/m)$. This definition makes sense for if $n = rn_1$, and $m = rm_1$, where r, n_1 and m_1 are in $\mathbb{Z}$, then

$$b(n/m) = ((bn_1)r)(1/r)(1/m_1) = b(n_1/m_1).$$

In the particular case where $m_1 = 1$, we see

$$b(n_1 r/r) = b(n_1/1) = bn_1$$

ensuring agreement with elements of $\mathbb{Z}$. Also the fact that $b(n/m) = b(1/m)n$ is easily proved.   □

Next comes the matter of showing V is a $\mathbb{Q}$-group. Toward this end we first show.

Proposition 12.6. *If a is in V, n and m in $\mathbb{Z}$ and r in $\mathbb{Z}\backslash\{0\}$, then*

$$(an + am)(1/r) = a(n/r) + a(m/r).$$

Proof. Suppose the expression holds for $r \geq 1$. On substituting $-a$ for a, while remembering an and am commute, it follows readily that the expression holds with r replaced by $-r$. To prove the result for $r \geq 1$, observe that

$$[a(1/r)n + a(1/r)m]r = a(1/r)n + a(1/r)m + \cdots + a(1/r)n + a(1/r)m,$$

where the sum of the $a(1/r)n + a(1/r)m$ is r-times. However, $a(1/r)n$ and $a(1/r)m$ additively commute and the right hand side becomes

$$a(1/r)nr + a(1/r)mr = an + am.$$

From this the proposition follows.   □

As indicated above we have:-

Proposition 12.7. *V is a $\mathbb{Q}$-group.*

Proof. The fact that if q_i, $i = 1, 2$, are in $\mathbb{Q}$ and a is in V, then $a(q_1 q_2) = (aq_1)q_2$ is not difficult to check. We must show $a(q_1 + q_2) = aq_1 + aq_2$. Suppose $q_i = n_i/m_i$, $i = 1, 2$, where n_i is in $\mathbb{Z}$ and m_i in $\mathbb{Z}\backslash\{0\}$. Now, by 12.6,

$$
\begin{aligned}
a[(n_1/m_1) + (n_1/m_2)] &= a(n_1 m_2 + n_2 m_1)(1/m_1 m_2) \\
&= (an_1 m_2 + an_2 m_1)(1/m_1 m_2) \\
&= (an_1 m_2)(1/m_1 m_2) + (an_2 m_1)(1/m_1 m_2) \\
&= aq_1 + aq_2.
\end{aligned}
$$

The proof is complete.   □

Proposition 12.2 and 12.3 extend to $\mathbb{Q}$ and of these, the 12.3 extension is required later.

Proposition 12.8. *If $a > 0$ is in V and $q_2 > q_1$ in $\mathbb{Q}$, then $aq_1 < aq_2$.*

Proof. For $i = 1, 2$, we have $q_i = n_i/m_i$ where n_i is in $\mathbb{Z}$ and m_i in $\{1, 2, \ldots\}$. Now $n_1 m_2 < n_2 m_1$ and, by 12.3, $bn_1 m_2 < bn_2 m_1$, for all $b > 0$ in V. However,

$$a(1/m_1 m_2)(m_1 m_2) = a > 0,$$

which can only happen if $a(1/m_1 m_2) > 0$. Replacing b by $a(1/m_1 m_2)$, we see that $aq_1 < aq_2$. The proposition is proved.   □

For $a > 0$ in V, it is important to have information as to how the elements of $a\mathbb{Q}$ are distributed in V. A first step in this direction is provided by:-

Proposition 12.9. *If $b > 0$ and $a > 0$ are in V, then there exists q in $\mathbb{Q}$ such that $0 < bq < a$.*

Proof. By proposition 12.4 there exists n in $\mathbb{Z}$ such that $an > b$. Clearly $n > 0$ and, by proposition 12.8

$$(an)(1/n) = a > b(1/n) > 0,$$

so that 12.9 holds with $q = 1/n$.   □

Proposition 12.9 has a useful extension.

Proposition 12.10. *If $d > c > 0$ are in V and $a > 0$ in V, then there exists q in $\mathbb{Q}$ such that aq is in (c,d).*

Proof. Suppose for all q in $\mathbb{Q}, aq$ is in $V\backslash(c,d)$. If g is the minimum of c and $d - c$, then clearly $g > 0$ and, by 12.9, there exists q_1 in $\mathbb{Q}$ such that aq_1 is in $(0,g)$. By 12.4, there exists m in $\mathbb{Z}$ such that $aq_1 m \geq d$. However, $aq_1 > 0$ so that $m \geq 1$ and inductively it can be shown $aq_1 m \leq c$. This is true for $m = 1$ by the nature of g. Assuming $m \geq 2$ and $aq_1(m-1) \leq c$, we see

$$aq_1 m = aq_1 + aq_1(m-1) \leq aq_1 + c < d - c + c = d,$$

implying (see the opening assumption) that $aq_1 m \leq c$. The contradiction obtained means proposition 12.10 holds true.   □

A useful extension of 12.10 is not difficult to establish. We state without proof.

Proposition 12.11. *If a, c and d are in V, where $a > 0$ and $d > c$, then there exists q in $\mathbb{Q}$ such that aq is in (c,d).*

The third step in proving (iii) of 11.1 is now undertaken. Thus $\mathbb{Q}$ (as above) is considered contained in $\mathbb{R}$ and we seek to define ar, a in V, r in $\mathbb{R}$, in such a way that on $\mathbb{Q}$ the definitions agree. Other properties of this 'action' will be required later but a full definition for all a in V is not essential. Most of what follows concerns only $a > 0$.

To carry out the program indicated some notation is introduced. For r in $\mathbb{R}$ we let L_r be the set of all q in $\mathbb{Q}$ such that $q < r$. Thus r is the least upper bound of L_r. Also for r in $\mathbb{R}$ we define U_r as all q in $\mathbb{Q}$ such that $q > r$. For a in V we let $L_r(a)$ be the union of all $(-\infty, aq)$ with q in L_r and $U_r(a)$ the union of all (aq, ∞) with q in U_r.

Proposition 12.12. *If q is in $\mathbb{Q}$ and r in $\mathbb{R}$ and $q \geq r$, then for $a > 0$ in V, and d in $L_r(a), d < aq$.*

Proof. Since d is in $(-\infty, aq_1)$, where q_1 is in L_r, we see $q_1 < r \leq q$. Thus, by 12.8, $d < aq_1 \leq aq$ and the result follows.   □

The next result is essential to further progress.

Proposition 12.13. *If $a > 0$ is in V and r in $\mathbb{R}$ and H is the closure of $L_r(a)$ in V, then $H\backslash L_r(a)$ contains a unique element.*

Proof. Since $L_r(a)$ is open and, by 12.12, $H \neq V$ we see from the connectedness of V, that $L_r(a)$ is a proper subset of H. Suppose b_i, $i = 1,2$, are two elements of $H \backslash L_r(a)$ with $b_1 < b_2$. Now $b_1 \geq aq$ for q in L_r, otherwise b_1 is in $L_r(a)$. Thus $(-\infty, b_1]$ contains $(-\infty, aq)$ (q in L_r) and $(-\infty, b_1] \supseteq L_r(a)$. However, $(-\infty, b_1]$ is closed and does not contain b_2 so the result holds. $\square$

Before we can actually define the single element of $H \backslash L_r(a)$ of 12.13 as ar we must check that this agrees with the previous definition using $\mathbb{Q}$.

Proposition 12.14. *If the conditions of 12.13 hold and r is in $\mathbb{Q}$, then the unique element of $H \backslash L_r(a)$ is ar.*

Proof. By 12.12, if d is in $L_r(a)$, then $d < ar$. Thus, being closed $(-\infty, ar]$ contains H. It follows that if b is the unique element of $H \backslash L_r(a)$, then either $b = ar$ or $b < ar$. If $b < ar$ we have, by 12.11, that there exists q in $\mathbb{Q}$ such that $b < aq < ar$. By 12.8, this can only happen if $q < r$ and b is in $(-\infty, aq)$ and in $L_r(a)$. A contradiction has been reached and 12.14 is proved. $\square$

In view of proposition 12.13 and 12.14, it makes complete sense to define $ar, a > 0$ in V and r in $\mathbb{R}$, as the unique element of $H \backslash L_r(a)$ (see 12.13).

To obtain further information we make use of $U_r(a)$.

Proposition 12.15. *If a, r and H are as in proposition 12.13, then $U_r(a) \cup H = V$ and $U_r(a) \cap H = \emptyset$.*

Proof. An element b of $U_r(a)$ is such that $b > aq_2$ where q_2 is in U_r. If b is in $L_r(a)$, then $b < aq_1$ where q_1 is in L_r. Thus $q_1 < q_2$ and, by 12.8, $b < aq_1 < aq_2 < b$. We conclude that $U_r(a) \cap L_r(a) = \emptyset$. Since $U_r(a)$ is open, $L_r(a)$ is contained in the closed subset $V \backslash U_r(a)$ of V and, therefore, $U_r(a) \cap H = \emptyset$. Now suppose c in V is not in $U_r(a) \cup H$. Since for $q < r$ in $\mathbb{Q}$, c is not in $(-\infty, aq)$ we have $(-\infty, aq) \subseteq (-\infty, c]$ and $L_r(a) \subseteq (-\infty, c]$. It follows that if $c < ar$, then H does not contain ar. Thus $c \geq ar$ and since c is not in $H, c > ar$. Now, by 12.11, (ar, c) contains an element aq_3 with q_3 in $\mathbb{Q}$. We have $aq_3 \neq aq_4$ for any q_4 in L_r. Also, by 12.14, $r \neq q_3$ so q_3 is in U_r. Thus since $aq_3 < c, c$ is in $U_r(a)$ and we have obtained a contradiction. The proof is complete. $\square$

Corollary 12.16. *The closure of $U_r(a)$ is $U_r(a) \cup \{ar\}$.*

Proof. By 12.15 and 12.13, $V \backslash L_r(a)$ is $U_r(a) \cup \{ar\}$. Also $U_r(a)$ is not closed, otherwise it is a proper open and closed subset of V. The corollary therefore holds. $\square$

Proposition 12.17. *If r is in $\mathbb{R}$ and $a > 0$ in V, then $a(-r) = -ar$.*

Proof. Since $U_r(a)$ is the union of all (aq, ∞) with q in $U_r, -U_r(a)$ is the union of all $(-\infty, aq_1)$ with q_1 in L_{-r} (note that $a(-q) = -aq$ by 12.7). Thus $-U_r(a) = L_{-r}(a)$. However, by 2.9, the map taking v in V to $v(-1)$ is a homeomorphism and therefore maps the closure of $U_r(a)$ to that of $L_{-r}(a)$. From the above corollary this closure is $L_{-r}(a) \cup \{-ar\}$. It now follows from the definition of $a(-r)$ and 12.13 that $a(-r) = -ar$. The proof of the corollary is complete. $\square$

The final part of this section is devoted to establishing the existence of an isomorphism of $(\mathbb{R},+)$ onto V. To this end we fix some $a > 0$ in V and define a map δ of $\mathbb{R}$ into V by setting $(r)\delta = ar$, for all r in $\mathbb{R}$. δ will in fact turn out to be such an isomorphism. The first stage of showing δ is a homomorphism requires the following result.

Proposition 12.18. *If r_i, $i = 1,2$, are in $\mathbb{R}$, then $ar_1 + ar_2 \geq a(r_1 + r_2)$.*

Proof. From proposition 12.13 and the definition of ar_i, $i = 1,2$, the ar_i cannot be contained in any $(-\infty, aq_i)$ with q_i in L_{r_i}. Thus $ar_i \geq aq_i$ for all q_i in L_{r_i}. It follows, by 12.7 and 12.1, that

$$ar_1 + ar_2 \geq aq_1 + aq_2 = a(q_1 + q_2)$$

for all q_i, $i = 1,2$, in L_{r_i}. It will now be shown that we can find q_1 and q_2 such that $q_1 + q_2 = q$, for any given q in $L_{r_1+r_2}$. For such a q we have $q < r_1 + r_2$ and with q_1 in $(q - r_2, r_1)$ and $q_2 = q - q_1$, we see $q - q_1 < r_2$ (i.e., q_2 is in L_{r_2}) and $q_1 + q_2 = q$. It follows that $ar_1 + ar_2$ is not in $(-\infty, aq)$ for all q in $L_{r_1+r_2}$ and consequently

$$(-\infty, aq) \subseteq (-\infty, ar_1 + ar_2].$$

Taking the union over all such q we see

$$L_{r_1+r_2}(a) \subseteq (-\infty, ar_1 + ar_2]$$

and because $(-\infty, ar_1 + ar_2]$ is closed the definition of $a(r_1 + r_2)$ (see 12.13) yields the fact that $ar_1 + ar_2 \geq a(r_1 + r_2)$. The proposition is entirely proved. □

Corollary 12.19. *The map δ is a homomorphism.*

Proof. If r_i, $i = 1,2$, are in $\mathbb{R}$, then

$$(r_1 + r_2)\delta = a(r_1 + r_2) \leq ar_1 + ar_2,$$

by 12.18. However,

$$a(r_1 + r_2 - r_2) = ar_1 \leq a(r_1 + r_2) + a(-r_2)$$

and, by 12.17, $ar_1 + ar_2 \leq a(r_1 + r_2)$. Thus we have

$$(r_1 + r_2)\delta = ar_1 + ar_2 = r_1\delta + r_2\delta$$

and the corollary is completely proved. □

The next step is to prove:-

Proposition 12.20. *The map δ is onto.*

Proof. Let b be in V and L_b be the set of all q in $\mathbb{Q}$ such that $aq < b$. With the h of 12.4 taken as $-b$ we obtain n in $\mathbb{Z}$ with $an > -b$, thereby concluding $-n$ is in L_b and $L_b \neq \emptyset$. Also there exists n_1 in $\mathbb{Z}$ with $an_1 > b$ and, by 12.8, n_1 is an upper bound of L_b. Thus L_b has a least upper bound r in $\mathbb{R}$. It will be shown $b = ar$. If $b < ar$, then there exists q_1 in L_r with $aq_1 > b$, since otherwise the closure of $L_r(a)$ does not contain ar. This means there exists q_2 in L_b with $q_2 > q_1$ (the least upper bound of L_b is r). This implies $aq_2 < b$ contrary to 12.8. We have $b \geq ar$. If $b > ar$, then there exists, by 12.11, q_3 in $\mathbb{Q}$ such that $b > aq_3 > ar$. Thus q_3 is in L_b and clearly there exists $q_4 > q_3$ in L_r. Now, by 12.8,

$ar \geq aq_4 > aq_3$ since otherwise ar is in $(-\infty, aq_4)$ contrary to the definition of ar. This is a contradiction, since $aq_3 > ar$. We conclude that $ar = b$ and $r\delta = b$. The proposition is entirely proved.   □

It is now shown that:-

Proposition 12.21. *The map δ is an isomorphism of $\mathbb{R}$ (i.e., $(\mathbb{R}, +)$) onto V (i.e., $(V, +)$).*

Proof. The fact that δ is a group homomorphism onto follows from 12.19 and 12.20. Suppose $r > 0$ in $\mathbb{R}$ is such that $ar = 0$. Let q in $\mathbb{Q}$ be in $(0, r)$. We have $ar \geq aq$ otherwise ar is in $L_r(a)$. Now $q > 0$ and, by 12.8, $aq > 0$ so that $ar \neq 0$ contrary to our assumption. Thus for all $r > 0, ar \neq 0$ and taking $r_1 < 0$ we see, by 12.17, $ar_1 = -a(-r_1) \neq 0$ so that there is no r_2 in $\mathbb{R}\backslash\{0\}$ for which $ar_2 = 0$. This means ker $\delta = \{0\}$ and the proof of 12.21 is complete.   □

It has been shown that (iii) of theorem 11.1 holds true with $\mu = \delta^{-1}$. Having reached this intermediate goal, the next section completes the proof.

13. FINISHING THE PROOF

The first half of this section completes the proof of 11.1. The second half covers a number of interesting results which have, due to one reason or another, not been included in previous sections.

In order to prove (iv) of 11.1 we shall be adopting notation of the previous section. In particular $a > 0$ is a fixed element of V and δ is the isomorphism of $(\mathbb{R}, +)$ onto $(V, +)$ (see the last section) defined by $r\delta = ar$, for all r in $\mathbb{R}$. The μ of (iii) of 11.1 was then taken as δ^{-1}. We also need a method of representing all r in $\mathbb{R}$ with $r > q_1$ and $r < q_2$, where q_i, $i = 1, 2$, are in $\mathbb{Q}$. It is desirable that this subset of $\mathbb{R}$ is not confused with similar subsets of V. Thus the indicated open interval of $\mathbb{R}$ will be denoted by $\lambda(q_1, q_2)$.

Proposition 13.1. *If q_i, $i = 1, 2$, are in $\mathbb{Q}$, then*

$$(\lambda(q_1, q_2))\delta = (aq_1, aq_2).$$

Proof. For r in $\lambda(q_1, q_2)$, we have $q_1 < r < q_2$ and ar is the unique element of $H\backslash L_r(a)$, where H is the closure of $L_r(a)$ (see the last section). It follows that $ar > aq_1$. A similar argument using 12.16 yields $ar < aq_2$. Thus $\lambda(q_1, q_2)$ is mapped by δ into (aq_1, aq_2). We must show δ maps $\lambda(q_1, q_2)$ onto (aq_1, aq_2). Suppose b in V is such that $aq_1 < b < aq_2$. Now $b = ar$, where r is in $\mathbb{R}$. Either $q_1 > r, q_1 = r$ or $q_1 < r$. If $q_1 > r$, then $(aq_1, \infty) \subseteq U_r(a)$, and, by 12.16, $ar < aq_1$. If $q_1 = r$, then $aq_1 = ar$, by 12.14, so we conclude $q_1 < r$. A similar argument shows $q_2 > r$. It has been shown that δ maps $\lambda(q_1, q_2)$ onto (aq_1, aq_2) and the proof is complete.   □

Corollary 13.2. *If V and N are as in 11.1, then the μ of (iii) is continuous.*

Proof. Any open set O of $\mathbb{R}$ is a union of open sets $\lambda(q_1, q_2)$ with q_i, $i = 1, 2$, coming from $\mathbb{Q}$. It follows readily from 13.1 that $O\delta$ is a union of (aq_1, aq_2) with q_i, $i = 1, 2$, coming from $\mathbb{Q}$. Thus $O\delta$ is open and, since $\delta = \mu^{-1}$, the continuity of μ follows.   □

The problem of showing μ^{-1} is continuous makes use of the local compactness of V. Before commencing this undertaking two straightforward propositions are given.

Proposition 13.3. *The V of* 11.1 *is Hausdorff.*

Proof. If c and d are in V with $c \neq d$, then assuming $c < d$ we have, by 12.11, that for some q in $\mathbb{Q}, c < aq < d$. This means $(-\infty, aq)$ and (aq, ∞) are disjoint open subsets of V respectively including c and d. The proof of 13.3 is complete.   □

A question needing answering is that of the compactness of V.

Proposition 13.4. *The V of* 11.1 *is not compact.*

Proof. Suppose V is compact. By 12.4, if $b \neq 0$ is in V, then the family $(-\infty, bn), n \in \mathbb{Z}$ is such that any h of V is in some such subset. The family $(-\infty, bn), n \in \mathbb{Z}$ therefore covers V. Thus there exists an integer $k \geq 1$ and $n_1, \ldots, n_k$, in $\mathbb{Z}$ such that $(-\infty, bn_i), i = 1, \ldots, k$, cover V. However, these subsets of V are such that for i_1 and i_2 in $\{1, \ldots, k\}$ either

$$(-\infty, bn_{i_1}) \subseteq (-\infty, bn_{i_2})$$

or the reverse inclusion holds. This in turn means there is some m in $\{n_1, \ldots, n_k\}$ with $V \subseteq (-\infty, bm)$. By 12.4, some r in $\mathbb{Z}$ is such that $br > bm$ yielding a contradiction. The proof of 13.4 is complete.   □

We are ready to use the locally compact assumption. The lemma now given is an important step in proving 11.1(iv).

Lemma 13.5. *If V as in* 11.1 *is locally compact and b is in V, then there exist a_i, $i = 1, 2$, in V with $a_1 < a_2$ and a compact subset Γ of V such that $(a_1, a_2) \subseteq \Gamma$, where b is in (a_1, a_2).*

Proof. Since V is locally compact a point of V has a compact neighbourhood Γ_1. If all elements of $\{v + \Gamma_1 : v \in V\}$ contain 0, then $\Gamma_1 = V$ contrary to 13.4. It follows, if necessary by translation, that Γ_1 can be taken as not containing 0. From §8 of [1], compact subspaces of a Hausdorff space are closed and, by 13.3, Γ_1 is closed. Now Γ_1 being a neighbourhood contains a non-empty open subset O of V. Also $O \neq \Gamma_1$ otherwise Γ_1 is a proper open and closed subset of V contrary to connectedness. By 2.5, $V \backslash O$ is Z-closed and $V \backslash O = Z(S)$, where S is a non-empty subset of N. Thus there exists γ in S, such that $(V \backslash O)\gamma = \{0\}$ and $O\gamma \neq \{0\}$. Now

$$V\gamma = [(V \backslash O) \cup \Gamma_1]\gamma = \{0\} \cup \Gamma_1\gamma$$

and since $O \neq \Gamma_1$, there exists d in $(V \backslash O) \cap \Gamma_1$. Because $d\gamma = 0$ we see $V\gamma = \Gamma_1\gamma$ and $V\gamma$ is the image under the continuous function induced by γ (see 2.9) of the compact subspace Γ_1. From §8 of [1] this means $V\gamma$ is compact and non-zero (we have $O\gamma \neq \{0\}$). If (D_1, D_2) is a disconnection of $V\gamma$, then $D_i\gamma^{-1}$, $i = 1, 2$, are, by 2.9, non-empty disjoint open and closed subsets of V with $D_1\gamma^{-1} \cup D_2\gamma^{-1} = V$. This is clearly contrary to the connectedness of V and therefore $V\gamma$ is connected. Now $V\gamma$ being non-zero and including 0, contains two distinct elements c_i, $i = 1, 2$. We may assume $c_1 < c_2$. If c in (c_1, c_2) is not in $V\gamma$, then $(-\infty, c) \cap V\gamma$ and $(c, \infty) \cap V\gamma$ contain c_1 and c_2 respectively and are a disconnection of $V\gamma$. As was proved above this would imply the disconnectedness of V. Consequently, (c_1, c_2) is contained in the compact subspace $V\gamma$ of V. By 12.11, (c_1, c_2) is non-empty. If d_1 is in $(c_1, c_2), a_i = b - d_1 + c_i$, for $i = 1, 2$, and $\Gamma = b - d_1 + V\gamma$, then Γ is clearly a compact subset of V and $a_1 < a_2$. Furthermore, $(a_1, a_2) \subseteq \Gamma$ and since (a_1, a_2) contains $b - d_1 + d_1$ the lemma is completely proved.   □

The next goal is the proof of (iv) of 11.1.

Theorem 13.6. *If in* 11.1, V *is locally compact, then (iv) holds.*

Proof. It was shown in 13.2 that μ is continuous. It remains to show that local compactness implies μ^{-1} is continuous. This amounts to showing $O\mu (= O(\mu^{-1})^{-1})$ is open whenever O is an open subset of V. Now, if O is a union of subsets (aq_1, aq_2) with q_i, $i = 1, 2$, coming from $\mathbb{Q}$, then by 13.1, $O\mu$ is a union of the $\lambda(q_1, q_2)$ and is clearly open. To prove the theorem it is therefore sufficient to show O is such a union. Thus it is required to prove that for b in O, there exist q_i, $i = 1, 2$, in $\mathbb{Q}$ with b in (aq_1, aq_2) such that $(aq_1, aq_2) \subseteq O$.

By 13.5, there exist a compact subspace Γ of V and a_i, $i = 1, 2$, in V with $a_1 < a_2$ such that b is in (a_1, a_2) and $(a_1, a_2) \subseteq \Gamma$. Now

$$(\Gamma \backslash O) \cap (-\infty, b] (= \Gamma_1 \text{ say})$$

is clearly closed in Γ and is therefore compact. Furthermore, the collection of all $(-\infty, c)$ where $c < b$, is an open cover of Γ_1. Indeed, if d is in Γ_1, then $d \le b$ and since b is in $O, d < b$. Therefore, 12.11 implies the existence of d_1 in (d, b) and the inclusion of d in $(-\infty, d_1)$ and, therefore, in the union of the collection $(-\infty, c), c < b$, of open subsets of V. Thus these open subsets are indeed a cover of Γ_1 and contain a finite subcover. We have shown that there exists an integer $k \ge 1$ and c_i, $i = 1, \ldots, k$, in V with $c_i < b$ such that $(-\infty, c_i)$, $i = 1, \ldots, k$, cover Γ_1. Clearly for all i_1 and i_2 in $\{1, \ldots, k\}$, either $(-\infty, c_{i_1}) \subseteq (-\infty, c_{i_2})$, or the reverse inclusion holds. This means all $(-\infty, c_i)$, i in $\{1, \ldots, k\}$ are contained in some $(-\infty, c_j)$ with j in $\{1, \ldots, k\}$. Therefore no elements of (c_j, b) are in Γ_1. However, if g is the maximum of c_j and a_1, then $g < b$ and clearly no elements of (g, b) are in Γ_1. since (g, b) is contained in $\Gamma \cap (-\infty, b]$, we conclude that $(g, b) \subseteq O$. By 12.11, there exists q_1 in $\mathbb{Q}$ such that aq_1 is in (g, b). Clearly $(aq_1, b) \subseteq O$. An entirely similar argument gives q_2 in $\mathbb{Q}$ with $aq_2 > b$ and $(b, aq_2) \subseteq O$. Since b is in O, it now follows that $(aq_1, aq_2) \subseteq O$, thereby establishing the desired result. The proof of the theorem is complete.   □

Theorem 11.1 now stands proved. If (iv) holds, then with respect to Z^t, V is simply $\mathbb{R}$. As explained, in this case, N is a primitive compatible subnon-ring of $C_0(\mathbb{R})$.

Before finishing this section there are a number of further results that need mentioning. In the light of what has gone before it is not difficult to supply proofs. However, their statement and a very brief indication of how to attempt a proof will be our present aim.

It is not difficult to show that a non-trivial direct decomposition of the N of 11.1, is *antisymmetric*. This means it is of the form $R_1 \oplus R_2$, where $(-1)R_1 = R_2$ and $(-1)R_2 = R_1$. There are several matters of interest that occur for a primary N-group, where N has antisymmetric direct decomposition. From theorem 9.6, it is readily deduced that:-

Theorem 13.7. *If V is a primary N-group and N has antisymmetric direct decomposition, then every non-zero N-subgroup of V is both open and closed.*

Another aspect of antisymmetric direct decomposition is that:-

Theorem 13.8. *If V is a primary N-group where N has ACCR and an antisymmetric direct decomposition, then V is finite.*

The proof of 13.8 is accomplished by showing V is necessarily Hausdorff and, then observing that, by 10.5, $c(V)$ must be zero and again, by 10.5, that V is finite.

The fact that many primary N-groups have the property that all non-zero N-subgroups are open (see 9.2) is of considerable interest. A more general notion is the existence of small open N-subgroups. A primary N-group will be said to have *small open N-subgroups* (have *SONS*), if every open neighbourhood of 0 contains an open N-subgroup. If V is such an N-group and $v \neq 0$ in $c(V)$, then $V \backslash \{v\}$ contains an open subgroup U of V which yields the disconnection $(c(V) \cap U, c(V) \backslash U)$ of $c(V)$. This means $c(V) = \{0\}$ and it follows that:-

Theorem 13.9. *If a primary N-group has SONS, then it is totally disconnected.*

According to proposition 2 of [1] the primary N-group V will, with respect to Z^t, be a topological group if for any neighbourhood Γ of 0 there exists an open subset O of V such that $O(-1) + O \subseteq \Gamma_1$. As a consequence of this fact we have:-

Theorem 13.10. *If a primary N-group has SONS, then with respect to Z^t it is a topological group.*

Theorem 13.10 tends to suggest there may be many situations where primary N-groups are, with respect to Z^t, topological groups. Indeed one case of this is provided by 13.7 and 4.8.

Corollary 13.11. *If V is a primary N-group without a minimal N-subgroup and N has antisymmetric direct decomposition, then V is a topological group (under Z^t).*

Another example where this happens is provided by 9.2. Again by 9.2, 6.2 and 4.4 we conclude.

Corollary 13.12. *If the primary N-group V is totally graded and has no minimal closed N-subgroups, then V is a topological group.*

However, it seems likely that many more examples of primary N-groups being topological groups can arise. It would be of considerable interest to have reasonably deep theorems yielding such conclusions.

No previous developments in the area of this paper have taken place. Indeed as far as I am aware nothing even similar has been developed (apart possibly from [5]). However, the topological techniques involved yield substantial theory. The author feels it is likely that many very meaningful results along the lines of those unearthed in this paper may hold true.

We are now ready to briefly provide proofs of results extracted from manuscripts.

14. Loose Ends

The proof of 1.1 follows easily from the fact that, if V is a compatible N-group with a submodule U such that $(U : V) = (0 : V)$, then $U \leq z(V)$. To prove this we take any u in U and α in N and an element β of N such that

$$(u + w)\alpha - w\alpha - u\alpha = w\beta,$$

for all w in V. Such a β exists by compatibility. Also because u is in U, β is in $(0 : V)$ and $w\beta = 0$, for all w in V. Clearly this can only mean $U \leq z(V)$.

Proposition 1.6 will be deduced as a corollary of the following theorem:-

Theorem 14.1. *If V is a 3-tame N-group and S a non-empty subset of V, then $C_V(S) = C_V(W)$ where W is the N-subgroup of V generated by S.*

Proof. Suppose it has been shown that for u in S, $uN \leq C_V(C_V(S))$. If this holds, then clearly $W \leq C_V(C_V(S))$ and $C_V(W) \geq C_V(S)$. since $C_V(W) \leq C_V(S)$, the result will then follow. Let α and β be in N and let

$$\gamma = (1+\alpha)\beta - \alpha\beta - \beta.$$

Now for w in $C_V(S)$, $(u+w)\gamma = u\gamma + w\gamma$. If v_i, $i = 1,2,3$, are any given elements of V, then the 3-tame assumption implies that there exists λ_1 in N such that

$$(v_i + (u+w)\alpha)\beta - (u+w)\alpha\beta - v_i\beta = v_i\lambda_1.$$

Furthermore, suppose λ_1 is chosen so that $v_1 = u+w$, $v_2 = u$ and $v_3 = w$. Now from the definition of $C_V(S)$ we have

$$(u+v)\lambda_1 = (u+v)\gamma = u\lambda_1 + w\lambda_1.$$

Thus

$$
\begin{aligned}
(u+w)\gamma &= (u+(u+w)\alpha)\beta - (u+w)\alpha\beta - u\beta \\
&+ (w+(u+w)\alpha)\beta - (u+w)\alpha\beta - w\beta.
\end{aligned}
$$

Again with v_i, $i = 1,2,3$, in V we can find λ_2 in N, such that

$$(u+v_i\alpha)\beta - v_i\alpha\beta - u\beta = v_i\lambda_2.$$

Furthermore, suppose λ_2 is chosen so that $v_1 = u+w$, $v_2 = u$ and $v_3 = w$. Since $(u+w)\lambda_2 = u\lambda_2 + w\lambda_2$ we have,

$$
\begin{aligned}
(u+(u+w)\alpha)\beta - (u+w)\alpha\beta - u\beta &= (u+u\alpha)\beta - u\alpha\beta - u\beta \\
&+ (u+w\alpha)\beta - w\alpha\beta - u\beta
\end{aligned}
$$

and a similar argument shows,

$$
\begin{aligned}
(w+(u+w)\alpha)\beta - (u+w)\alpha\beta - w\beta &= (w+u\alpha)\beta - u\alpha\beta - w\beta \\
&+ (w+w\alpha)\beta - w\alpha\beta - w\beta.
\end{aligned}
$$

We therefore conclude (see the expansion of $(u+w)\gamma$) that

$$
\begin{aligned}
(u+w)\gamma &= u\gamma + (u+w\alpha)\beta - w\alpha\beta - u\beta \\
&+ (w+u\alpha)\beta - u\alpha\beta - w\beta + w\gamma.
\end{aligned}
$$

However, $w\alpha$ is in $C_V(S)$ and therefore,

$$(u+w\alpha)\beta - w\alpha\beta - u\beta = 0.$$

Thus $(w+u\alpha)\beta = w\beta + u\alpha\beta$, for all w in $C_V(S)$ and α and β in N. It follows that $uN \leq C_V(S)$ and the theorem is completely proved. $\square$

Corollary 14.2. *Proposition 1.6 holds true.*

Proof. If V is primary and $u \neq 0$ in V, then by 14.1, $C_V(u) = C_V(uN)$ and $C_V(uN) \cap uN$ is clearly a ring module. Since $uN \neq \{0\}$, this can only mean $C_V(uN) = \{0\}$. The corollary holds. $\square$

The final result needing proof is 1.7. Here in the primary N-group V, we have two non-zero elements v_i, $i = 1, 2$, such that $(0 : v_1) \leq (0 : v_2)$ and it is required to show that $v_1 = v_2$. First, it is easy to establish, that an N-homomorphism μ of $v_1 N$ onto $v_2 N$ can be defined by setting $(v_1 \alpha)\mu = v_2 \alpha$ for all α in N. Now for w_1 in $v_1 N$ and β in N we have $(w_1 + v)\beta - w_1\beta = v\gamma$, for some γ in N independent of v in V. Thus putting $v = w_2\mu$ with w_2 in $v_1 N$ we see

$$(w_1 + w_2\mu)\beta - w_1\beta = w_2\gamma\mu = (w_1\mu + w_2\mu)\beta - w_1\mu\beta,$$

for all w_i, $i = 1, 2$, in $v_1 N$ and β in N. Now taking w_2 as $-w_1 + w_2$ gives

$$(w_1 - w_1\mu + w_2\mu)\beta - w_1\beta = w_2\mu\beta - w_1\mu\beta.$$

Also taking $w_2 = 0$ we see $w_1(1 - \mu)\beta = -w_1\mu\beta + w_1\beta$, so that

$$(w_1(1 - \mu) + w_2\mu)\beta = w_2\mu\beta + w_1(1 - \mu)\beta.$$

Now $(v_1 N)\mu = v_2 N$, so that it follows readily, the centralizer of $w_1(1 - \mu)$ in V contains $v_2 N (\neq \{0\})$. By the above corollary this can only mean $w_1(1 - \mu) = 0$ and, in particular, $v_1 = v_1\mu = v_2$. Thus 1.7 holds true and the end of this paper has been reached.

In conclusion, I would like to thank Olita Moala for her effort in reproducing this material.

REFERENCES

[1] P. J. Higgins, *An Introduction to Topological Groups*, Camb. Univ. Press, London, (1974).

[2] S. D. Scott, *Compatible Near-rings with Minimal Condition on Ideals*, manuscript (1995).

[3] S. D.Scott, *N-solubility and N-nilpotency in Tame N-groups*, Alg. Coll. **5** (1998), 425-448.

[4] S. D.Scott, *Primary and Semiprimary N-groups*, manuscript (1997).

[5] S. D.Scott, *Primitive Compatible Near-rings*, manuscript (1994).

[6] S. D. Scott, *The Structure of Ω-Groups*, in Nearrings, Nearfields and K-Loops, 47–137. Klu. Acad. Pub, Netherlands (1997).

[7] S. D. Scott, *Zero Sets-Consequences for Primitive Near-rings*, Proc. Edin. Math. Soc. **25** (1982), 55–63.

DEPARTMENT OF MATHEMATICS, UNIVERSITY OF AUCKLAND AUCKLAND, NEW ZEALAND. E-MAIL: stuscott@math.auckland.ac.nz

ON THE RADICALS OF COMPOSITION NEAR-RINGS

STEFAN VELDSMAN

ABSTRACT. Let α be a Hoehnke radical in the variety of near-rings. Using α we then define a corresponding radical for a composition near-ring C. This is done via the foundation of C (that is, the constant part of the composition.

A *composition near-ring* C is a quadruple $C = (C, +, \cdot, \circ)$ where both $C_1 := (C, +, \cdot)$ and $C_2 := (C, +, \circ)$ are near-rings which satisfies $(xy) \circ z = (x \circ z)(y \circ z)$ for all $x, y, z \in C$. If C_1 is a ring, then C is called a *composition ring*. The *foundation* of a composition near-ring C, denoted by $Found(C)$ or K_C (or just K if there is no reason for confusion) is the subset $K_C = \{x \in C | x \circ 0 = x\}$. The two standard examples of composition near-rings (or rings) are $(K[x], +, \cdot, \circ)$ and $(K^K, +, \cdot, \circ)$ where the former is the polynomials over the near-ring (respectively ring) K and the latter is the set of all functions from K into K. Both these composition near-rings have foundation K.

Composition rings have been studied by many authors, see for example Adler [1], Speegle [8] and Pilz [6] together with their references. Composition near-rings have not yet received their due attention with only limited contributions, see for example Petersen [5].

For a radical α and a ring K, much effort over many years has gone into determining the relationship between the radical $\alpha(K)$ of K and the radical of the ring $(K[x], +, \cdot)$ and the near-ring $(K[x], +, \circ)$, particularly for K a commutative ring with $1 \in K$. In this regard, one may consult, for example, Amitsur [2], Mlitz [4], Pilz [6] or Kautschitsch [3].

In this note we will present a procedure for defining a radical for a composition near-ring. The underlying principle of our approach is that the radical of a composition near-ring should reflect the "radicalness" of the foundation of the composition near-ring. Hence our proposal is really only sensible for composition near-rings with non-zero foundation. In any case, $(K[x], +, \cdot, \circ)$ is really our motivating force.

Of course, radicals for composition rings have been defined, but these were always "modulo" the underlying ring structure. By this is meant that the radical of the composition ring $(C, +, \cdot, \circ)$ depends mainly on the structure of the near-ring $(C, +, \circ)$. In fact, one may have $ab = 0$ for all $a, b \in C$ without this pathological behaviour being reflected in the radical of the composition ring C.

In the sequel, C will always be a composition near-ring with foundation K. Clearly K is a subcomposition near-ring of C, but usually when we refer to the foundation of C, we mean the near-ring $(K, +, \cdot)$. So, for example, an ideal of K will mean an ideal of the near-ring $(K, +, \cdot)$. A subset I of C may be an ideal of C_1, denoted by $I \lhd C_1$, or an ideal

---

1991 *Mathematics Subject Classification.* 16Y30, 16N80, 16S36 .

Y. Fong et al. (eds.), *Near-Rings and Near-Fields*, 198–201.
© 2001 *Kluwer Academic Publishers. Printed in the Netherlands.*

of C_2, denoted by $I \lhd C_2$. In case I is both an ideal of C_1 and C_2, then we say I is an ideal of C and it will be denoted by $I \lhd C$. In such a case, many authors would call I a full ideal of C, but we try to be consistent with the general algebraic nomenclature that ideals should be the kernels of homomorphisms in the specific variety. For subsets $S, T \subseteq C$, $(S : T)_{C_2}$ denotes the subset $(S : T)_{C_2} = \{x \in C \mid x \circ T \subseteq S\}$.

Let $C(C) = \{x \in C \mid \text{For all } J \lhd K, k \in K, j \in J, x \circ (k + j) - x \circ k \in J\}$. The elements of $C(C)$ are called the compatible elements of C. It can be verified that $C(C)$ is a subcomposition near-ring of C with foundation K. If $C = C(C)$, then C is called *compatible*. These notions, at least with respect to the near-ring structure C_2, have been studied extensively, see Pilz [6] and its references.

An ideal J of the near-ring $K = Found(C)$ is an C-ideal of K if $c \circ (k + j) - c \circ k \in J$ for all $c \in C, k \in K, j \in J$. Every ideal of K is an $C(C)$-ideal of K. Our interest in these ideals are mainly because of:

Proposition 1. *Let* $J \lhd K$. *Then the following are equivalent*

(a) J *is an* C-ideal of K.
(b) $(J : K)_{C_2}$ *is an ideal of* C.
(c) $J = \langle J \rangle_C \cap K$ *where* $\langle J \rangle_C$ *denotes the ideal of* C *generated by* J.

Proof. $(a) \Rightarrow (b)$ is routine.

$(b) \Rightarrow (c)$ If $(J : K)_{C_2}$ is an ideal of C, then $\langle J \rangle_C \subseteq (J : K)_{C_2}$ since $J \subseteq (J : K)_{C_2}$. Thus $\langle J \rangle_C \cap K \subseteq (J : K)_{C_2} \cap K = J$. Clearly $J \subseteq \langle J \rangle_C \cap K$ and so $J = \langle J \rangle_C \cap K$.

$(c) \Rightarrow (a)$ Suppose $J = \langle J \rangle_C \cap K$. For $x \in C, k \in K$ and $j \in J, x \circ (k + j) - x \circ k \in \langle J \rangle_C \cap K = J$; hence J is an C-ideal of K. □

Let $\mathcal{M}$ be a class of near-rings and let α be the corresponding Hoehnke radical, i.e. for every near-ring N, $\alpha(N) = \cap(I \lhd N \mid N/I \in \mathcal{M})$. As is well-known, the semisimple class of α, denoted by $S\alpha = \{N \mid \alpha(N) = 0\}$ is just the subdirect closure of $\mathcal{M}$ (i.e. any near-ring in $S\alpha$ is a subdirect sum of near-rings from $\mathcal{M}$). Let $\mathcal{M}^* = \{$composition near-rings $C \mid \alpha(K_C) = 0\}$. Thus $\mathcal{M}^*$ is the class of all composition near-rings whose foundation is a subdirect sum of near-rings from $\mathcal{M}$. Let α^* be the Hoehnke radical for composition near-rings determined by $\mathcal{M}^*$, i.e. $\alpha^*(C) = \cap(I \lhd C \mid C/I \in \mathcal{M}^*)$. Since $Found(C/I) = (K_C + I)/I \cong K_C/(K_C \cap I)$, $\alpha^*(C) = \cap(I \lhd C \mid K_C/(K_C \cap I) \in S\alpha)$.

If a composition near-ring D is a subdirect sum, say D is a subdirect sum of $D/I_\gamma, \gamma \in \Lambda$, then $Found(D)$ is a subdirect sum of $Found(D/I_\gamma), \gamma \in \Lambda$. Hence

$$K_C / (K_C \cap \alpha^*(C)) \cong (K_C + \alpha^*(C)) / \alpha^*(C) = Found (C/\alpha^*(C))$$

which is isomorphic to a subdirect sum of the near-rings $Found(C/I_\gamma), \gamma \in \Lambda$, where $\alpha^*(C) = \bigcap_{\gamma \in \Lambda} (I_\gamma \lhd C \mid C/I_\gamma \in \mathcal{M}^*)$. Since $Found(C/I_\gamma) \in S\alpha$ for all α, we have

$$K_C / (K_C \cap \alpha^*(C)) \in S\alpha$$

and so $\alpha(K_C) \subseteq K_C \cap \alpha^*(C)$. For a compatible composition near-ring C, we have equality:

Proposition 2. *Let* $\mathcal{M}$, α *and* α^* *be as above. Let* C *be a compatible composition near-ring. Then* $\alpha(K_C) = K_C \cap \alpha^*(C)$.

Proof. Let $J \lhd K_C$ with $K_C/J \in \mathcal{M}$. By Proposition 1 we have $J = \langle J \rangle_C \cap K$ and since $\text{Found}(C/\langle J \rangle_C) = (K_C + \langle J \rangle_C)/\langle J \rangle_C \cong K_C/(K_C \cap \langle J \rangle_C) = K_C/J \in \mathcal{M} \subseteq S\alpha$, we have $C/\langle J \rangle_C \in \mathcal{M}^*$. Hence $\alpha^*(C) \subseteq \langle J \rangle_C$. Then $\alpha^*(C) \cap K_C \subseteq \langle J \rangle_C \cap K_C = J$ for all $J \lhd K_C$ with $K_C/J \in \mathcal{M}$. We conclude that $\alpha^*(C) \cap K_C \subseteq \alpha(K_C)$. $\square$

Although this result gives a relationship between $\alpha^*(C)$ and $\alpha(K_C)$, we want to determine $\alpha^*(C)$ explicitly. For this we need some preparations. The next result is part of the folklore of the general radical theory - it is recorded here for the ease of reference.

Lemma 3. *Let ρ be a Hoehnke radical for composition near-rings. For any composition near-ring C and $I \lhd C$ with $I \subseteq \rho(C)$, $\rho(C/I) = \rho(C)/I$.*

Lemma 4. *Let ρ be a Hoehnke radical for composition near-rings. For a composition near-ring C, let $J := \rho(C) \cap K_C$. Then $\rho(C) = \langle J \rangle_C$ for all C if and only if ρ satisfies: For any composition near-ring D, if $\rho(D) \neq 0$, then $\rho(D) \cap K_D \neq 0$.*

Proof. Suppose the condition is satisfied. Since $\rho(C) \lhd C$ and $J \subseteq \rho(C)$, $\langle J \rangle_C \subseteq \rho(C)$. By Lemma 3 we have

$$\rho(C/\langle J \rangle_C) \cap \text{Found}(C/\langle J \rangle_C) = (\rho(C)/\langle J \rangle_C) \cap (K_C + \langle J \rangle_C)/\langle J \rangle_C$$
$$= ((\rho(C) \cap K_C) + \langle J \rangle_C)/\langle J \rangle_C$$
$$= (J + \langle J \rangle_C)/\langle J \rangle_C = 0.$$

By the assumption $\rho(C/\langle J \rangle_C) = 0$; thus $\rho(C) \subseteq \langle J \rangle_C \subseteq \rho(C)$.

Conversely, suppose $\rho(C) = \langle J \rangle_C$ for all composition near-rings C where $J = \rho(C) \cap K_C$. Let D be a composition near-ring with $\rho(D) \neq 0$. If $\rho(D) \cap K_D = 0$, then $J' := \rho(D) \cap K_D = 0$. Thus $0 = \langle J' \rangle_D = \rho(D)$; a contradiction. $\square$

Lemma 5. *Let $\mathcal{M}$, α and α^* be as above. Then $\alpha^*(D) \neq 0$ implies $\alpha^*(D) \cap K_D \neq 0$ for any composition near-ring D.*

Proof. Suppose $\alpha^*(D) \neq 0$ but $\alpha^*(D) \cap K_D = 0$. Now $\alpha(K_D) \subseteq K_D \cap \alpha^*(D) = 0$ and so $D \in \mathcal{M}^*$ which gives the contradiction $\alpha^*(D) = 0$. $\square$

Theorem 6. *Let $\mathcal{M}$ be a class of near-rings and let α be the corresponding Hoehnke radical. Let $\mathcal{M}^*$ be the class of all composition near-rings C with $\alpha(K_C) = 0$ and let α^* be the Hoehnke radical determined by $\mathcal{M}^*$. Then $\alpha^*(C) = \langle \alpha(K) \rangle_C$ for any compatible composition near-ring C with $\text{Found}(C) = K$.*

Proof. By Proposition 2, $\alpha(K) = \alpha^*(C) \cap K$ and by Lemmas 5 and 4, $\alpha^*(C) = \langle \alpha(K) \rangle_C$. $\square$

We note that for a compatible composition near-ring C, $\alpha^*(C) \subseteq (\alpha(K) : K)_{C_2}$ and in general this inclusion is strict as the next example will show.

Example 7. Let $\mathbb{Z}_4 = \{0, 1, 2, 3\}$ be the ring of integers mod 4. Here we have $C(\mathbb{Z}_4) = \{f : \mathbb{Z}_4 \to \mathbb{Z}_4 \mid f(x+2) - f(x) \in \{0, 2\}$ for all $x \in \mathbb{Z}_4\}$ and in fact, if $\mathcal{P}(\mathbb{Z}_4)$ denotes all the polynomial functions on $\mathbb{Z}_4$, we have $C(\mathbb{Z}_4) = \mathcal{P}(\mathbb{Z}_4) = \{a_0 + a_1 x + a_2 x^2 + a_3 x^3 \mid a_0, a_1 \in \mathbb{Z}_4, a_2, a_3 \in \{0, 1\}\}$. Thus the composition near-ring (actually composition ring) $C := \mathcal{P}(\mathbb{Z}_4)$ is compatible with foundation $\mathbb{Z}_4$.

Let $\mathcal{M}$ be the class of all 2-primitive near-rings. Then α is the $\mathcal{J}_2$-radical and for $\mathbb{Z}_4$, $\alpha(\mathbb{Z}_4) = \{0,2\}$ (i.e. just the Jacobson radical of $\mathbb{Z}_4$ = nil radical of $\mathbb{Z}_4$). Now $\alpha^*(C) = \langle\{0,2\}\rangle = \{0, 2, 2x, 2+2x\}$ which is strictly contained in

$$(\alpha(\mathbb{Z}_4) : \mathbb{Z}_4) = \{f \in C \mid f(\mathbb{Z}_4) \subseteq \{0,2\}\}$$

$$= \{a_0 + a_1 x + a_2 x^2 + a_3 x^3 \mid a_0 \in \{0,2\} \text{ and } a_1 + a_2 + a_3 \in \{0,2\}\}.$$

For example, $x^2 + x^3 \in (\alpha(\mathbb{Z}_4) : \mathbb{Z}_4)$, but $x^2 + x^3$ is not in $\alpha^*(C)$.

As mentioned earlier, our approach to radicals for composition near-rings is really only sensible for those with non-zero foundation. If $Found(C) = 0$, then C is trivially compatible and $\alpha^*(C) = \langle\alpha(K_C)\rangle = 0$ which may not be desirable (e.g. if C has $a \circ b = 0$ for all $a, b, \in C$). We conclude with a few consequences of our present theory. Below $\mathcal{M}$, α and α^* are as above and C is a compatible composition near-ring with $Found(C) = K$.

(a) $\alpha^*(C) = 0$ if and only if $\alpha(K) = 0$. Thus, if $\alpha^*(C) = 0$, then C is a composition near-ring with $Found(C)$ a subdirect sum of near-rings from $\mathcal{M}$.

(b) $\alpha(K) = K$ if and only if $\alpha^*(C) = \langle K \rangle_C$. (Indeed, if $\alpha^*(C) = \langle K \rangle_C$, then $\alpha(K) = \alpha^*(C) \cap K = \langle K \rangle_C \cap K = K$.) and $\langle K \rangle_C = C$.

(c) If $K = C$, then $K = C_1$ and $\alpha^*(C) = \alpha^*(C) \cap K = \alpha(K) = \alpha(C_1)$.

(d) For any $x \in C$, $(C \circ x, +, \cdot)$ is a subnear-ring of C_1 as well as a homomorphic image of C_1 via the map $\gamma_x : C_1 \to C \circ x$ defined by $\gamma_x(C) = C \circ x$. Since Hoehnke radicals are preserved by homomorphisms, $\alpha(C_1) \circ x \subseteq \alpha(C \circ x)$ for all $x \in C$. Then $\alpha(C_1) \subseteq (\alpha(K) : K)_{C_2}$; hence if $\alpha(C_1) = C_1$, $K = C \circ K = \alpha(C_1) \circ K \subseteq \alpha(K)$, i.e. $\alpha(K) = K$ and so $\alpha^*(C) = \langle K \rangle_C$ (by (b) above).

REFERENCES

[1] I. Adler. Composition Rings, *Duke Math. J.* **29** (1962), 607–623.

[2] S.A. Amitsur. Radicals of Polynomial Rings, *Canad. J. Math.* **8** (1956), 355–361.

[3] H. Kautschitsch. Maximal ideals in the near-ring of polynomials, Radical Theory, Proc. 1st Conf. Eger 1982, *Colloq. Math. Soc. J. Bolyai* **38** (1985), 183–193.

[4] R. Mlitz. Ein Radikal für universale Algebren und sein Anwendung auf Polynomringe mit Komposition, *Monatsh. Math.* **75** (1971), 144–152.

[5] Q.N. Petersen. *Composition near-rings.* M.Sc Treatise, University of Port Elizabeth, 1995.

[6] G. Pilz. *Near-rings.* North-Holland, Amsterdam. Second, revised edition, 1983.

[7] G. Pilz and Yong-Sian So. Near-rings of polynomials and polynomial functions, *J. Austral. Math. Soc.* (Series A) **29** (1980), 61–70.

[8] A. De Bruyn Speegle. *Sandwich composition rings.* Ph. D. Dissertation, Texas A & M University, 1997.

DEPARTMENT OF MATHEMATICS, UNIVERSITY OF PORT ELIZABETH, SOUTH AFRICA.
EMAIL: maassv@upe.ac.za

9 780792 367062